Lecture Notes in Management Engineering

Series Editor

Adolfo López-Paredes, INSISOC, University of Valladolid, Valladolid, Spain

This book series provides a means for the dissemination of current theoretical and applied research in the areas of Industrial Engineering and Engineering Management. The latest methodological and computational advances that can be widely applied by both researchers and practitioners to solve new and classical problems in industries and organizations contribute to a growing source of publications written for and by our readership.

The aim of this book series is to facilitate the dissemination of current research in the following topics:

- Strategy and Entrepreneurship
- Operations Research, Modelling and Simulation
- Logistics, Production and Information Systems
- Quality Management
- Product Management
- Sustainability and Ecoefficiency
- Industrial Marketing and Consumer Behavior
- Knowledge and Project Management
- Risk Management
- Service Systems
- Healthcare Management
- Human Factors and Ergonomics
- Emergencies and Disaster Management
- Education

More information about this series at https://link.springer.com/bookseries/11786

N. C. Saha · Anup K. Ghosh · Meenakshi Garg ·
Susmita Dey Sadhu

Food Packaging

Materials, Techniques and Environmental Issues

With Contributions by Deepak Manchanda and Subramoni Chidambar

N. C. Saha
Foundation for Innovative Packaging
and Sustainability
Mumbai, Maharashtra, India

Meenakshi Garg
Department of Food Technology
Bhaskaracharya College of Applied Science
New Delhi, India

Anup K. Ghosh
Department of Materials Science
and Engineering
Indian Institute of Technology Delhi
New Delhi, India

Susmita Dey Sadhu
Department of Polymer Sciences
Bhaskaracharya College of Applied Science
New Delhi, India

ISSN 2198-0772　　　　　　　　ISSN 2198-0780　(electronic)
Lecture Notes in Management and Industrial Engineering
ISBN 978-981-16-4235-7　　　ISBN 978-981-16-4233-3　(eBook)
https://doi.org/10.1007/978-981-16-4233-3

© The Editor(s) (if applicable) and The Author(s), under exclusive license to Springer Nature Singapore Pte Ltd. 2022
This work is subject to copyright. All rights are solely and exclusively licensed by the Publisher, whether the whole or part of the material is concerned, specifically the rights of translation, reprinting, reuse of illustrations, recitation, broadcasting, reproduction on microfilms or in any other physical way, and transmission or information storage and retrieval, electronic adaptation, computer software, or by similar or dissimilar methodology now known or hereafter developed.
The use of general descriptive names, registered names, trademarks, service marks, etc. in this publication does not imply, even in the absence of a specific statement, that such names are exempt from the relevant protective laws and regulations and therefore free for general use.
The publisher, the authors and the editors are safe to assume that the advice and information in this book are believed to be true and accurate at the date of publication. Neither the publisher nor the authors or the editors give a warranty, expressed or implied, with respect to the material contained herein or for any errors or omissions that may have been made. The publisher remains neutral with regard to jurisdictional claims in published maps and institutional affiliations.

This Springer imprint is published by the registered company Springer Nature Singapore Pte Ltd.
The registered company address is: 152 Beach Road, #21-01/04 Gateway East, Singapore 189721, Singapore

Foreword

Packaging is an essential part of getting goods of all types from their point of production to their point of use. Packaging is particularly important for perishable and sensitive products such as foods, which require protection from contamination by microorganisms that can pose serious health hazards, as well as benefitting greatly from packaging systems that reduce food waste by extending shelf life, protecting against physical damage, etc.

At the same time, there is great concern about the impacts of use of packaging on the overall sustainability of our environment and resources, especially when it comes to disposal of used packaging. Proper design of packaging systems can do a great deal to make these systems for efficient and more sustainable.

This book sets out to discuss the basic functions of packaging, especially as they pertain to foods. It describes the major types of packaging materials and how they are combined into packaging systems, examining the advantages and tradeoffs of various types of packaging designs, including flexible, semi-rigid and rigid packaging systems. Special requirements for fresh and processed foods are discussed, along with the design process and how to evaluate packaging performance.

It also offers a discussion of environmental issues as they pertain to packaging, with a special focus on sustainability and on life cycle assessment as a tool for evaluation of sustainability.

Proper design of packaging systems, based on thorough knowledge about the field, can have a significant impact on enhancing delivery of safe and healthy food to people around the world, while reducing adverse environmental impacts. This book seeks to contribute to that goal.

Susan E. Selke, Ph.D.
Professor Emeritus and former Director
School of Packaging
Michigan State University
Michigan, USA

Preface

Food is one of the basic and critical needs of human being for survival and is very much necessary for growth, physical ability and good health. The human health is completely dependent on the nutrition that is received through assimilation of food. It is the most important complex natured commodity which do require to avail the advantages of the dynamics of packaging and its relevant innovative techniques. A food package is to be designed to provide many functions like preservation, protection, presentation, communication, logistics, legislation, product-package compatibility and consumers satisfaction. In addition, growing consumers demand for convenience features, enhanced shelf life, safety and hygienic aspects, coupled with environmental regulation, are also needed to be considered. In order to cater the demand of modern consumers and to meet the market requirement, the selection of suitable packaging materials, its techniques and an effective design of packages for food products have really become a very challenging task.

In fact, food packaging is always considered as a complex and complicated subject as compared to the packaging of any other commodity. Because, food itself are available in different category like basic foods, fresh fruits and vegetables, spices and condiments, processed foods and interestingly, the critical factors responsible for spoilage of foods of different category varies from each other. Due to this, the requirement of packaging does also vary with different categories of foods. This has further created a great demand for different innovative packaging materials with adequate functional properties and techniques to enhance the shelf life of food products.

Over the years, a number of packaging materials, techniques and technologies have been developed to meet the market requirement and to satisfy the consumers demand. And, packaging will always remain an important tool for marketing and food distribution. The development of innovative technologies like aseptic packaging, retort packaging, MAP and CAP system, smart and intelligent packaging is capable to increase the acceptability of packaged food by the modern consumers, and thus, these are capable to change the entire marketing system by way of providing longer shelf life of food products, reduction of food wastage, RFID for identification, innovative techniques like ripening and spoilage indicator, etc. These techniques and technologies will also improve the market acceptability of food in packaged condition for domestic and export market.

At the same time, the demand of packaged foods has also increased substantially which has led to the increase in consumption of packaging materials resulting into the accumulation of post-consumers packaging wastes. The increase of packaging wastes has created a great concern to environmental pollution due to littering, and hence, collection, segregation and recycling of post-consumer packaging wastes are considered as the "need of the hour." The study of life cycle analysis (LCA) of packaging materials is also considered to be an important tool for framing the policies towards the management of post-consumer packaging wastes.

Considering all these important aspects, the co-authors have taken a collective initiative to gather all relevant information related to "food packaging" and to publish this book by Springer Nature Publisher with the objective of highlighting

1. the importance of packaging, concept of food packaging, the history of packaging, evolution of packaging.
2. the latest development of different packaging materials like flexible, semi-rigid and rigid packaging materials, manufacturing process, properties and their application.
3. the developments and latest trends in packaging techniques for fresh produces and processed food products.
4. the importance of package design and its impact on the marketing of packaged food products.
5. the environmental issues by emphasizing the concept of sustainability and life cycle analysis (LCA) of different packaging materials with case study.

All these aspects are covered in ten chapters which are mainly contributed by four co-authors: Prof. N. C. Saha, Prof. A. K. Ghosh, Dr. Meenakshi Garg and Dr. Sushmita Sadhudey. In addition, two invited authors, Mr. Deepak Manchanda and Mr. S. Chidamber who have got over 40 years rich experience in packaging materials and design, each of them has also contributed one chapter in their areas of expertise to meet the objectives of this book.

We are also indeed grateful to Mr. Vijay K Sood, Past Chairman, Indian Institute of Packaging, who has got over 4 decades of rich experience in food packaging as a part of his long associations with Nestle India Limited for the implementation of innovative packaging materials and systems for a variety of food products. His constant inspiration and guidance have motivated us to write this book.

It is believed that the knowledge shared by all the authors will fulfill the long-felt need of updated information to meet the requirement of "food packaging." It is hoped that this book would be reference guide and help the students of food technology and packaging courses. Moreover, this book would also help the industry to update their knowledge in packaging materials, techniques and technologies and environmental issues related to food packaging.

We are extremely thankful to Prof. Susan K. Selke, Professor Emeritus and Former Director, School of Packaging, Michigan State University, USA, for contributing her "Foreword" for setting the tune of the book.

We would also like to express our sincere thanks to Springer Nature for showing their keen interest to publish this book.

I am confident that this book would be a valuable asset to the students community, packaging industry, R&D institutions and all related organization in India and overseas who are directly or indirectly involved to packaging of food products.

Mumbai, Maharashtra, India
Prof. Dr. N. C. Saha
Chairman
Foundation for Innovative Packaging
and Sustainability

Contents

1 **Food Packaging: Concepts and Its Significance** 1
   N. C. Saha
   1.1 Introduction 1
   1.2 History of Packaging 6
   1.3 Concepts of Food Packaging 10
   1.4 Functions of Packaging 18
   1.5 Components of Packaging 20
   1.6 Packaging—Its Linkage to Labeling, Branding
       and Marketing 23
   1.7 Packaging in the Nineteenth Century 29
   1.8 Evolution of Modern Food Packaging 36
   References 44

2 **Flexible Packaging Material—Manufacturing Processes
   and Its Application** 47
   N. C. Saha, Anup K. Ghosh, Meenakshi Garg,
   and Susmita Dey Sadhu
   2.1 Introduction 47
   2.2 Paper 49
       2.2.1 Manufacturing 50
       2.2.2 Types of Paper 51
   2.3 Plastic Films and Laminates 54
       2.3.1 Materials 55
       2.3.2 Polyethylene 55
       2.3.3 Low-Density Polyethylene (LDPE) 58
       2.3.4 High-Density Polyethylene (HDPE) 58
       2.3.5 Linear Low-Density Polyethylene (LLDPE) 59
       2.3.6 Polypropylene (PP) 59
       2.3.7 Polyethylene Terephthalate (PET) 60
       2.3.8 Nylon (PA6, PA66, PA11) 61
       2.3.9 Ethylene Vinyl Alcohol (EVOH) 62

		2.3.10	Cellulose-Based Materials	63
		2.3.11	Manufacturing Process	64
	2.4	Aluminum Foil		71
		2.4.1	Foil is Made by Two Methods	71
		2.4.2	Properties	72
	2.5	Plastic Woven Sack		73
		2.5.1	Materials	74
		2.5.2	Properties	74
		2.5.3	Manufacturing	75
		2.5.4	Applications	76
	2.6	Jute—As Packaging Medium		78
		2.6.1	Use of Jute Fibers	78
		2.6.2	Manufacturing Process of Jute Fabric	79
		2.6.3	Terminology Related to Jute Fabrics	80
		2.6.4	Benefits of Jute Packaging	82
		2.6.5	Limitation of Jute Fabrics	83
	2.7	Leno Bag		84
		2.7.1	Advantages of Leno Bags	84
		2.7.2	The Process of Manufacturing	84
		2.7.3	Application	86
	References			86
3	**Semi-rigid Materials—Manufacturing Processes and Its Application**			89
	N. C. Saha, Anup K. Ghosh, Meenakshi Garg, and Susmita Dey Sadhu			
	3.1	Introduction		89
	3.2	Paperboard		89
		3.2.1	White Board	90
		3.2.2	Solid Board	90
		3.2.3	Chipboard	91
		3.2.4	Fiber Board	91
		3.2.5	Ovenable Board	91
	3.3	Folding Carton		91
		3.3.1	Properties of Folding Carton	92
		3.3.2	Application of Folding Carton	92
	3.4	Lined Carton		92
		3.4.1	There Are Different Types of Lined Board	93
		3.4.2	Properties of Lined Carton	93
	3.5	Paper-Based Multilayer Composite Carton		94
		3.5.1	Advantages	94
		3.5.2	Materials Used	94
		3.5.3	Types of Composite Cartons	96
	3.6	Bag in Box		96
		3.6.1	Advantages	96

		3.6.2	Applications	97
3.7	Composite Containers			97
		3.7.1	Composite Containers Are Made from Two Methods	98
		3.7.2	Advantages	98
		3.7.3	Properties	98
		3.7.4	Application	98
3.8	Multilayer Squeezable Tube			99
		3.8.1	Manufacturing Process	100
		3.8.2	Types of Tube	101
		3.8.3	Applications	104
3.9	Thermoformed Container			104
3.10	Thermoforming Process			105
		3.10.1	Advantages of Thermoforming	105
		3.10.2	Disadvantages of Thermoforming	106
		3.10.3	Material Selection	106
		3.10.4	Process Factors	107
		3.10.5	Quality Control	107
References				111

4 Rigid Packaging Materials-Manufacturing Processes and Its Application ... 113
N. C. Saha, Anup K. Ghosh, Meenakshi Garg, and Susmita Dey Sadhu

4.1	Introduction			113
4.2	Metal Containers			113
		4.2.1	Tinplate	114
		4.2.2	Coating/Lacquering	115
		4.2.3	Manufacturing of Three-Piece Cans	116
		4.2.4	Aluminum	118
		4.2.5	Two-Piece Can Manufacturing Process	118
		4.2.6	Properties of Metals Which Make Them Suitable for Food Packaging	118
		4.2.7	Applications	120
4.3	Glass Container			121
		4.3.1	Manufacturing of Glass	122
		4.3.2	Types of Coatings on Glass	123
		4.3.3	Properties of Glass	123
		4.3.4	Applications	125
4.4	Plastics Containers			126
		4.4.1	Production of Rigid Plastic Containers	126
		4.4.2	Thin-Wall Injection Molding	133
		4.4.3	Blow Molding	133
		4.4.4	In Extrusion Blow Molding (EBM)	137
4.5	Plastic Crates			139

		4.5.1	Criteria for the Selection of Crates	140
		4.5.2	Materials Used for the Manufacturing	140
		4.5.3	Manufacturing Process	141
		4.5.4	Type of Plastic Crates	141
		4.5.5	Disadvantages of Plastic Crates	145
		4.5.6	Applications	145
	4.6	Wooden Crates		145
		4.6.1	Material Used	145
		4.6.2	Design of Wooden Crates	148
		4.6.3	Advantages of Wooden Crates Are	150
		4.6.4	Disadvantages of Wooden Crates Are	151
	4.7	Corrugated Fiber Board Boxes		151
		4.7.1	Components of Corrugated Fiber Board	153
		4.7.2	Manufacturing Process of Corrugated Fiber Board Box	155
		4.7.3	Design of Corrugated Fiber Board Boxes	157
		4.7.4	Advantages of Corrugated Fiber Board Boxes	158
		4.7.5	Applications	162
	References			162
5	**Packaging Techniques for Fresh Fruits and Vegetables**			165
	N. C. Saha, Meenakshi Garg, and Susmita Dey Sadhu			
	5.1	Introduction		165
	5.2	Breathable Films		167
		5.2.1	Manufacturing Process	168
		5.2.2	Types of Breathable Films	169
	5.3	Molded Pulp Trays		170
		5.3.1	Manufacturing Process	170
		5.3.2	Properties of Molded Pulp Packaging	171
		5.3.3	Applications	171
	5.4	Expanded Polystyrene (EPS) Containers		172
		5.4.1	Manufacturing Process	172
		5.4.2	Properties	173
		5.4.3	Applications of Expanded Polystyrene	174
	5.5	Modified Atmosphere Packaging (MAP)		174
		5.5.1	Role of Gases in MAP	175
		5.5.2	Classification of MAP	175
		5.5.3	Principle of MAP Technique	176
		5.5.4	Techniques of MAP	177
		5.5.5	Advantages of MAP Technique	178
		5.5.6	Limitation MAP Technique	179
		5.5.7	Application of MAP	179
		5.5.8	Packaging Material Used for MAP	180
		5.5.9	Application of MAP Technique	180
	5.6	Controlled Atmosphere Packaging (CAP)		181

		5.6.1	Advantages of CA Technique	181
		5.6.2	Disadvantages of CA Technique	182
		5.6.3	Application	182
	5.7	Active Packaging		183
		5.7.1	Principles of Active Packaging	184
		5.7.2	Active Packaging Principles and Its Application to Perishables	184
		5.7.3	Different Types of Active Agents	185
	5.8	Intelligent Packaging		187
		5.8.1	Quality or Freshness Indicators	187
		5.8.2	Time–Temperature Indicators	187
		5.8.3	Self-adhesive Labels	188
		5.8.4	RFID	188
	5.9	Antimicrobial Packing		188
		5.9.1	Microbial Agents Are of Different Types	189
		5.9.2	There Are Different Methods of Incorporating Antimicrobial Agents into the Film	189
		5.9.3	Properties of Antimicrobial Packing	189
		5.9.4	Antimicrobial Package Material Can Be Classified into Two Types	190
	References			190
6	**Packaging Techniques for Processed Food Products**			193
	N. C. Saha and Meenakshi Garg			
	6.1	Introduction		193
	6.2	Canning		195
		6.2.1	Methods of Canning	196
		6.2.2	Material Used for Canning	196
		6.2.3	Application	197
	6.3	Aseptic Packaging		198
		6.3.1	Chemical Treatments	200
		6.3.2	Packaging Materials Used	201
		6.3.3	Main Features of Aseptic System	202
		6.3.4	Advantages of Aseptic Packages of Food Products	203
		6.3.5	Application	203
	6.4	Retort Pouch		204
		6.4.1	Characteristics of Materials Used for Retort Pouch	204
		6.4.2	Advantages of Retort Packaging Technique	205
		6.4.3	Disadvantages of Retort Packaging Technique	206
		6.4.4	Application	206
	6.5	Ultra-High-Temperature Processing (UHT)		207
		6.5.1	Advantages of UHT Technique	209
		6.5.2	Disadvantages of UHT Technique	209
		6.5.3	Application	209
	6.6	High-pressure Processing		210

		6.6.1	HPP Systems Are of Two Types	210
		6.6.2	Foods Suitable for HPP	210
	6.7	Individual Quick Freezing (IQF)		212
		6.7.1	Benefits	212
		6.7.2	Methods	213
	6.8	Freeze-Drying		214
		6.8.1	Packaging Requirement	216
		6.8.2	Advantages of Freeze-Drying Technique	216
		6.8.3	Disadvantages of Freeze-Drying Technique	216
		6.8.4	Application	217
	6.9	Microwave Heating		217
		6.9.1	Packaging Materials Used for Microwave Heating	218
		6.9.2	Advantages of Microwave Oven	218
		6.9.3	Disadvantages of Microwave Oven	219
		6.9.4	Applications	219
	6.10	Irradiation		220
		6.10.1	Packaging Material	220
		6.10.2	Advantages of Irradiation Technique	220
		6.10.3	Disadvantages of Food Irradiation	221
		6.10.4	Application	221
	6.11	Pulse Electric Field		222
		6.11.1	Advantages	222
		6.11.2	Disadvantages	222
		6.11.3	Applications	223
	6.12	UV Sterilization		223
		6.12.1	Advantages	223
		6.12.2	Disadvantages	223
	References			224
7	**Food Packaging Design—Its Concept and Application**			227
	Deepak Manchanda			
	7.1	Introduction: Significance of Design for Packaged Food		227
		7.1.1	First Moment of Truth	228
	7.2	Role of Package Design as a Marketing Tool		229
		7.2.1	Disruption and Differentiation	230
		7.2.2	Convenience and Desirability	230
		7.2.3	Communication and Branding	230
		7.2.4	Brand Integrity and Consumer Safety	230
		7.2.5	Process Feasibility	231
		7.2.6	Supply Chain and Logistics	231
		7.2.7	Commercial Viability	231
		7.2.8	Sustainability and CSR	231
		7.2.9	Special Needs	232
	7.3	Branding and Integrated Brand Identity		232
	7.4	Design and Sensation Transference		234

		7.4.1	Packaging is the Future of Finishing	235
		7.4.2	Quality and Perception—Neck Finish	235
	7.5	Pack Structure Design		236
	7.6	Package Structure Design Process		237
	7.7	Product and Pack Compatibility for Design Decisions		238
	7.8	Graphic Design Significance and Process		239
		7.8.1	Graphic Design Process	239
		7.8.2	Visual Hierarchy	240
	7.9	Evolution of Print Processes		242
	7.10	Types of Food and Typical Pack Formats		245
	7.11	Design Guidelines		247
	7.12	Statutory Requirements		249
	7.13	Brand Launch, Pilot Production and Pre-testing		253
		7.13.1	Finish or Finesse	254
	7.14	Consumer Connect and Digitalization		254
	7.15	Challenge of Online Retail		255
	7.16	Supply Chain and Secondary Packaging		255
	7.17	Visualizing the Future		256
	7.18	Examples of Modern Packaging Design		257
	References/Suggested Reading			259
8	**Testing and Quality Evaluation of Packaging Materials and Packages**			**261**
	N. C. Saha			
	8.1	Introduction		261
	8.2	Paper and Paper Board		262
		8.2.1	Physical Tests	262
		8.2.2	Mechanical Tests	264
		8.2.3	Chemical Tests	268
		8.2.4	Optical Tests	270
	8.3	Corrugated Fibre Board		271
	8.4	Corrugated Fibre Board Box		273
		8.4.1	Box Compression Strength (BCT)	273
		8.4.2	Drop Test	274
		8.4.3	Vibration Test	275
		8.4.4	Stack Load Test	275
		8.4.5	Inclined Impact Test	275
		8.4.6	Rolling Test	276
		8.4.7	Water Spray Test	276
	8.5	Plastic Films and Laminates		277
		8.5.1	Physical Tests	277
		8.5.2	Mechanical Tests	280
		8.5.3	Optical Tests	285
		8.5.4	Phsico-chemical Properties	285
	8.6	Plastic Containers		286

	8.7	Metal Containers	289
	8.7.1	Physical Properties	289
	8.7.2	Mechanical Properties	290
	8.7.3	Chemical Properties	291
	8.7.4	Testing of Tinplate Containers or Cans	292
	8.8	Glass Containers	294
	8.9	Composite Containers	298
	8.9.1	Weight of Empty Container	299
	8.9.2	Adhesive Bonding	299
	8.9.3	Body Wall Thickness	299
	8.9.4	Drop Test	299
	8.10	Plastic Sqeezable Tube	300
	8.10.1	Physical Tests	300
	8.10.2	Mechanical Tests	301
	8.10.3	Physico-chemical Test	302
	8.11	Plastic Pouches	302
	8.11.1	Plastic Pouch Compression Test	303
	8.11.2	Static Compression Test	303
	8.11.3	Pouch Burst Test	304
	8.11.4	Vacuum Leakage Test (ASTM F2338-09(2013))	304
	8.11.5	Drop Test of Pouches	305
	References		305

9 Sustainable and Green Packaging-Environmental Issues 309
S. Chidambar

9.1	Introduction	309
9.2	Defining Sustainability	309
9.3	Origins of Sustainability	310
9.4	Addressing the Concerns	311
9.5	Some Guidelines	313
9.6	Hierarchy of Options	314
9.7	Evaluating Sustainability	315
9.8	Debunking Some Myths	316
9.9	Biodegradability	317
9.10	Some New Concepts on Resource Utilisation	318
9.11	Linear Economies	319
9.12	Circular Economies	320
9.13	Thinking Out of the Box	321
9.14	Greenwashing	322
9.15	Statutory Mandates	322
9.16	Final Recommendations	323
References		324

10	**Life Cycle Analysis of Packaging Material**		325
	Anup K. Ghosh and Susmita Dey Sadhu		
	10.1	Life Cycle Analysis (LCA)	325
	10.2	Definition	327
	10.3	Methodology	327
	10.4	Life Cycle Impact Assessment	329
	10.5	Carbon Footprint	331
	10.6	Significance of LCA Study in Packaging Material	331
		10.6.1 Plastic Pouches Versus Glass Bottles in Milk Packaging	333
		10.6.2 Plastics Versus Tin for Oil Packaging	335
		10.6.3 PP-HDPE Woven Sacs Versus Jute/Paper Sacks	335
	10.7	Conclusion	338
	References		340

Appendix: Relevant Indian Standards of Packaging Materials for Food Products ... 341

Index ... 349

Authors and Contributors

About the Authors

Prof. N. C. Saha is an internationally acclaimed Professor with over 34 years of rich experience in the field of Packaging Science and Technology. He was the former Director of the Indian Institute of Packaging, Government of India. He obtained Master's in Food Technology from Central Food Technological Research Institute, Mysore, and Ph.D. in Packaging Management from North Maharashtra University, India. He is an alumni of School of Packaging, Michigan State University, USA, as UNDP fellow and University of Tehran for Advanced Program in Productivity Concepts & Tools as a fellow of Asian Productivity Organization. He has published two books, two chapters in packaging, 15 research papers, over 450 articles in packaging and acquired one patent in "Packaging Container made of Plastic laminated Tube for liquid jaggery" to enhance shelf life for 180 days. Currently, he is Technical Advisor (Packaging), India Biomass Project initiated by the Government of Netherlands; Chairman, five Sectional Committees—Bureau of Indian Standards (BIS), Government of India; Founder Chairman, Foundation for Innovative Packaging and Sustainability; Visiting Professor, Indian Institute of Technology Delhi, among others. His research interest is in the areas of Food Packaging.

Prof. Anup K. Ghosh is Distinguished and Renowned Professor in the field of Polymer Science & Engineering at Indian Institute of Technology (IIT) Delhi, India, and Fellow of National Academy of Sciences, India. Professor Ghosh has over 30 years of research and teaching experience and has 12 patents to his credit. He is known for his accomplishments in the field of Polymer Processing and Rheology, having significantly contributed in the areas of Reactive Processing of Polymer Blends and Alloys, 3D Printing of Polymers, Polymer Packaging and Microcellular Processing of Polymeric Materials. Professor Ghosh obtained his M.Tech. degree in Chemical Engineering from the Indian Institute of Technology, Kanpur, India (1982), and a Ph.D. in Chemical Engineering from the State University of New York at Buffalo, NY, USA (1986). He has supervised 30 Ph.D. thesis and over 100 M.Tech. thesis. or his innovative development of polymeric orthotic knee joints for loco motor disabled people, with successful field trials on 1000 patients, he was conferred the National Award (2015) by the Government of India.

Dr. Meenakshi Garg is currently a faculty member at the Department of Food Technology, Bhaskaracharya College of Applied Sciences, University of Delhi, India. She has 20 years of experience in research and academics. Dr. Garg holds M.Sc. and Ph.D. degree from Chaudhary Charan Singh Haryana Agriculture University, India. Her research interest includes food processing, food packaging and food science and nutrition. Her research findings have appeared in many international and national peer-reviewed journals, published five books chapters and filed two patents. She is a Life Member of the Association of Food Scientists and Technologists (India), Indian Dietetic Association, Green Chemistry and Asian Polymer Association.

Dr. Susmita Dey Sadhu is currently a faculty member in the Department of Polymer Science, Bhaskaracharaya College of Applied Sciences, University of Delhi, India. She completed her Bachelor's and Master's degree in Chemistry from Burdwan University, India and completed her Ph.D. in Rubber Technology from Indian Institute of Technology Kharagpur, India. She is also actively engaged in research activities in the areas of polymer nanotechnology, polymers in packaging and polymer composites. Dr. Sadhu has several publications in reputed international and national journals. Her interest includes polymer composites, packaging and polymer characterization.

Contributors

Deepak Manchanda is a Formerly Head of Packaging at reputed consumer brand companies—Oriflame-Silver Oak; Dabur and Ranbaxy. Initially he worked with Metal Box India, later with integrated brand development consultancy, Autumn Design and Firstouch Solutions. He was a consultant for Jindal Polymers and Michelman. He was closely associated with Indian Institute of Packaging (IIP), Packaging South Asia magazine and industry conferences. Presently, he is Chief Consultant (Packaging Design) at Foundation for Innovative Packaging and Sustainability (FIPS), a Non-Govt, Non Profit Organization, registered under Section 8(1) of the Companies Act, 2013 in India.

Subramoni Chidambar is one of India's leading consultants to the Packaging/Plastics/Printing industries, has over 53 years' experience including top-level assignments with several leading companies. Pioneered the development of many hi-tech products/applications over five decades. Founding editor of Packaging South Asia. Papers presented at several international seminars and writings widely published across the world. Visiting faculty since 1970 at IIP/UNDP/Universities/Management Institutes

Chapter 1
Food Packaging: Concepts and Its Significance

N. C. Saha

1.1 Introduction

Indian economy is based on agriculture where two-thirds of Indian work force still earn their livelihood directly or indirectly from agriculture. Indian agriculture sector accounts for 18% of India's gross domestic product (GDP) and provides employment to 50% of the countries workforce. India is blessed with diversified agro-climatic condition which is led to produce a large variety of agricultural crops like rice. wheat, maize, millet, pulses, cash crops like sugarcane, tobacco, cotton, jute, oil seeds, plantation crops like tea, coffee, spices, coconut, rubber and a wide variety of horticultural crops like fruits and vegetables [1]. Moreover, the constant development of agricultural science has enabled to make the country to become world's largest producer of pulses, rice, wheat, spices and spice products which is required to feed the humans of second largest populated country of the world. The large production of commodity has also made it possible on the availability of huge quantity of produce. This has further created an opportunity for the processing of food products which can be stored for longer duration for the consumption by the human being during off-season.

Packaging of food products has become one of the most exciting areas in the world of retailing, multidisciplinary and multi-industry.

In fact, the fastest growth of Indian economy is largely effected by the socioeconomic changes like globalization, urbanization, liberalization of trade policies, rise in middle-class population, increase in literacy, growth in number of working women, rise in consumerism, increase in number of nuclear families with dual income, change in lifestyle and booming of retail market. A number of mega trends resulting into the changes into socioeconomic scenario of the country are as follows [2]:

(a) **Globalization**

- World would become a "Single Global Village."
- Reduction of trade barrier.

(b) **Liberalization**

- Encourages to foreign investors
- Booming of international trade
- Easy availability of foreign goods
- Emphasis to global competition.

(c) **Urbanization**

Conversion of more rural areas into urban society as the urban/middle class income group represents an enormous potential for innovative and value added goods. Almost 20% of rural control 2/3 of all assets in rural India—almost similar consumption like urban counterparts. And thus, the urban lower-income groups are a major source of business for daily consumption items which creates an opportunity area for "value of money" brands. Due to this, the rural India is the "CHALLENGE OF DISTRIBUTION—THE MARKET FOR FUTURE."

(d) **Increase in Middle-Class Population**

According to recent study by McKinsey report reflected that there would be dramatic growth in 10 folds of Indian middle-class population and the household income will be jumped into triple fold, and thus, the purchasing power of middle-class consumer will be quadruple. A large number of middle class population and increase of higher purchasing power.

(e) **Growth of Literacy**

Leads to increase in human health consciousness and safety which has resulted into great demand of branded packaged food products.

(f) **Growth in Number of Working Women**

- Causes to lack of time at kitchen
- More demand for ready-to-eat foods
- Frequent dining in outside restaurant.

(g) **Rise in Consumerism**

- Change in consumer attitude
- More demands of goods in smaller packs
- More awareness about packaged goods to get right quantity and quality.

(h) **Increase in Youth Earning Group**

- More younger earning population due to the opening of "Call Centers"
- Increase in brand consciousness
- More acceptances for new concepts
- More aspiration for change
- Surplus income with younger generation
- Tendency to foreign visits.

(i) **More nuclear families with dual income**

- Seventy-five million household called "consuming classes" with annual income between USD 2500 and USD 20,000
- One million household in creamy layer with high income and get easier financing options.

(j) **Change in food habits due to surplus income among the middle-class population leads to**

- More incidence of eating away from home
- Emergence of more Italian and Chinese restaurants
- Increase in purchasing of packed foods.

(k) **Booming of retail market causes to**

- Opening of shopping malls and supermarket
- Entrance of global players into retail market
- Availability of goods in consumer package
- Shopping becomes easier and customer friendly.

(l) **Change in Lifestyle**

- Global lifestyle due to media exposure and overseas travel
- Demand on convenience in life
- Look at ME (LAM) becomes the most important criteria.

Considering the recent socioeconomic changes and the requirements of consumers, the marketing trend has also been changed in a big way. In the twenty-first century, "marketing" has become an integral part of human life. The marketing process is applicable not only to selling of any commodities but also to various kinds of services like usage of mobile communication system, credit card facility, e-commerce, e-banking, e-insurance, etc., which are availed by the modern citizen. Even the hitherto considered the noblest professions of education, medical care and legal services are now offered in highly attractive and presentable forms. The documents need high-quality presentation packaging. The stationery is no more functional wrapping; it is a high-design packaging now [3].

Over a period of time, the concept of packaging has undergone an enormous change. It is no more restricted to only the containment and carrier of goods from one place to another; rather, it serves many purposes like preservation, protection, presentation, communication, marketing and safety of food products to be packed. In fact, it is no more considered as a common sense subject; rather, it is now well established as a combination of science, art and technology. The subject has become more complex in nature in order to satisfy the requirements of modern consumers.

In the retail world, food packaging is considered to be the most exciting areas due to the availability of variety of foods in the market which require multidisciplinary and involved to multi-industry. In fact, this industry has crossed all kinds of boundaries. In general, packaging for food products needs to provide three basic functions, i.e., preservation, protection and presentation. But, over a period of time, the role of packaging has become more complex and complicated in nature due to fast-changing scenario of the modern society. In fact, packaging is a multidisciplinary area which requires the knowledge of physics, chemistry, mathematics, chemical engineering, mechanical engineering, social science, economics and material science [4].

In addition, constant growth and development of food technology has also created a great demand for the requirement of innovative packaging materials as well as technology which would meet the requirement to enhance the shelf life of fresh produces and processed food products.

A number of innovative packaging techniques and technologies, namely canning, retort packaging, aseptic packaging, vacuum packaging, vacuum and gas flushing, modified atmosphere packaging (map), controlled atmosphere packaging (CAP), active packaging, intelligent packaging, etc., are developed to enhance the shelf life of food products. In addition, the presentation aspect of packaging does also require innovative package; structural design in terms of suitable packaging materials, shape, size, capacity and packaging formats with adequate functional properties and graphic design with attractive graphics and printing to enhance the eye appeal and also to make the package "standout" on the shelves. Due to this, printing technology and design are also covered under packaging. Apart from science, engineering and technological aspects, management and logistics are considered to be very important to have proper inventory control of packaging materials in the store, distribution network and efficient supply chain management.

Packaging is not merely wrapping the product by means of any materials; rather, it is a very dynamic natured, complex and complicated system with more scientific approaches. It is normally defined as a coordinated system of production of goods with all safety aspects, storage, distribution, logistics and effective supply chain and delivery to the end user [5].

It is not involved to consumption of packaging materials and its disposal; rather, it is responsible for recovery, reuse or recycle of the materials in compliance to environmental regulation so that there will not be any burden to the society due to the accumulation of post-consumer packaging wastes and thus create unhealthy environment. In fact, packaging is responsible to maximize the sales value and thus profit. As the consumer would first touch and feel the package, it is termed as **"First moment of truth"** to the consumer. It is an essential tool for product identification, consumer satisfaction and business promotion. In fact, it is also considered to be the heart and soul of merchandising. Packaging also play an important role to formulate the business strategy for the growth and development of business as well as its performance. In fact, packaging becomes an integral part for any business decision as it provides the important functions like interaction with the customer by means of statutory and promotional labeling for information, marking for identification, attractive graphics, printing, decoration to enhance eye appeal to the consumer for maximizing the sales. Considering the complexity of modern consumer behavior, the future packaging has to be more innovative with less resources, maximum level of functional properties, hygienic and safe, high degree of performance with optimum cost to satisfy the consumer. In short, packaging would be an important component in the entire value chain [6].

According to Smithers PIRA on The Future of Global packaging to 2018, the total value of packaging market in India was USD 27.7 billion in 2018 with an annual growth rate of 10.7% and it was expected that the same would continue till 2021. In India, there are a number of megatrends which lead to the increase in demand of packaging. The trends are globalization, rapid growth of urbanization, growing middle class population, change in lifestyle, convenience, lack of time for cooking leads to increase in demand of packaged foods, demography, safety and sustainability. The largest category of India's flexible plastic packaging market is polyethylene film with 75% volume share. In 2012, its sales recorded growth of 2.5%, reaching almost 2 million tons. In 2019, the growth rate was very limited due to the governments notification on the banning of single-use plastic-based sachet and pouches, but ultimately, its consumption has grown in other area like agriculture. In case of rigid plastics, the total demand was 1.1 million tons in 2013 which has grown to 2.3 million tons in 2018; mainly, the consumption was for polypropylene (PP) and polyethylene terephthalate (PET). Rigid plastics, mainly PET, have been growing for all the packaged consumer goods category including of packaged drinking water and beverages, whereas in few areas like home care, personal and beauty care product categories, the fastest growth of consumption is observed for high-density polyethylene (HDPE) bottles [7].

In addition, the demand of rigid plastics are growing due to certain important factors like easy availability, user-friendliness and adaptability. Indian market has also experienced about burgeoning demand for corrugated fiber board boxes over the last 5 years. The growth is mainly due to the increase of packaged consumer goods where corrugated fiber boxes are used as transport packaging materials. The growth of middle-class population in India is leading to higher demand for processed foods and beverages, and in turn, packaging materials is one of the primary growth drivers for the domestic metal packaging industry. The per capita consumption of metal packaging is very less in India as compared to global market. Moreover, the growth of consumption of metal packaging for processed foods and beverages is noticeable which is mainly due to fast-growing trend of middle-class population and growth in retail chain.

The Indian packaging industry is considered to be the fastest growing segment for consumer goods, mainly for processed food products. It is reported by the Federation of Indian Chambers of Commerce and Industry (FICCI) that the increase of per capita income, steady growth of urbanization and rise in the number of working women have resulted into the increase of expenditure for processed food products. The penetration of online retail market is less in India due to the strong presence of traditional grocery market in India. However, it is reported that Indian consumers do incur an expenses of almost 60% of the total expenses toward food and groceries. Moreover, it is also expected that food delivery and service industry will be growing fast in the years to come. Due to this, there would be an "increased demand" of "Take away food," but the greatest challenge would be to deliver the "ready-to-eat" food in packaged condition in hygienic, safe and free from all kinds of pathogen to ensure that packaged food products would not create any harmful affect to human health [8].

1.2 History of Packaging

It is very difficult to trace the precise beginning of packaging, but there are anthropological evidences that primitive humans used some crude forms of packaging to wrap up their animal and plant food to store in caves before settling as an agrarian society. During that time, they used to form social groups and traversed distances to cover geographical migrations in search of animals and the seasonal plant foods. The primitive humans normally used to search the food sources around and quite often experienced an extremely nomadic existence without property accumulation. Under these circumstances, they could carry only what was possible on their backs properly packed and secured. In ancient times, there was not necessary of packaging as the food used to be produced locally and then consumed. But as the primitive people require to contain the food and also to carry, a need was created for a device. This necessity has created a need for a container or a package. The availability of natural materials like tree leaves, bamboo, lotus leaves, palm leaves, gourds, Coconut Shells, shells and animal skin etc. has helped the primitive people to use these materials as Containers. However, a need was arisen to contain, protect and transport of food due to continuous growing of civilization where packaging became an essential even for primitive people [9].

In 5000 BC, there is an evidence that some plants and animals had been domesticated and the forage or hunt was still important. However, the availability of food supply in a given vicinity was reasonable. This evolutionary stage had resultant into the formation of the support larger social groups and small tribal villages. The villagers used to store and transport available foods in different types of containers made of natural resource materials. During that time, few villages with access quantity of different resources used to do trading with their neighbors to meet the demand of transport containers. It is believed that a number of packages like buckets and bags made from materials of plants or animal origin were considered as primitive packaging and hollow logs were replaced by wood boxes. The evolution of the development of different shaped containers made from river bank and then allowed to dry was also made available under the category of primitive packaging. It is assumed that few Neolithic genius might have tried to place the clay containers into fire and made available for wet product resulting into birth of pottery and ceramic wares [10].

There is also an evidence that in the Mediterranean Coast, salt blocks were used by Phoenician sailors to protect their fire from wind on the sandy and thus discovered a hard and inert substance which could remain in fire. In the modern day of Iraq, probably around 2500 B.C, the glass beads and figures were made in Mesopotamia and in about 1500 B.C, the earliest hollow glass objects appeared in Mesopotamia and Egypt. Further, a hot strands of glass were used to wrap around a core of clay and dung and then roll it against a smooth flattened surface to smother the strand

1.2 History of Packaging

lines, cooled the glass and the core was dug out to form the ancient Egyptian glass containers [11].

Historically, there have been only five different types of packaging materials, namely

- Cloth
- Paper
- Glass
- Metal
- Plastics.

These packaging materials were used by ancient people and even now, the same materials are also being used by modern society in the twenty-first century. However, it is experienced that even today, the application of Cloth is restricted to the packaging of seeds only. At the angle of historical perspective, these traditional five packaging materials really makes quite interesting [12].

- **Cloth**

Cloth and paper are made from cellulosic materials and considered to be the oldest forms of packaging materials. Both the materials are flexible in nature. The package made of flexible material require very less quantity to contain and hold the product as compared to either rigid or semi-rigid packaging formats. In fact, China is the first country who started to use flexible materials by using sheets of treated mulberry bark to wrap food as early as the first or second century B.C. [13].

- **Paper**

In USA, the first paper mill was built near Philadelphia in 1690. During that time, handmade paper used to make from parchment and rags, but these materials were having limited in supply and expensive. However, in the following centuries, the refund techniques for paper making process were developed by Chinese. These techniques of paper making process were slowly moved into other parts of the world, mainly Asia and Europe. The paper is defined as a thin-layer sheet of cellulose where cellulose fibers are normally derived from plant. In the earlier days, the Linen cloth was used to be made from cellulosic materials derived from flax. However, subsequently, old linen rags were also used as a source of fibers [14].

After almost 94 years, lithography technology was invented by Alois Senfelder in Munich which has enabled to manufacture printed label with black and white color. Subsequently, this particular technology became popular for the printing on the surface of glass bottles, metal boxes and paper board boxes. However, in 1837, the color printing or chromolithography was invented and became popular. Mr. Thomas Gilpin installed the first paper making cylinder machine, and thus, the paper making technology was spread over many countries. In 1850, paper making process was started by using wood pulp, and thus, the technology enabled to make the paper in flexible form. Mr. Francis Wolle had invented first paper bag making machine in 1852, and thus, paper bag become popular for packaging application. In fact, even in the twenty-first century, the technology of making paper bag has been modernized, and these bags are well accepted for packaging application.

- **Glass**

The glass is one of the oldest materials which is still popular even in twenty-first century for packaging application. Most importantly, the raw materials used for making glass are available in plenty as important materials are sand, lime stone and soda. The press-molded cups and bowels were first developed in 1200 B.C. In seventeenth-century, the technology was further evolved and blow glass techniques was made possible for making irregular shaped container. Later on, glass as packaging materials for number of products like medicine, alcoholic beverages, processed foods became popular. In nineteenth century [15].

- **Metal**

Metal is also another old material next to glass. This technique was invented in 1200 AD in Bohemia. The application of tin plate for making tin can and tin foil was found to be more economical option. However, the complete scenario in terms of popularity of metal container was changed by the innovation of canning process in early 1800 by Nicholas Appert who had sterilized the food by means of thermal processing techniques to increase the shelf life of processed foods for longer duration. Later on, the continued innovations in metal packaging has enabled to develop tin box for packaging of bakery products, confectionery and tobacco items. Precisely, tin box was developed in 1830 and later on, soft metal tubes were first introduced to the market in 1841. However, there was a constraint of tinplate in terms of rigidity until alloy was developed. This development has enabled to manufacture thinner gauge materials. Later on, coating techniques has made this tin plate as most versatile packaging materials in terms of high mechanical properties, thermal resistance, opaqueness, barrier properties against the ingress of moisture and oxygen gas and penetration of light. All these properties have become very useful to preserve the food products for longer duration [16].

- **Plastics**

In nineteenth century, plastics was invented during American Civil War, and these materials were mainly used for military and wartime use. In fact, this is considered to be most youngest material for packaging application. It is believed that Mr. John Wesley and Mr. Isaiah Smith Hyatt, two brothers from New York, did experiments for several years with "celluloid," and invented this material. Over a period of time, a number of research work across the world has been able to developed number of plastics materials in flexible as well as sheet by means of different plastic processing techniques. In this twenty-first century, plastics has become the most popular versatile packaging materials in terms of its numerous advantages and to develop different types, shaped, molded containers for packaging application [17].

Labels and Trademarks
The first labels and trademarks are found in the ancient wine flasks and pottery containers where the business cartels advised their *companies* fellow traders to label their seals on their merchandise when sailing in common ships. Now, it has become a common practice that the package containing of any goods where the labels are printed on the outer surface of the package to provide all information related to the product inside the package including ingredients, nutritional requirement and brand names and marking as symbols for easy identification of the goods.

The history of packaging indicates that the label was first used in England in 1660. During that period, a phrase about "**let the buyer beware**" became very popular. The labels were used to be affixed on the package to distinguish the inferior quality, impure products with superior one so that customers would not be cheated. Keeping with this objective, honest merchants who were unhappy with this kind of deception began to mark on the package with this phrase to make an identification and also to alert the potential customer. In 1866, the official trade mark was used for the first time by Smith Brothers for their cough drops which was ultimately marketed in glass jars. This new idea was well accepted by the customer as a "brand" during that period. In the same line, Eagle–Arwill chemical paint company was awarded the first US trademark in 1870.

Today, trade mark of any company is used to enhance the image of company, and it facilitates the customer for brand royalty [14].

The evidences stated that the concept packaging was realized by primitive people during the nomadic period itself. However, the development and application of different types of packaging materials are stated below in a chronological order in Table 1.1 [18].

It is quite interesting to learn about the journey of packaging over the last 1000 years. The main factor for the development was due to social changes which

Table 1.1 Packaging materials in ancient period

Period	Materials	Items
100,000 years ago	Natural material	Animal skins, horns, bamboo, large gourds, leaves, etc.
20,000 years ago	Natural materials (modified)	woven baskets, leather bottles
8000 years ago	Developed in the Middle East	Ceramics, amphorae, beakers, etc.
5000 years ago	Developed in Egyptian tombs	Wood, barrels, boxes, crates. Wooden boxes
3500 years ago	Potter's wheel was invented	Mass production of Ceramics
3000 years ago	Developed glass blowing technique	Glass Vessels
2000 years ago	Invention of true paper by Cai Lung, a Chinese in 105 AD	Paper and cellulose fibers

occurred over the last 1000 years. Moreover, all the development was occurred based on the necessity, and demand was arisen by the primitive people. In the seventeenth century, there was plagues and sicknesses which affected the poor disproportionately resulting into death and revolt of peasants, civil war. But in 1795, Nicholas Appert invented the canning system to preserve the food for longer duration and he earned 12,000 Francs from Napoleon for this invention. This invention has led for the setting up of canned food Factory by Brian Donkin at Dart ford.

In eighteenth and nineteenth century, the Industrial Revolution resulting into the growth of middle-class population has led to the development of packaging. In 1930, a wonderful plastics material like polyethylene (PE) was discovered. Later on, the Second World War created a great demand on the insulation material used for the magnetron used in radar and eventually used as a component for atomic bomb. In 1940, the polymer scientists across the world decided to explore the possibility for the invention of new polymeric materials for harnessing to the war, and subsequently, the versatile nature of this material was realized, and thus, the material was adapted. However, post Second World War, another polymeric material like polypropylene was invented. This material has gradually replaced other basic packaging materials like tin, aluminium, nickel, etc.

The development in packaging since 1000 year is given in chorological order and is listed in Table 1.2 [19].

Few photographs of packaging materials developed during seventeenth centuries are given in Figs. 1.1, 1.2, 1.3 and 1.4.

1.3 Concepts of Food Packaging

According to dictionary meaning, the word "Food" is defined as "any substances taken into and assimilated by a plant or animal to keep it alive and enable it to grow and repair tissue and promote nourishment." It is universally accepted fact that food is one of the basic and critical needs of human being for survival and is even listed before clothing and shelter. Food is very much necessary for survival, growth, physical ability and good health. In fact, the physical and mental functioning of human body is dependent on the nutrition it receives through assimilation of food [20]. The necessity of food has been known to humans from the beginning of times, and therefore, primitive mankind onwards the hunters major part of their time to search food to meet their necessity. Food packaging is considered to be most complicated subject as different types of food would require different of packaging materials, packaging system and machinery. In fact, one cannot think today's world without packaging from the point of production to the point of end user. Packaging provides the suitability for transportation of food products and protects them from physical and environmental hazards (Figs. 1.5, 1.6 and 1.7).

In general, packaging is considered to be a combination of science, art and technology. It is also termed as a container which is responsible to contain the product to preserve, store, handle, transport, sale and use. Packaging is also defined as an integral

1.3 Concepts of Food Packaging

Table 1.2 Types of packaging materials developed during seventeenth and eighteenth century

Year	Types of packaging materials developed
In 7000 B.C	Beginning of glassmaking started as an offshoot of pottery
In 1500 B.C	Glass making process was first industrialized in Egypt
In 1200 B.C	Development of glass cups and bowls by pressing into mold
In 300 B.C	Phoenicians invented blow-pipe and then into round containers
In 1035	In Cairo, paper was used for wrapping purpose
In 1040	Pi Cheng, Chinese invented movable type printing machine
In 1100	In Europe, paper was received for the first time from China
In 1250	To prevent armor against corrosion, tinplate was developed in Bohemia
In 1310	The paper making process was first introduced in London
In 1450	Guttenberg's printing press was developed
In 1630	"Grocer's paper bags" was first recorded
In 1690	The paper making technology was arrived to Germantown, Pennsylvania, USA
In 1700	Egham, UK-based glass factory produced 3 million bottles per year. This was the first commercial glass production by this company
In 1750	For general purpose, jute sack was used
In 1764	Selling snuff in metal canisters began by London tobacconists, as another type of "rigid packaging"
In 1788	Food items were first stored in wooden barrel as boxes, barrels and jute sacks were found to be heavy in weight and inconvenient to carry. Moreover, these packages were not found to be suitable for storage of food items as the items were easily contaminated by insects, mice and rats
In 1794	A new design of "lay down" bottle for mineral water was developed by Schweppes with an idea to keep the cork wet, and thus gas-tight
In 1795	Nicolas Appert was offered a prize of 12,000 Francs by Napoleon for the development of canning technique to preserve foods
In early 1800	In France, the importance of metal containers for safe preservation of food was realized by overcoming the misconception about its hazardous effect
In 1806	The machine for making paper in reel form was patented by Mr. Bryan Donkin
In 1807	A new material like aluminum was invented by Humphrey Davy
In 1809	The sterilization technique of food in sealed container was first developed by Nicholas Appert, a Parisian chef and confectioner, to preserve food for longer period of time
In 1810	The sealed cylindrical "can" was patented by Peter Durand of Britain
In 1817	After a gap of 200 years for the invention of paper by Chinese, England produced the first paperboard carton, often called a cardboard box
In 1820	At Dartford, Kent, the first food canning factory was set up
In 1825	Charlie Smith developed the first carbon dioxide-pressurized dispensing container and the devise of "Regency Portable Fountain" soda water
In 1825	The first tiny pellet of aluminum was produced by Danish scientist

(continued)

Table 1.2 (continued)

Year	Types of packaging materials developed
In 1831	The first distilled form of Styrene was produced from balsam tree
In 1835	Vinyl chloride was discovered for further development of rubber chemistry
In 1840	The canning of food was first experimented in Sydney
In 1841	The artist first used the collapsible, soft metal tubes for packaging of paints
In 1844	The paper bags were developed and first commercial paper bags manufactured in Bristol, England
In 1848	Tin cans were invented for packaging and transportation of condensed milk, butter and meat in safe condition
In 1850	Australia started the export of surplus food in tins to Europe. During that period, card board-based corrugated box were also appeared
In 1850	Decorative fancy paperboard boxes were in use
In 1852	The bag making machine was invented by Francis Wolle in the USA. Francis Wolle invented
In 1855	Ingot of aluminum was first commercially produced
In 1855	The Can Opener was designed by Mr. Robert Yeates in UK
In 1860	First synthetic plastic "parkesine" was produced by Parkes in UK
In 1860	Paperboard tubes described
In 1862	First fruit jam factory made tinned jam in South Australia
In 1866	Development of first hand blown glass bottles
In 1867	The process for deriving useful cellulose fiber from wood pulp was developed
In 1868	The tissue paper for wrapping was first produced by Samuel Ramsden, paper mill in Melbourne, Australia
In 1870	Glued paper sacks with gusseted design to produce different types of paper bags was first developed
In 1870	Robert Gair developed the folding cartons for first time in USA
In 1871	Single-faced Corrugated board was first developed by Albert Jones at New York, USA, and also got patented
In 1879	The paper bag making machine was patented by Margaret Knight in USA
In 1880	Kellogg, USA, used paperboard cartons on large scale for first time
In 1889	The automatic rotary bottle making machine was developed to enhance the process of glass containers, and the same was patented
In 1890	Collapsible metal tube was used for first time for packaging of tooth paste
In 1890	The first automatic rotary bottle making machine was invented by Michael Owens
In 1894	The first double-faced corrugated boxes was produced by Thompson and Norris to prevent scratches during transportation. This boxes played an important role for the transportation of goods
In 1900	All sizes and shapes of glass containers were produced for consumer products

Fig. 1.1 Wooden barrels. Bottle *Source* Brody AL, Marsh KS (eds) (1997) The Wiley Encyclopedia of Packaging, 2nd edn. John Wiley & Sons, New York

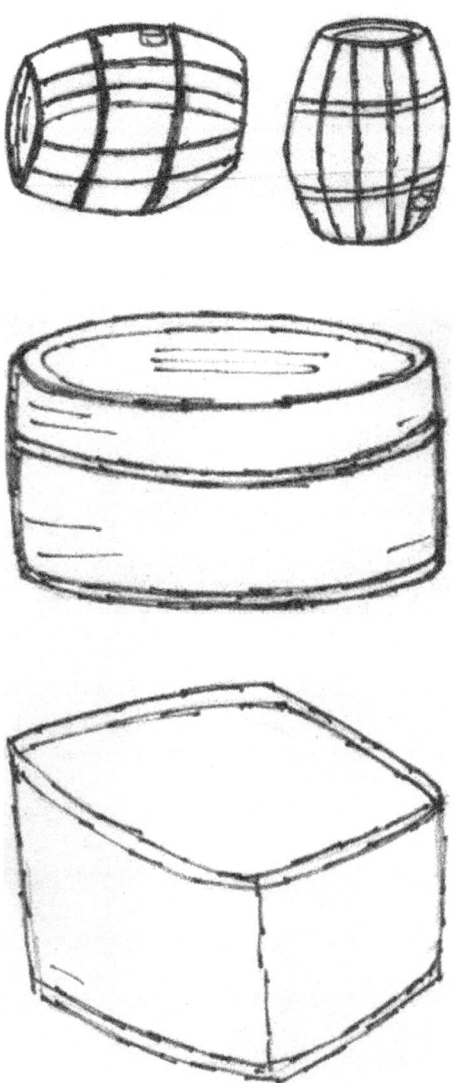

Fig. 1.2 Metal containers. Bottle *Source* Brody AL, Marsh KS (eds) (1997) The Wiley Encyclopedia of Packaging, 2nd edn. John Wiley & Sons, New York

Fig. 1.3 Tinned jam. Bottle *Source* Brody AL, Marsh KS (eds) (1997) The Wiley Encyclopedia of Packaging, 2nd edn. John Wiley & Sons, New York

part of production, distribution, logistics and supply chain system. It is considered as the last output of production and the first input for marketing. Packaging is always considered as the tool or instrument for marketing. Packaging is also termed as the coordinated system for preparing the goods for storage, handling, effective distribution and just-in-time delivery to the destination. In other words, packaging is called as "need-based technology." It is also believed that it is a complicated, complex, dynamic system to cater the demand of customer and meet the market requirement [21].

Fig. 1.4 Blown glass. Bottle *Source* Brody AL, Marsh KS (eds) (1997) The Wiley Encyclopedia of Packaging, 2nd edn. John Wiley & Sons, New York

Fig. 1.5 Hexagonal-shaped corrugated box. *Source* Modern Food Packaging Book, Indian Institute of Packaging

It is defined as "coordinated system of preparing goods for safe, efficient and cost-effective transport, distribution, storage, retailing, consumption and recovery, reuse or disposal combined with maximizing consumer value, sales and hence profit." In short, it is very clear that packaging is the driver for brand promotion and sales where package design plays an important role to attract the customer at the point of purchase (POP).

Traditional concept of packaging as a medium to meet the purpose of storage, protection and transportation of food products has undergone a big change now as more and more features are incorporated in a package to enhance its attraction to the consumers. Packaging has become a resource utility and facilitator in the overall distribution and marketing of food products.

Fig. 1.6 Plastic crates for fresh vegetables. *Source* Modern Food Packaging Book, Indian Institute of Packaging

Fig. 1.7 Metal container. *Source* Modern Food Packaging Book, Indian Institute of Packaging

Food packaging is also considered as the linkage between the food production and marketing of the same. It is termed as the last output of production and the first input of marketing. It is also termed as the "**first moment of truth**" to the consumer. Packaging is an essential tool for protection, distribution, marketing of food products and consumer satisfaction. It is the heart and soul of merchandising of fresh as well as processed foods [22].

Food packaging is not a common sense subject; rather, one has to understand about the nature of product and also the critical factors which are responsible for the spoilage of food during handling, storage and transportation. Food packaging is defined as "a system by which a food product would reach from the manufacturing point to the consumer's hand in safe and sound condition at an affordable cost with freshness and eye appeal at the consumption level [23].

In food packaging, the stiff market competition would always give rise to an opportunity for innovation but at the same time, there is a need to have proper regulation in order to overcome any kind of hindrance for the implementation of innovation to increase the economic growth and development. First, regulation that restricts firms' ability to compete effectively through their packaging may impede the competitive process. Second, unfettered firms that deliberately free-ride on and undermine the intellectual property investments of others can distort competition to the detriment of consumers. The first of these, regarding regulatory requirements on the nature of communications, can take the form of either mandating specific information or requiring the removal of information. Inappropriate regulatory action can detrimentally impact competition and confuse consumers.

Packaging in terms of both volume and value is a large and important business in the UK and across the world. It is one of the most prominent and visible sectors which reaches society in everyday life through different mediums. The nature of packaging is such that it is intertwined with all industries, both large and small. It cannot exist on its own, but only as an integral part of the food and drink, personal care, pharmaceuticals or chemicals industries, to name just a few.

Packaging is also necessary in a world dominated by marketing-based economics which mainly rely on it for the safe delivery of profitable products. The importance of properly designed packaging lies in the fact that it is necessary to protect the product from damage and deterioration. At the same time, it must also provide identification, attractive presentation and also meet the appropriate environmental criteria. It is not an activity in isolation from others but a combined socioeconomic responsibility consisting of preservation, protection, presentation, distribution, safety, being eco-friendly and also being economical. In addition, it plays a great role for the promotion of marketability of any goods including food products combining branding with advertising [24].

In other words, packaging is considered to be the tool or instrument for marketing. In the past, the functional role of food packaging to preserve and secure the product was taken for granted by the general community. This latest attribute is now a patent quality of the packaging. Food product security is now a major global issue for all companies involved to packaging supply chain.

1.3 Concepts of Food Packaging

Fig. 1.8 Modern retail store. *Source* Retail4growth.com

From a customer perspective, the value of food packaging is evaluated on its contributions as to

- The condition of food products on the receipt by the customer.
- The promotional and informative value of the package.
- The role of packaging in physical distribution at all times in the marketing channel.
- Environmental and ecological aspects of packaging.

The emergence of super bazar and malls has ushered in an era of modern self-service system. They have mass display and well-designed packages to attract attention. Through their silent sales talk, the sales turn over increases. Due to this, packaging itself is a device of sales promotion [25] (Fig. 1.8).

The packaging of any product is its most powerful spokesman in a store environment. It offers a level of interactivity with the consumer like no other form or media can match.

The packaging operation has now converted with more automation. Due to this, the productivity has grown in big way, and it can happen now very easily more quickly and efficiently than ever before. Packaging materials used in consumer package are generally lighter in weight, uses less materials and is easier to open, dispose from, reseal, store and dispense. Packaging is now an essential component in our daily life. It is fundamental to the way modern commerce is organized. Packaging is everywhere, and it surrounds every space of our daily life commencing from morning beverage and breakfast and moves through the use of toiletries, laptop/computer at office, till dinner and retiring to bed. Without packaging, materials handling would be messy, inefficient and costly and modern consumer marketing would be difficult. In short, packaging of food products need detailed understanding in terms of product package compatibility, quality of packaging materials and its safety aspects.

1.4 Functions of Packaging

Packaging has many roles beyond protection, preservation and presentation. Notably, packaging offers brand owners the possibility to communicate with consumers through distinctive designs and on-pack communication in the form of logos, graphics, images, colors, messages and product information. This represents an important medium for marketing communication and an important battleground for the intense rivalry evident in most FMCG markets where brands compete for the attention of consumers. Such competition is both immediate in nature, i.e., how existing packaged products compete with each other, as well as dynamic in the sense of the process by which new products enter the market and existing ones adapt and improve through innovation and new product development. Both aspects of competition are vitally important to a well-functioning market and for economic progress for the public good.

The role of packaging is critical to the commercial success of both consumer and industrial products in that it protects the product, provides information about the product and provides tamper-evidence for the product. Additionally, in the case of fast-moving consumer goods, it also markets the product and is typically seen by consumers as an intrinsic element of the product, not merely as a means of transport, protection and storage but as means of use and as a means of identifying the product.

Indeed, packaging plays a vital role for both consumers and producers in the functioning of the market economy. Packaging acts as an important market signal to reduce the information barriers between producers and consumers of the product, by providing different information (advertising, branding, reputation, trials, warranties and other signals to reduce asymmetric information). Moreover, the role of packaging in reducing information imbalance between producer and consumer becomes vital when the consumer cannot determine the attributes of a product prior to purchase, for example ice cream, ready-to-eat meals, washing powders and other FMCG products. This role of packaging allows efficient and effective transactions in business and competition.

The functions a package is expected to perform will depend on the end-use requirements and the requirement to suit the packaging line/system. End-use performance is important because defects and limitations will adversely affect customer satisfaction resulting in reduction in sales. End-use package functions include

- Display
- Ease of opening
- Convenience
- Dispensing, reuse and
- Quantity level.

There are two basic functions of packaging which are given below [26]:

(i) **Marketing**: Package promotes the product through attractive graphics and also attracts attention to the customers. Packaging also acts as a silent salesman.

1.4 Functions of Packaging

However, marketing functions can be elaborated further by considering the following six important activities like

- **Presentation**—the packages should have the attributes of presentation by way of attractive structural design in terms of proper selection of packaging materials, proper shape, size, capacity and functional properties as well as graphic design in terms of be attractive art work, font, letter size and printing to improve upon its visual appearance.
- **Communication**—The package provides all information by means of proper labeling and marking like ingredients, name and address of manufacturer, gross and net weight, date of manufacturing, best before, batch/lot no, MRP, vegetarian or non-vegetarian, etc.
- **Branding**—The package also promotes the brands of any goods which are packed within the package.
- **Promotional information**—The package would provide the promotional information like "Buy one get on free," gift coupon inside the package," etc.
- **Sustainability**—The package will provide the information by putting marking the logo of "Recyclability."
- **Identification**—The package would also facilitate for proper distribution or channel cooperation among distributor, whole sellers, etc.

(ii) *Logistics*: A means of ensuring safe delivery of goods to the ultimate destination in safe and sound condition at an affordable cost. However, logistics functions can be elaborated by considering nine important activities, which are as follows:

- **Containment**—Products must be contained in the package till reaches to the destination.
- **Protection**—To protect the contents of the package from damage or loss due to any kind of hazards like mechanical hazards, chemical hazards and biological hazards.
- **Preservation**—The package must preserve the products which are within the package for longer duration by considering the important properties of packaging materials like physical, mechanical, chemical and physicochemical properties.
- **Apportionment**—To reduce the output from industrial production to a manageable, desirable "consumer" size that is translating the large output of manufacturing into smaller qualities of greater use of customers.
- **Segregation**—The package would also enable to segregate the goods easily.
- **Unitization**—To permit primary packages to be unitized into secondary packages (e.g., by placing inside the corrugated fiber board boxes), the secondary packages are reinforced by means of straps and placed onto stretch wrapped pallet and ultimately into a container in order to reduce multiple handling.
- **Convenience**—To allow the products to be used conveniently by using convenient features into the packages.

- **Storage**—The package would also provide to store the goods in godown by properly utilizing the space like palletization and racking, etc. This would also depend on the structural design of package.
- **Disposal**—The package should be enabled to dispose of easily. However, this would depend on the selection of right type of packaging materials which would be in compliance to environmental regulation.

However, in any fast moving consumer goods (FMCG) company, the packaging functions are described by considering the following six elements:

- Protection
- Preservation
- Presentation
- Price
- Promotion
- Profit.

1.5 Components of Packaging

In general, packaging components is considered to be either assembled part of a package, manufactured in India or abroad, including of any interior or exterior blocking, bracing, cushioning, weather proofing, exterior strappings, coating, closures, labels, inks and varnishes, dyes and pigments, adhesives, stabilizers or any other additives. The packaging components are broadly classified into the following six categories [27].

1. Unit pack
2. Intermediate pack
3. Outer or shipping container
4. Inner packaging components
5. Closures
6. External reinforcements.

1. **Unit Pack**

Unit pack or package is basically the package which comes in contact with the product. In general, unit packages are normally used for one-time application. However, depending on the capacity in terms of volume or weight, the unit package can be for multiple use. This package should be able to provide all its necessary three important functions like protection, preservation and presentation. The unit package would provide protection to the goods against all kinds of mechanical hazards which might occur during handing, storage and transportation.

In addition, the unit package would also provide the preservation to the food products either to maintain its desired shelf life or it would also enhance shelf life by

1.5 Components of Packaging

preventing the package along with the product from environmental hazards, biological hazards and also physicochemical properties like barrier properties of the packaging materials which will be used to make the package or the individual package against the permeation of moisture content and oxygen gases from outside to inside the package to protect the product for any kind of chemical reaction like oxidative or hydrolytic rancidity for fatty foods in order to avoid the quality deterioration.

Quality deterioration can be due to a number of factors such as ingress of moisture, light, microorganisms, gases like oxygen, etc. Hence, the selection of packaging material for unit package is very critical and the package should have adequate functional properties to meet the basic requirements of any package. The unit package should also have additional features like security features, convenience features like easy opening and closing and reusability while it is used as retail package.

The unit package would also help to attract the modern consumers through attractive graphics and printing to enhance the eye appeal.

Some of the common types of unit packages are given below:

(i) Metal containers such as deep-drawn aluminum containers, tinplate containers and tin-free steel containers, etc.
(ii) Composite containers made from paperboard body with metal or plastic ends.
(iii) Glass containers—carboys, bottle, jar, ampoules, etc.
(iv) Plastic containers like bottles, jars, jerry cans, drums, barrels, etc.
(v) Plastic squeezable laminated or coextruded tube.
(vi) Aluminum collapsible tube.
(vii) Plastic-based flexible laminated pouch or sachets.
(viii) Paper-based multilayered laminated composite cartons made of paperboards, plastic and aluminum foils.
(ix) Plastic-based flexible single-layered blow-n-fill package for liquid medicine.
(x) Ready-to-serve package made from E-fluted corrugated carton.
(xi) Duplex board-based cartons for pastry, cakes, etc.
(xii) Cloth bag for seeds.

1. Intermediate pack

The main purpose of utilization of Intermediate packs are as follows:

(a) Unitization
(b) Marketing requirement
(c) Display value.

A number of packaging materials are converted into intermediate package to serve the purpose. However, the most important types of intermediate packages are given below:

(i) Box made of paperboards such as graybeard, duplex board, E-fluted corrugated board, etc.
(ii) Flexible pouches made of either single-layered plastics materials or different combinations of laminated structure.

- (iii) Shrink films.
- (iv) Stretch films.
- (v) Plastic woven fabric-based bag.

2. **Outer or shipping containers.**

The shipping containers are mainly used to perform the following functions:

- (a) To protect the contents from all kinds of hazards such as mechanical, chemical, biological and environmental hazards
- (b) To provide for easy handling, storing and transportation
- (c) To provide for easy identification
- (d) To facilitate for easy segregation

There are number of packaging options which are used as shipping containers. The important types of shipping containers are as follows:

- (i) Timber/plywood cases/crates
- (ii) Wooden barrels
- (iii) Wooden box
- (iv) Plywood box
- (v) Steel and aluminum drums
- (vi) Plastic drums made either from HDPE or LDPE
- (vii) Plastic woven sack made either from high density polyethylene (HDPE) or polypropylene (PP)
- (viii) Flexible intermediate bulk containers (FIBC)
- (ix) Metal intermediate bulk containers (IBC)
- (x) Plastic intermediate bulk containers (IBC)
- (xi) Fiber board drums
- (xii) Solid fiber board box
- (xiii) Corrugated fiberboard and combination board boxes
- (xiv) Jute bag or sacks
- (xv) Multiwall paper sack
- (xvi) Flexible plastic-based single-layered or laminated sack.

3. **Inner Packaging Components**

Inner packaging components are used within the package (outer container or unit pack) for protection/barrier purposes. Inner packaging components can be in various forms and some of them are indicated below:

- (i) Positioning devices and supports
- (ii) Separators made of CFB or paper board
- (iii) Clearance blocks/pads
- (iv) Suspension devices
- (v) Cushioning materials like shredded paper, wood wools, bubble film, expanded polyethylene sheets (EPE), expanded polystyrene (EPS), CFB pad, separators, etc.

1.5 Components of Packaging

(vi) Barrier materials to protect from dust, oils, water vapors, etc.
(vii) Labels
(viii) Instruction booklets (ix) Paper-based angle board.

4. **Closures**

A closure is the weakest point in any package. Today, different types of caps and closures are used depending upon the type of package. Some of the material used for the purpose of caps and closures of packages are given below:

(i) Nails for wooden box
(ii) Metal-based crown caps, lug caps for glass bottles
(iii) Metal-based roll-on pilfer-proof caps (ROPP) for glass and plastic containers
(iv) Different types of plastic caps like flip top, ROPP for glass, plastic containers
(v) Different types of plastic caps for aluminum collapsible tube or plastic-based laminated tube
(vi) Self-adhesive pressure-sensitive plastic tapes for CFB boxes
(vii) Adhesives for closing CFB box
(viii) Metal closures for metal containers.

5. **External Reinforcements**

The main purpose of using external reinforcements is as follows:

(a) To improve stacking strength
(b) To avoid bursting in case of failure
(c) To increase weight carrying capacity
(d) To protect corners/edges.

In order to achieve the purpose, the following action should be taken.

(i) Providing girth-wise battens in wooden containers
(ii) Steel/wire strapping in girth direction for wooden containers
(iii) Providing corrugated hoops to the cylindrical drums
(iv) Using steel corners for wooden containers
(v) Nylon/polypropylene straps for corrugated fiber board boxes.

1.6 Packaging—Its Linkage to Labeling, Branding and Marketing

Packaging drives the customer to get attracted and helps them to make up their mind to buy the product. In other words, packaging of any consumer goods makes the **"first impression to the consumer"** [28]. The consumer package of food products need to perform many functions like contain the product, protect the protect and also the package from all kinds of mechanical hazards which might occur during transit, preserve the food product against chemical and biological hazards and thus, enhance the shelf life of packaged food. Moreover, the package will have attractive graphics and printing to enhance eye appeal to attract the consumer and thus to increase the sales turn over.

Packaging and labeling do provide more than protection and identification of the products. They play a vital role in developing the image and brand of the product within the target market. In fact, failing to pay attention to the design of packaging and labeling aspect which can decrease the visibility and attractiveness of the products leads to devastating for sales. In other words, branding of any products is promoted only through the labels which are normally printed on the surface of package, and thus, labeling becomes an integral part of packaging. Hence, packaging is an essential component for the identification of any products in marketing. Packaging also enhances the appearance of the label including brands for promoting the product. And labeling plays an important role in marketing. In addition, labeling also helps to provide all the information including statutory about a product to the prospective customer. In short, packaging, labeling and branding are interlinked to each other and all these are in combined form and act as facilitator for effective marketing.

What Is Label

In general, label can be defined as either a piece of paper, fabric or plastic film or may be of a tin plate where the information about the product inside the package will be printed. However, the constant growth and development of automation has enabled to make the label in altogether different manner. Nowadays, the label is generally printed either on the outer surface of a package made of paper board carton or it could be online reverse printed on plastic-based laminated film which would be converted into pouches.

The main purpose of putting a label on the package is to provide the information about the product packed inside the package. The information would be consisting of brand name, statutory information like ingredients, nutritional requirement, name and address of the manufacturer, or packer, date of manufacture, best before (for food products), expiry date (for pharmaceuticals) maximum retail price (MRP), etc. In addition, there could also be certain optional labels like "Buy one, GET one Free," Gift Coupon inside the package, etc. However, there are certain labels or claims like Sudh Ghee or Pure Ghee where the word "SUDH" or "PURE" is not permitted as per the prescribed norms mentioned in The Food Safety and Standards Authority of India (FSSAI) under the Ministry of Food and Consumer Affairs, Govt. of India. Similarly, the terms like "family pack" or "Jumbo pack" are not permitted rather the package need to declare the gross weight as well as net weight.

Labeling and Its Functions

In general, labeling is defined as a process in which all kinds of information including statutory are printed on the package in the form of a label in order to provide the information and also facilitate to easy identification of the product by the customer. The package could be either of consumer package or bulk package. In case of bulk or transport package, the label could be either printed on the surface or affix a sticker. However, labels are always printed on the surface of the consumer package. The main purpose of labeling is to provide detailed information about the product inside the package in terms of quantity and quality, instruction for the use of product and storage condition of products. In case of consumer package, the information would include name and address of the manufacturer or marketing Company or agents if the manufacturer and marketer are different, brand name, batch/lot number, date of manufacturing, best before (in case of food products) and expiry date (mainly for pharmaceutical products), gross weight/net weight, maximum retail price (MRP), ingredients and nutritional information (food products) and chemical names with dosage (pharmaceutical products). In addition, the label of consumer package would also carry the barcode and marking on the package for the purpose of identification. A brown circle is to indicate the presence of non-vegetarian ingredients in the food item, while a green circle in a white square would indicate as vegetarian food. In fact, these information on the package are the statutory requirement as per the Food Safety and Standards Authority of India (FSSAI), Ministry of Health and family Welfare, Government of India. In short, a proper labeling of any product would facilitate the customer or consumer to take a proper buying decision at "Point of Purchase" (POP).

Requirement of Labeling for Export Consignment

Labeling is also considered to be very important for export consignment of food products. In fact, every label of any bulk package for export should be having the following information:

- Information related to the legal requirement of any particular importing country
- Instructions about the storage condition of the food products
- Dimensions of the outer package
- Gross/net/tare weight of the bulk package
- The number of unit packages stored inside the bulk package and quantity of each package.
- Instructions for the use of the product
- Country of origin
- Name and address of the manufacturer
- Lot number of the consignment
- Date of manufacture and date of its expiry
- Description of the goods with brand name.

Functions of Labeling

The important functions of labeling are as follows:

- To provide easy identification of the products
- To promote the brands of any products
- To facilitate the consumer to know about date of manufacturing and date of expiry of food products
- To attract the customer by communicating the promotional scheme like "Buy one and get one free" and "Gift Coupon inside the package," etc.
- To convey the nutritional information of any food product

- To protect the customer for not buying any fake or duplicate products
- To allow the customer to take a correct buying decision about any product
- To identify the package through printed barcode on the labels
- To identify the brands of any packaging.

Types of Label

Depending upon the performance, the labels can be broadly classified into the following two categories [29]:

(a) Removal type for temporary label
(b) Non-removal type for permanent label.

(a) **Removal type for temporary label**: This type of labels is applicable for temporary use. The label may have adhesive that would provide adequate bond strength to keep them fasten with the surface till the labels are no longer needed or they may use "Static Cling" to keep them attached to the surface of the package.

In this case, the labels are made with removal adhesive with light adhesive bond strength which would be enough to hold a label in place while it is needed, but at the same time, the adhesive will allow to remove the label quickly and cleanly without leaving any residual adhesive on the surface.

(b) **Non-removal type for permanent label**: This type of labels is applicable for permanent use. The label will have permanent adhesive with strong adhesive strength so that the label will be attached to the surface firmly and it will also not allow to remove the label easily.

In fact, the adhesive bond strength and the performance of permanent label will depends on the following important factors:

- **Type of Surface**: The different surfaces will have different degree of smoothness from flat surfaces which will not have any peaks and troughs to rough. In fact, uneven surfaces will have more number of peaks and troughs. In general, stronger bonds could be achieved on flat surfaces as compared to curved surfaces.
- **Time**: The label with self-adhesive will "set" over time. In other words, the adhesives should have good performance for "initial tack" as well as "ultimate tack." Initial tack means the initial adhesive bond when a label is applied, and an ultimate tack would be achieved once the adhesives has fully strengthen and set.
- **Temperature**: Temperature plays an important role for adhesion performance. Adhesion performance will be faster at higher temperature and the adhesion process will be slow down at lower temperature. But, it is always recommended that different types of adhesive will have different recommended temperature, and hence, it is advisable that labels should be applied at extreme high as well as extreme low temperature.

1.6 Packaging—Its Linkage to Labeling, Branding and Marketing 27

- **Adhesive strength**: The adhesive strength will be varying with the type of adhesives used for the manufacturing of labels. Adhesive strength ranges from high tack label with extremely high strong adhesion strength to low initial tack which would allow the label to remove easily.

However, the labels are also available in different types depending upon its manufacturing process and application. Few of the important labels are given below:

- **Shrink Label**: This kind of labels are used in unit package, mainly in glass or plastic bottle where a plastic sleeve is dropped over the container and then shrunk by means of heat. Poly vinyl chloride (PVC), polypropylene and poly ethylene terephthalate (PET) are normally used.
- **Pressure-Sensitive (PS) Label**: This kind of label is manufactured, pre-coated with a pressure-sensitive adhesive and then adhered to a silicone-coated release paper. While the label is applied, the release paper acts as carrier that requires a minimum pressure to get the adhesion bond on the surface.
- **Heat transfer label**: This is popularly known with its trade name like Therimage. The label is manufactured by using thermoplastic printing inks for printing the desired design on a carrier sheet. The ink and carrier sheet are brought into contact with the container. Heat is applied to melt the ink and then transfer it to the object.
- **Bar coded label**: A label which would carry a carry a bar-code symbol that can be applied to a product or a container.
- **In-mold Label**: This type of label has got a special application for blow-molded or injection-molded containers. The labeling of in-mold is done where a pre-cut label coated with a heat-sealable adhesives is placed into the mold cavity before the plastic is induced. During the molding process, adhesive is activated by the heat of the plastic melt, causing the label to adhere to the plastic container as it is being formed.
- **Spot-glue label**: In this case, a label is attached to a container in such a manner that glue will be applied in few spots only.
- **Cut-in-place (CIP) label**: This kind of label is used for special application. A plastic film label is normally cut from the parent stock in an in-mold-labeling operation. The label material can be made significantly thinner than that used for a pre-cut label.
- **Sleeve label**: The sleeve label is made of decorated plastic sleeve which fits over and on plastic bottles. In general, these types of labels are made from polyethylene or polyvinyl chloride (PVC) which are shrinkable in nature. The sleeve is normally put on and over the bottles and then shrunk by means of heat to conform to a shape that will hold them in place.
- **Insert label**: This kind of label is placed between the seal and cover or inside the packages of dry products such as egg cartons.
- **Neck label**: A label that is used around the neck of a bottle or container is called as neck label.
- **Die-cut label**: The die-cut-labels are pressure-sensitive labels mounted on a carrier web which is die-cut to the level of the carrier sheet surface. This kind of labels

are die-cut in a die-cutting machine from irregular shaped label stock and then made into desired shape of the container.

The self service system in the retail outlets where the packages do always act as "Silent salesman" and then get inspired to buy the product. In short, the package stimulates the customer to take buying decision.

Moreover, the packaged goods provide the customer to get all the relevant information like date of manufacturing, "best before" lot no./batch no., name and address of the manufacturer, gross weight, net weight and maximum retail price (MRP) which are printed prominently on the packages. Hence, packages act as vehicle for communication about the product. It is reported by the market research study that about 15,000 new products are introduced every year in the consumer market. As a result, the companies face a difficult situation for making them accepted by the consumers [30].

The emergence of retail marketing has resulted in the introduction of thousands of new products with varied brands and has resulted in a "marketing war" leading to a stiff competition among the established and new brands.

Moreover, the new brands/products do also face stiff competition with the existing brands. Hence, he retailer has to adopt a mechanism by which the new brands or new products get attracted the customer which has become the most challenging task today.

In this critical situation, "packaging" plays a crucial role in the selling of customer goods by convincing the modern customer of its unique features like specific shape, size, printing format, color combination and graphic design.

In many cases, a particular type of package design of a particular brand always registers in the mind of customers, which compels them to look for the same goods to buy even at a higher price. In another situation, several products of the same type but with different brands are placed on a particular shelf of a retail outlet, which create difficult situation for the customer to take a final decision on the purchase. In such a situation, only a package with its elaborate printing and graphic design would make an impact among the numerous brands and thus help the customer to finally make the buying decision [31].

A numerous number of new branded food products are introduced every year. All these new brands have to compete with all established brands which would be placed in the same shelf of retail outlet. These products get a chance to position themselves onto the shelves of retail outlets after a lot of market research, promotion and distribution.

The final objective is how the products will move out of the shelves and get accepted by the customer so that the goods would enter into the shopping baskets where packaging is the only mechanism which stimulates the customer to buy the product.

Due to this fact, the role of packaging is now much beyond of its basic functions, i.e., preservation, protection and presentation. Today, it performs a more complicated and complex role. In fact, packaging, today and more so tomorrow, is an indispensable vehicle to holistically deliver goods to the consumer.

1.6 Packaging—Its Linkage to Labeling, Branding and Marketing

In a situation, where a number of new non-popular brands of any products have to compete with other established brands in the same retail outlet, packaging plays a significant role to make the difference by attracting customers and thus stimulate customers to purchase even non-popular brands. In fact, the attraction of a particular package depends on many factors. It could be due to the specific shape and size of the package or the printing or graphics, etc. The final objective is to compel the customer to go near a particular package and establish some relationship with the package as if the package speaks to the customer—"touch, feel and buy." In this situation, a package definitely acts as a silent salesman. Considering all these features, packaging is no more considered as a component of product element in the traditional marketing mix; rather today, packaging has established its own identity in the entire marketing scenario. As a result, packaging itself acts as an individual variable to achieve the success of target market [32].

Today, packaging also plays an important role for business growth and development. In the modern marketing scenario, packaging plays the leading role to satisfy all the relevant activities like proper storage, effective supply chain management system and timely delivery from the production point to the consumer's end.

1.7 Packaging in the Nineteenth Century

It is observed that there has been a revolution in the packaging industry during the nineteenth century. The packaging industry has been able to evolve to develop different types of formats of packaging by covering the entire range of packaging materials like flexible, semi-rigid, rigid and ancillary packaging materials. Due to this, a number of new types of packaging have been introduced to make available for a wide range of packaging materials, systems, machinery and ancillary materials to the consumers. This has not only helped the consumers to get a variety of packaging options, but at the same time, this development has broadened the horizon of packaging industry.

The important development of packaging occurred during nineteenth century which is given in Table 1.3 in chorological order [33]:

In fact, industrialization had occurred during the end of the nineteenth century in Europe and UK which has further expanded the opportunities across the border and made the packaging industry to become global. This has also enhanced the demand of goods by the overseas customers. This has further necessitated the requirement of innovative packaging materials, systems and technology for variety of products which could travel for a longer distances by providing all the required functions of packaging like protection to the package from all kinds of mechanical hazards which would occur during handling, storage and transportation and to preserve the food and pharmaceutical products by avoiding all kinds of biological and chemical hazards which would deteriorate the quality of goods during transit. The innovations of packaging materials are mentioned in chronological order since 1900:

Table 1.3 Developments in packaging during the nineteenth century

Year	Development
1903	Corrugated boxes first produced on a large scale
1903	First fully automatic glass bottle-blowing machine, Toledo, USA
1909	Bakelite invented by Belgian/American, Leo Baekeland
1911	Wax-coated liquid-tight cartons for milk described (first produced in USA, in 1915)
1912	First aseptic food packaging system (for cream) described in Scandinavia
1920	Staudinger in Germany produced the first artificial macromolecules (polymers)
1922	Development of crimped-on hermetic seal for metal cans renders soldering obsolete and makes high speed food production feasible
1930	First beer cans, with conical tops and crown cork developed
1931	Liquid-tight cartons (the Perga system) commercialized
1933	Bottles made from molded fiber described (FESA) in France
1933	Polythene isolated by Swallow & Perin of ICI
1935	Cellophane, the first transparent film, produced in the UK
1937	Molded pulp egg trays designed (Hartman), patented in Europe
1940	Liquid cartons first used for milk on large scale in Chicago
1941	Goodhue and Sullivan in the USA produced the first practical aerosol container
1948	Denis Gabor in Hungary discovers the holographic effect
1950	First two-piece (impact extruded) aluminum beer "can" produced in Switzerland
1951	Original "Tetra Pak" designed
1956	Flip-top carton launched
1963	Pull-tab easy open end for beer cans
1964	DWI aluminum two-piece "can" for beer developed
1970	Two-piece "can" in steel for beer launched
1972	Stay-on easy open tab for beer cans
1972	First "widget" beer-foaming device manufactured for Guinness
1980	Micro-flute corrugated board launched

(i) *In 1900*

The boxes ad cartons made of corrugated card board was made by pasting of layers of paper together (Fig. 1.9).

(ii) *In 1910*

Waxed paper was used to wrap sweets.

(iii) *In 1914*

In Sydney, Australia, J. Fledging and Co. produced cardboard shipping containers by providing an alternative to wooden crates (Fig. 1.10).

1.7 Packaging in the Nineteenth Century

Fig. 1.9 Paper-based corrugated box. *Source* Brody AL, Marsh KS (eds) (1997) The Wiley Encyclopedia of Packaging, 2nd edn. John Wiley & Sons, New York

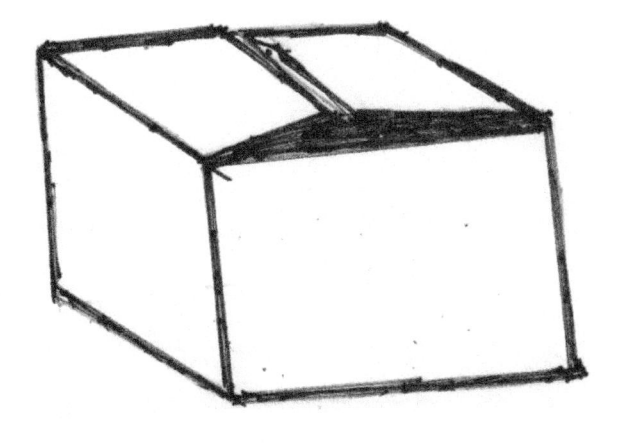

Fig. 1.10 Shipping carton. *Source* Brody AL, Marsh KS (eds) (1997) The Wiley Encyclopedia of Packaging, 2nd edn. John Wiley & Sons, New York

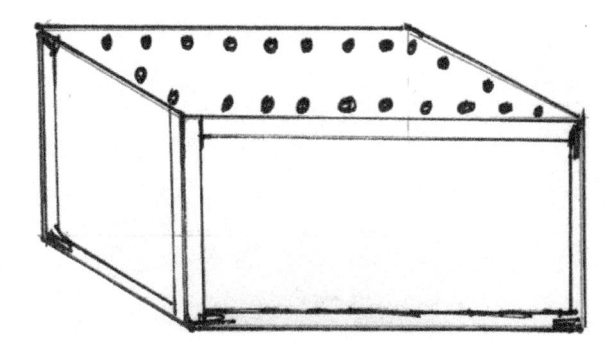

(iv) *In 1915*

Glass bottle made of machine made were used for milk (Fig. 1.11).

(v) *In 1917*

First milk carton was invented.

(vi) *In 1920*

Multiwalled paper sacks were produced for packaging cement and other bulk products. Incidentally, paper bags became the standard for retail packaging (Fig. 1.12).

(vii) *In 1928*

Heat sealed wax paper used to package bread was introduced for the future.

(viii) *In 1930*

Open-topped cylindrical cans used by Heinz to can spaghetti (Fig. 1.13).

Fig. 1.11 Glass bottle. *Source* Brody AL, Marsh KS (eds) (1997) The Wiley Encyclopedia of Packaging, 2nd edn. John Wiley & Sons, New York

Fig. 1.12 Paper bags. *Source* Brody AL, Marsh KS (eds) (1997) The Wiley Encyclopedia of Packaging, 2nd edn. John Wiley & Sons, New York

Fig. 1.13 Open top cylindrical can. *Source* Brody AL, Marsh KS (eds) (1997) The Wiley Encyclopedia of Packaging, 2nd edn. John Wiley & Sons, New York

(ix) *In 1940*

Steel drums used for shipping food, fuel etc. during World War II (Fig. 1.14).

(x) *In 1950*

Aluminum toothpaste tubes and aerosol cans were developed.

(xi) *In 1958*

In Australia, the first tetrahedron-shaped carton was made which could be formed, filled and sealed in one operation (Fig. 1.15).

(xii) *In 1960–1970*

Paper bag for shopping was introduced and became popular in supermarket (Fig. 1.16).

(xiii) *In 1961*

Single-trip glass bottles were introduced (Fig. 1.17).

Fig. 1.14 Steel drum. *Source* Brody AL, Marsh KS (eds) (1997) The Wiley Encyclopedia of Packaging, 2nd edn. John Wiley & Sons, New York

Fig. 1.15 Tetrahedron-shaped carton. *Source* Brody AL, Marsh KS (eds) (1997) The Wiley Encyclopedia of Packaging, 2nd edn. John Wiley & Sons, New York

1.7 Packaging in the Nineteenth Century

Fig. 1.16 Paper bag for shopping. *Source* Brody AL, Marsh KS (eds) (1997) The Wiley Encyclopedia of Packaging, 2nd edn. John Wiley & Sons, New York

(xiv) *In 1967*

Plastic margarine tubs came to the market.

(xv) *In 1968*

Milk cartons introduced into Australia (Fig. 1.18).

(xvi) *In 1969*

Comalco's aluminum drink cans was introduced (Fig. 1.19).

(xvii) *In 1969*

PET soft drink bottles, aseptic packages for long-life juice and milk cartons, microwave food packs, cling wrap, vacuum metalized papers, blister packs, laminates, Styrofoam trays and bar codes on packaging (Fig. 1.20).

(xviii) *In 1980*

HDPE plastic milk and juice bottles, small pouches for dishwashing detergent concentrates, trigger pumps for spray cleaners, ring pull for aluminum/steel cans, finger pumps for soap and cosmetics were introduced.

1990s

Cheer pack (like a milk carton without the paper core), stand-up plastic pouch packs for detergents and other liquids, synthetic wine cork, foil laminates for potato chips and other moisture sensitive foods, pouring spouts introduced for milk and juice cartons, transparent plastic "cans" to replace steel for fruit, pull top lid for steel cans [34].

Fig. 1.17 Glass bottle. *Source* Brody AL, Marsh KS (eds) (1997) The Wiley Encyclopedia of Packaging, 2nd edn. John Wiley & Sons, New York

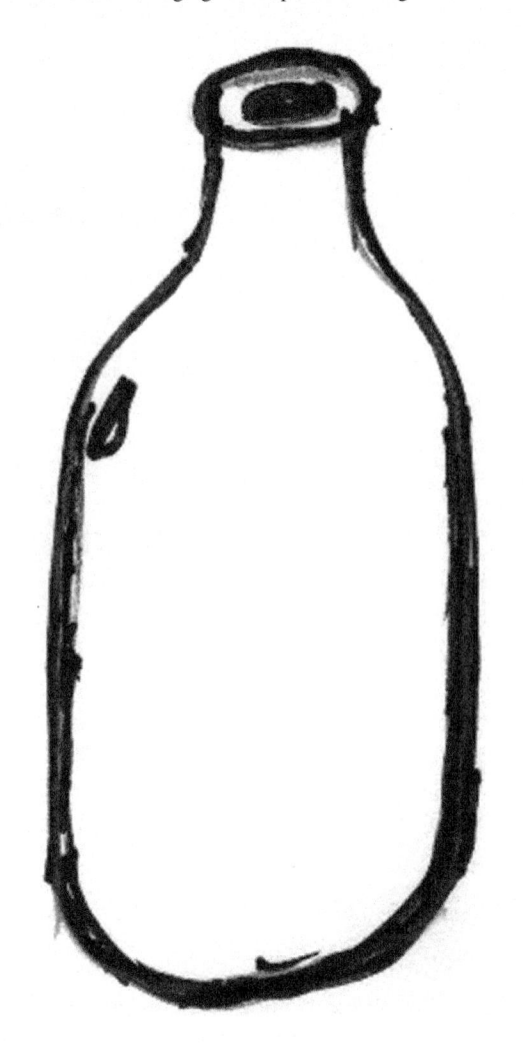

1.8 Evolution of Modern Food Packaging

The evolution of modern packaging is mainly due to the socioeconomic changes that occurred recently in the modern society in order to meet the requirements of modern consumers and to satisfy their needs. A rapid change is witnessed in the packaging industry during the first two decades of the twenty-first century. The increase of population with the change of demographic descriptors of the societies across the globe and social habits is also found to be changed.

More number of nuclear family with limited members has become very common. In addition, the number of working women has also increased as both the married are educated and engaged in jobs. This has also increased the purchasing power of any family. These kinds of socioeconomic changes have created also high demand for ready-to-eat packaged food, and the "convenience factor" has become highly

Fig. 1.18 Milk cartons.
Source Brody AL, Marsh KS (eds) (1997) The Wiley Encyclopedia of Packaging, 2nd edn. John Wiley & Sons, New York

preferable by the modern consumers. Meanwhile, the packaging technology has also been developed to cater the demands of the modern consumers.

Innovations in terms of newer packaging materials as an alternatives with higher functional properties at optimum packaging cost, newer conversions technologies with higher productivity and thus to reduce energy consumption as well as cost, highly sophisticated precision-based packaging machineries for high-speed operation to increase the scale of production and also the testing equipment's for accurate measurements for effective quality control measures.

The importance of packaging to a modern industrial society is most evident when we examine the food packaging sector. The organic natured food is sourced either from animal or from plant origin. Due to this fact, the fresh foods have got one of the important characteristics that the food products are perishable in nature and the food products have got a very limited shelf life. Besides, most of the foods are geography and season specific. For example, potatoes and apples are grown in temperate climate of having soil in high altitude area, whereas mango, jackfruit, etc., are grown in tropical climate. Due to this, there is a need to have proper storage, packaging and transportation techniques to deliver the fresh produce. At the same time, the modern consumers are demanding for "**ready-to-eat foods**" with convenient features and also for longer shelf life [35]. In fact, packaging is considered to be one of the **farm-to-consumer** food delivery systems. It is an integral part of food manufacturers and provides a link between the manufacturers and the consumers. In

Fig. 1.19 Aluminum drink can. *Source* Brody AL, Marsh KS (eds) (1997) The Wiley Encyclopedia of Packaging, 2nd edn. John Wiley & Sons, New York

today's modern scenario, packaging reflects the status symbol of the society and also about the consumer's habit and behavior. The packaging needs to have good mechanical strength, light in weight, easy to handle, convenience, attractive graphics and printing, safe to meet the demand of modern consumer [36]. The food industry is the largest user of packaging. Appropriate packaging helps to minimize wastage, promote hygiene and safety. In modern India, a food commodity and the consumers today make demands on the packaging which besides preservation, protection, safety and hygiene include consumer's convenience in product purchase, adherence to legislation and above all environmental issues. Adequate packaging has profound effect on both the pattern of food consumption and the amount of food consumed.

In fact, the evolution of packaging is witnessed mainly for food products either for fresh produce or for processed food products. For food products, the packaging is considered as an integral part of food processing, preservation and distribution of food. In addition to providing protection and sanitation, proper packaging also provides convenience, economy and visual appeal. Packaging is considered to be one of the most important mechanisms by which wastage of food can be reduced, and at

1.8 Evolution of Modern Food Packaging

Fig. 1.20 PET bottle.
Source Brody AL, Marsh KS (eds) (1997) The Wiley Encyclopedia of Packaging, 2nd edn. John Wiley & Sons, New York

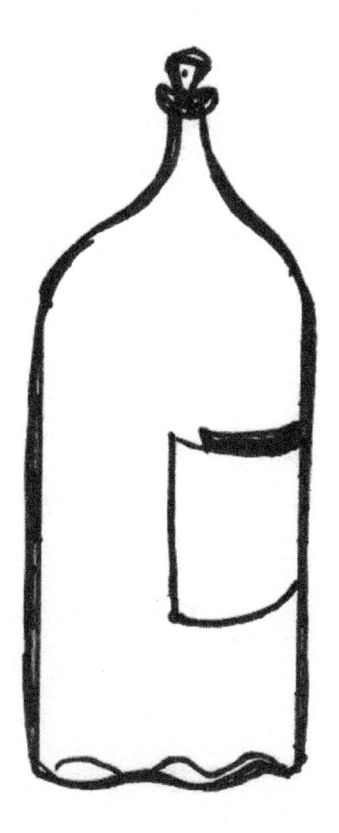

the same time, the surplus food produced during peak season could be processed and packed for wider distribution to meet off-season demand for seasonal foods [37].

In the modern packaging era, great revolutions of innovations in "Food Packaging" have been observed. A number of new techniques and technologies are employed for fresh and processed food products in order to enhance the shelf life. At the same time, a great change of market trends has come up as well. Few of them are as listed here [38]:

- Growing consumer awareness about the significance of packaging.
- Increase in consumer convenience.
- Increase in purchasing power of the modern consumer.
- Product branding and quality are desired by the consumer.
- Value addition to the products is considered in a significant manner.
- Product manufacturers are giving more thrust to exports.

On account of the changes in market trends, a shift of trends in packaging is also observed and preferred by modern consumers such as:

- Loose selling to packaged goods
- Bulk to retail consumer packs

- Conventional to newer packaging system
- Long life to optimum life of products
- Ease in handling and convenience
- Ease in production and distribution systems.

Considering the market trends and preferences of consumers, a number of new types of packaging materials and systems have been adopted in the modern market to satisfy the customer requirements. The evolution of packaging materials in the twenty-first century in comparison with (traditional/current) is as given in Table 1.4.

Table 1.4 Different types of packaging materials

Traditional trend		Current trend	
3-piece tin plate Cans		2-piece tin plate Can with easy open end	
Glass bottles for milk (Fragile)		Glass bottle for water. (Toughened)	
Earthenware pots for liquid		Plastic Jerri can for liquid	
Jute sack		Plastic Leno bag	
Wooden containers		Corrugated fiber board box	
Composite can		Leak-proof composite can	

(continued)

1.8 Evolution of Modern Food Packaging

Table 1.4 (continued)

Traditional trend		Current trend	
Folding Carton		Aseptic paper board carton	
Metal drum		Composite drum	
Glass bottle for beverages		Boat-shaped aseptic packs for beverages	
Plastic Pouch for edible Oil		Bag-in-box	
Injection molded plastic bottle		Stretch blow molded Plastic containers	
Hot foil stamping		HeatTransfer (Therimage)	
Reinforcement by plastic straps		Reinforcement with nylon strap	

(continued)

Table 1.4 (continued)

Traditional trend		Current trend	
Barcode		Quick response (QR) code	
Wooden pallet		Slip sheet	
Breathing film for fresh fruits		MAP package for fresh fruits	
PET bottle for edible oil		Active packaging (with oxygen scavenger)	
Single-layered plastic film-based pouch		Plastic-based flexible multilayered pouch for aseptic packaging	
Squeezable Single-layered plastic bottle		Squeezable coextruded plastic Bottle	
Plastic container		Micro-ovenable plastic container (C-PET)	

(continued)

1.8 Evolution of Modern Food Packaging

Table 1.4 (continued)

Traditional trend		Current trend	
Aluminum collapsible tube		Squeezable plastic laminated tubes	
Blister package		Temper-evident easy opening blister package	
Canned food in metal container		Retort package	

Similarly, there is an enormous change in the development of packaging systems and machinery during the twenty-first century which are being used. Few of the important packaging machineries are as given in Table 1.5

Packaging is a need-based technology, and "Innovation in Packaging" is a continuous process. As a result, a number of innovative packages with different features

Table 1.5 Development of packaging systems and machinery

Packaging system	Changing technology
Form fill seal machine	Fully automatic system with vacuumizing and gas flushing
Aseptic packaging	Aseptic packaging system with suitable packaging materials
Twist wrapping machinery	From low-speed to high-speed machineries to cater the organized sector
Pillow pack machinery	Shift from twist wrap to pillow pack machineries having good overprinting devices
Biscuit wrapping machineries	Shift from indigenous machineries fin seal wrapping machinery
Carton sealing machine	Automatic cartooning machinery
Hand-operated stretch wrapping machine	Automatic stretch wrapping machinery
Semi-automatic thermoformed machinery	Automatic thermoformed machinery with lid sealing
Offset/rotogravure printing	Digital printing
Pedal operated heat sealing machine	Automatic machine for heat sealing
Semi-automatic shrink wrapping machine	Automatic shrink packaging machine

such as convenient features with easy opening devices, usage of eco-friendly packaging materials to comply with environmental regulations, attractive graphics to make more eye appeal, "state-of-the-art" printing to make the packages more acceptable, etc., are being developed and introduced in the supermarket to meet the customer requirements and to satisfy the modern customers. The innovative packages are developed by using different types of packaging materials in different forms and format for specific application.

It is fact that packaging technology is driven by modern consumers. Due to this, the innovations in packaging in terms of newer materials with higher functional properties at optimum cost, newer packaging system with higher productivity, newer package design with various consumer-friendly features and alternative packaging materials in compliance to environmental regulations will be introduced at a constant pace to the market to meet demands of modern society. At the same time, it is projected that packaging will become the most acceptable technology by the society in years to come.

References

1. Annual report (2018–19) p 33, published by Ministry of Food processing industry, Government of India
2. Carl olsmats (2004) Packaging industry—a global overview. Technical publication of world packaging conference, India, PP-2, Oct 2004
3. Saha NC (2011) Packaging-the 5th P of marketing. Packaging India magazine, vol 43, April–May, p 7, Indian Institute of Packaging
4. A book on Packaging & distribution management, p-3, published by Indian Institute of Materials Management, 2001
5. Fundamentals of Packaging Technology, 2nd edn., Institute of Packaging Professionals (IOPP), USA, p 4. Ibid, pp 1–4
6. Saha NC (2010) Packaging enhancing product value through innovations. Published in Packaging India, p 3, Aug–Sept
7. The Future of Global Packaging to 2018, Smithers PIRA, UK, p 311
8. https://www.reportlinker.com/p05867471/?utm_source=GNW Packaging Industry in India-Growth, Trends, Forecasts (2020–2025)
9. Walter Saroka (1995) Fundamental of packaging technology, 5th edn. Institute of Packaging Professionals, p 4
10. https://medium.com/digital-packaging-experiences/the-evolution-of-packaging-57259054792d
11. http://www.historypackaging.com
12. http://article.unipack.ru/eng/17231/ahistoryofpackaging/. Accessed on 28th Nov 2008. Berger KR (2002), Op. Cit, p 7
13. Kenneth R. Berger reviewed by B. Welt (2002) A Brief History of Packaging. Agricultural and Biological Engineering Department, Institute of Food and Agricultural Sciences, University of Florida, Gainesville, p 4
14. http://www.historyofpackaging.com/. Accessed on 3 June 2006
15. Kenneth R. Berger (2002) Op.Cit. p-7
16. Hook P, Heimlich JE (1998) A history of packaging, Ohio State University, USA
17. Hook P, Heimlich JE (1998) Op.Cit. p-6

References

18. Brody AL, Marsh KS (eds) (1997) The Wiley Encyclopedia of Packaging, 2nd edn. Wiley, New York
19. http://www.historypackaging/en.encyclopedia/. Accessed on 30 June 2007
20. Narayanan PV, Dodi MC (1998) Indian food sector and packaging—an overview. In: Modern food packaging. Published by Indian Institute of Packaging
21. Saha NC (2011) Sustainable packaging—an urgent need for tomorrow. Packaging India magazine, Oct–Nov, vol 44, no 4. https://www.historyofpackaging.com
22. Saha NC (2010) Packaging enhancing product value through innovations. Packaging India, p 3, Aug–Sept, p 5
23. Packaging Technology Educational volumes (2001) Set 1, Indian Institute of Packaging, p 2
24. Narayanan PV (1994) Packaging-tool for brand Positioning, vol 27, no 2. Packaging India, June–July, Indian Institute of Packaging
25. Sherlekar SA (2008) Marketing management, 14th edn. Himalaya Publishing House, p 8
26. Evans JR, Berman B (2000) Marketing management, India edition, p 16
27. Packaging Technology educational, vol 1. Published by Indian Institute of Packaging, p 33
28. An article on "Perfect Packaging", published in The Hindustan Times, 8 Nov 2008
29. https://www.packaging-labelling.com/articles/importance-of-labelling-in-marketing
30. Hines JAR (2009) Packaging as a marketing tool, p 4, Packaging Diva Journal. Packaging Research: Background to Packaging, 2008, Institute of Packaging Professionals
31. https://www.yourarticlelibrary.com/marketing/product/product-branding-packaging-and-labelling/49054
32. Saha NC (2011) Packaging—5th P of Marketing, Op.Cit. P-5&6
33. https://www.thehistoryofpackaging/en.wikipedia/anon. Accessed on 30 July 2007
34. Packaging Research: Background to Packaging, 2008, Institute of Packaging Professionals
35. Narayanan PV, Dordi MC (1998) Indian food sector and packaging—an overview. Modern food packaging, a reference book, p 1, Indian Institute of Packaging
36. Saha NC (2011) "Preface" an "handbook on packaging of processed food products", solutions for micro and small scale industries, initiated by Ministry of Food Processing Industry, Government of India, Indian Institute of Packaging
37. Packaging Technology educational, vol 4, published by Indian Institute of Packaging, p 252, 2001
38. Saha NC (1999) Current trends in packaging of fresh and processed food products. Technical booklet of national conference, PICUP, Lucknow

Chapter 2
Flexible Packaging Material—Manufacturing Processes and Its Application

N. C. Saha, Anup K. Ghosh, Meenakshi Garg, and Susmita Dey Sadhu

2.1 Introduction

The main role of packaging material is to preserve the food products from any kind of chemical and microbial spoilage to increase the shelf life, provide protection against any kind of physical and mechanical damages during storage, handling and transportation and also to present the packaged food with attractive printing and graphics to enhance eye appeal to the customer.

Over a period of time, a wide variety of packaging materials have been developed with higher functional properties for their application into the packaging of food products. However, these packaging materials are different from one to another in terms of physical nature, mechanical strength as well as physicochemical properties. Due to this, the selection of a right type of packaging materials for a particular type of food products is always a great challenge to the food processing industries. It is a fact that the packaging industry is always under pressure by different stakeholders (producers, retailers and consumers) who have different priorities and do not always perceive that the packaging contributes as value addition to the product besides its traditional functions like containment, preservation, protection, preservation, providing information and also act as marketing tool or instrument [1].

In general, the concepts have always been related to an inert material, acting as a "passive" barrier between the foodstuff and therefore the outside environment, also avoiding the migration of harmful substances from the packaging to the food. The right selection of the packaging material plays an important role in maintaining product quality and freshness during distribution and storage. There are different packaging materials which have been traditionally used such as glass, paper, paperboard, aluminum, foil and laminates, tinplate, steel and plastics (both rigid and flexible) [2, 3]. The choice of a package depends mainly on the performance of the used packaging materials. This chapter introduces the flexible, rigid and semirigid packaging materials. The purpose of this chapter is to provide basic knowledge about food packaging requirements and the choice of material for food processing.

Requirements of food Packaging [4]:

While designing or selecting a suitable packaging material for food, there are following factors which need to be considered:

1. Physical protection—The chosen packaging material must protect the product enclosed from any mechanical or thermal shock or any other physical impact like compression, etc.
2. Preservation of product—In preserving a food product, there are two more factors other than physical protection, namely chemical and microbiological safety. To prevent chemical deterioration of products, the packaging must minimize the effect of change in atmosphere or exposure to gases, moisture, light, etc. On the other hand, to safeguard the product against any microbiological deterioration, a packaging material must provide barrier against microorganisms—pathogenic and spoilage, insects, rodents and potential agents which can cause spoilage or can spread disease through food. A suitable packaging material must also control factors such as aging and ripening.
3. Containment and tamper-evident—The packaging material should hold or contain the food product without affecting the shape or any property to enhance the handling or transportation of product. In addition to this, the packaging material containing a product must have some safety feature like some seals or caps which can indicate any kind of tampering or theft if done [5, 6].
4. Stability—There are many preservation or processing techniques which are operated after the packaging of the food. For example, we must be knowing about the effect of any kind of treatment on the quality of packaging material, and thereby choosing a suitable packaging material which can provide best quality product with minimal changes in its properties is a cumbersome task. For example—decrease in oxygen permeability of PET/PVDC/PE due to irradiation (ionizing radiation).
5. Convenience and cost—Other than protection and stability of the packaging material, the convenience in handling the package, its distribution, display, opening, closing, etc., is also considered as an important factor while selecting the packaging material for the product. Along with convenience, cost of the packaging material plays a decisive role during its selection. Because the cost of packaging directly affects the cost of the final packaged product.
6. Marketing—The marketing of a product is necessary to inform consumers and to encourage them to buy the product. For this, the labeling and the eye-catching design of the packaging plays an important role. Packaging should be so appealing that it attracts the consumer over other similar products of competitive brands and provides great customer satisfaction.
7. Legal compliances—Choose a packaging material that is legally compliant with the food safety and standard regulation of the country. The packaging should not violate the pre-existing copyrights and patent policies.

2.1 Introduction

Table 2.1 Properties of rigid, flexible and semirigid packaging

Property	Rigid	Flexible	Semirigid
Weight	Heavy in weight	Light	Light
Size	Used for bigger goods	For small packs	Used for small to medium pack
Storage	Require bigger space	60% less requirement	Require less space than rigid
Energy	Require more energy	40% lower energy	Less energy
Resealing	Not possible	Resealed	resealed
Re use	Can be reused	Not possible	Not possible
Disposal	Difficult to dispose	Easy	Easy

With packed foods, any package components that may contact the food are considered food contact materials. Packaging material that is mainly used by both small-scale and medium-scale food processors for products may be rigid or flexible. Market has a good share of rigid material and is used for packaging of fruits, vegetables and for other crushable items which include tin, glass, pottery, cans, wood, plastic bottles, etc. Flexible material uses less amount of material and energy as compared to rigid material which includes paper, plastic films, cloth aluminum foil, etc. Semirigid material is used to protect the product from harsh environment like humidity, temperature, etc., includes plastics, cardboard, etc. Table 2.1 shows the properties of rigid, flexible and semi rigid packaging material.

Flexible packaging material is light in weight, available in bags or pouches. It may be sealed using heat or pressure. Stand-up pouches with a zip lock, laminated tubes and vacuum pouches are the examples of flexible material. As the name suggests, flexible material can be customized easily. Manufacturing cost of flexible material is less, and they provide minimal protection against compression or perforation. It is used widely in food packaging because of its protective properties. Retort pouches are getting popularity as they are easy to use and are made up of multiple layers of polymers and from aluminum foil. This multilayer concept of different materials provides chemical and thermal shielding to the products from inside. Package offers an air-tight barrier, prevents exposure to light and heat and is temperature tolerant to maintain freshness of food for a long time. A good example is the development of paper boat pouch, used for beverages. It has broken the conventional trend of using rigid packaging for beverages.

2.2 Paper

Paper and paperboards are used in food packaging from old time at least from the seventeenth century, and its usage increased in the later part of the nineteenth century. Paper is defined as "all matted or felted sheet of fiber, formed on wire screen from water Suspension". Basic raw material of paper is cellulose, and its sources are the pulp made from wood, bamboo, bagasse, cotton, rice or wheat stalk, etc. It can be porous, waterproof, translucent, colored and soft in nature [7].

Paper and paperboard are non-specific terms that can be related to either material caliper (thickness) or grammage (basis weight). The International Standards Organization (ISO) states that the material weighing more than 250 gsm (gms per square meter) shall be known as paper board. In other words, materials more than 300 μm (micron) will be referred to as paper board. The properties are dependent on variables, i.e., sources of fibers, extraction process, the machine used and the treatment given for finished products.

2.2.1 Manufacturing

Raw material—Wood is used as a raw material for the pulp and paper industry, and it comes from the forest. Other material which is used for papermaking comes from sawmills, recycled newspaper and cloth and vegetable matter. Coniferous trees are used preferably for papermaking because these species of trees contain longer cellulose fibers which help in making strong paper. The wood from these trees is called "softwood". Wood from deciduous trees (leafy trees such as poplar and elm) is called "hardwood". In India, mostly paper is made from sugarcane waste called bagasse. Straw, bamboo and hemp jute fibers are also used for papermaking. Different steps in manufacturing of paper are discussed below.

Step-1 Striped logs are chipped into small pieces and recover the constituent of wood cells with no fiber forming ability.

Step-2 Pulp is the fibrous material. It separates cellulose fibers from soft or hard wood, fibrous crops and waste paper material like newspaper or rags. Individual fibers are produced after dissolving lignin. During papermaking process, these individual fibers can be reformed into paper sheet. The fibers obtained after the pulping treatment are termed as pulp. It is the basic raw material for the papermaking process. The process by which pulp is made is called pulping. There are three different types of pulping process, i.e., mechanical, chemical, semi-chemical or any of these in combination.

(a) Mechanical pulping—Water soluble impurities are removed leaving behind lignin. Paper made from this pulp does not generally contain a brightness and strength when compared to paper obtained from chemical pulp.
(b) Chemical pulping—The chemical pulping process produces the strongest pulp. This is basically an alkaline cooking at high temperature (160–1700 °C) and pressure with sodium hydroxide (NaOH) being the primary cooking chemical. Sodium sulfide (Na_2S) is added afterward to maintain the pH about 12. Sodium hydroxide contributes to the reduction in size of lignin molecules.

(c) Semi-chemical pulping—Pulp yield is 65–80%. This pulp is used for base paper, CFB, duplex boards, etc. Two processes are used in semi-chemical pulping.

   (i) Cold soda process: Wood chips are treated with cold sodium hydroxide (NaOH).
   (ii) Neutral sulfite semi-chemical process (NSSC):
   Sodium sulfite (Na_2SO_3) liquor buffered with sodium carbonate is used to cook the wood chips.

Step-3 Bleaching treatment is applied to improve the whiteness of pulp made of either Chemical or mechanical pulping process. In lignin, the color of pulp is decided by chromophoric groups. Chlorine, hydrogen per oxide or chlorine dioxide are usually used as a bleaching agent to remove the color.

Step-4 Surface area of fiber is increased by beating process. This process increases the water holding capacity of paper and develops additional bonding. Refining process is used to improve physical properties of finished paper and is almost similar to beating process.

Step-5 In this stage, mechanical treatment is given, just after pulping and beating, to convert fibrous material which is called stock into a paper sheet on paper machine.

Papermaking process uses three different types of machines: Fourdrinier machine, cylinder machine and twin wire formers. During the paper forming process, after sheet-making process, the fibrous material which contains about 99% water is usually passed through rollers or wire mesh to remove all the extra water which helps in forming the paper web.

Final treatment is given to paper according to the requirements of industry, which includes calendering, supercalendering, sizing, laminating, impregnating, etc.

Step-6 Pressure is applied in calendaring process to reorient the surface of fiber, and it also smoothens the surface of paper. After calendaring process, paper is called machine finished paper. In supercalendering process, extra pressure and moisture are added as compared to calendring process.

Step-7 Sizing is the process of coating paper with starch, casein, alum, etc., to improve its appearance, barrier properties and strength (Fig. 2.1).

2.2.2 Types of Paper

2.2.2.1 Paper

Plain paper has poor barrier properties against moisture, and it does not show any heat sealable properties also. Hence, it cannot be used for long time storage of products. Before using it as a primary packaging for food, it should be either coated or laminated with waxes, resins, lacquers, etc. So that its functional and protective properties can be improved. Properties and application of different types of paper are discussed below.

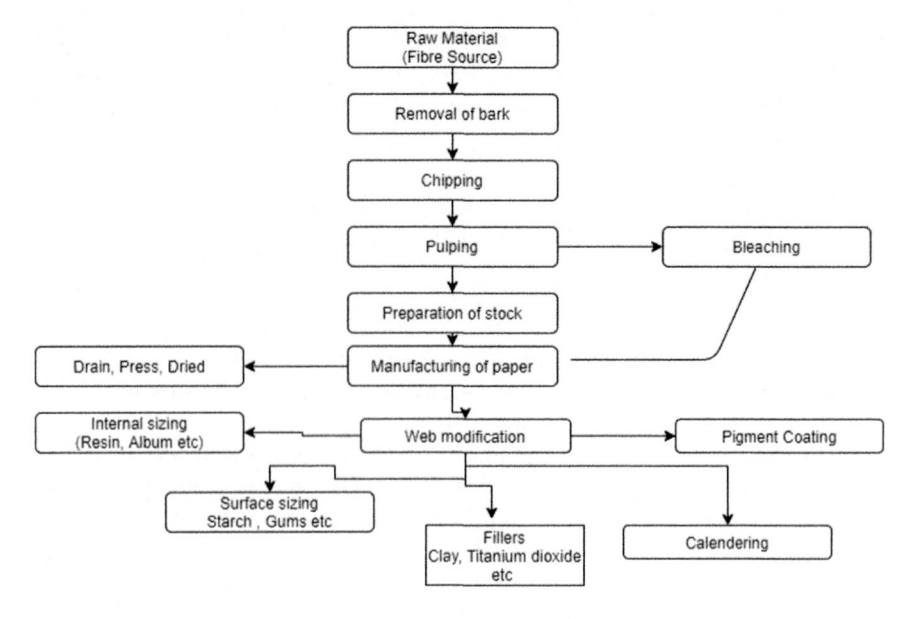

Fig. 2.1 Shows the various stages of paper manufacturing process

2.2.2.2 Kraft Paper

It is brown color paper made by chemical pulping process. High strength properties like tensile strength, good tearing resistance and folding endurance kraft paper are available in several forms: natural brown, unbleached, heavy duty and bleached white. The natural kraft is the strongest of all paper.

Application: It is commonly used for bags and wrapping. It is also used to package flour, sugar and dried fruits and vegetables.

2.2.2.3 Tissue Paper

It is made from strong cellulose fibers with 12–25 gsm, lightweight paper. It could be white or colored. It has toughness, even formation, finish and antitarnishing properties.

Application: It is used for wrapping of fruits like apples to avoid bruising.

2.2.2.4 Sulfite Paper

Sulfite paper is weaker and lighter in weight than kraft paper. It is usually glazed to improve its appearance, wet strength and oil resistance. For higher print quality, it can be coated. It is widely used in making laminates.

Application: It is used to pack confectionery items like biscuits and confectionary in small bags.

2.2.2.5 Sack Paper

It is made from unbleached sulfate wood pulp. It has high strength in both direction, i.e., MD and CD, and it has high burst factor.

Application: Mostly used to produce multiwall paper sack for the packaging of tea and other food products.

2.2.2.6 Wax Paper

It is coated with dry or wet wax to provide water vapor proof and greaseproof qualities.

Application: It is used for wrapping bread, biscuits, etc.

2.2.2.7 Greaseproof Paper

Greaseproof paper is manufactured by a process known as beating. In this process, the cellulose fibers undergo hydration for longer period than normal which makes the fibers gelatinous, and it can be broke up. These fibers are packed tightly so that its surface becomes resistant to fats and oils but not to wet agents.

Application: Greaseproof paper is used to wrap candy, cookies, bars, bhujia, roasted seeds, snack foods, etc. Greaseproof paper has replaced the use of plastic.

Glassine is a type of greaseproof paper in which extreme hydration provides glossy and smooth finish. It is supercalendered, transparent and impermeable to water vapor and smooth in texture. It can be coated with different types of varnishes or lacquers, etc. It is highly resistant to air and water vapor permeability and also has good aroma barrier properties.

Application: It is used as a liner for many types of products like biscuits, cooking fats, fast foods and baked goods. It is also used as protective wrapper for packed food stuff, tea and tobacco products. It also attracts printing because of its smooth surface.

2.2.2.8 Parchment Paper or Vegetable Parchment

This paper is a cellulose composite that is processed to achieve additional properties like non-stickiness, grease resistance and resistance to humidity. Acid-treated pulp (passed through a sulfuric acid bath) is used for making of parchment paper. The acid makes changes in the cellulose and makes it smoother and impermeable to water and fat, which gives strength, while it is wet. It does not provide a good barrier properties against water and air. It is also not heat sealable. Made from Chemical wood pulp and rags. It is not waterproof unless coatings are given. It is odorless and tasteless.

Application: It is used as wrapping paper for meat and fish packaging, butter, margarine and liner for fresh vegetable packaging.

2.2.2.9 Paper Laminates

Kraft or sulfite pulp is used to make paper laminates. It may be coated or uncoated depending upon the use. They can be laminated with plastic or aluminum metal to improve various barrier properties. For example, paper can be laminated with polyethylene and aluminum to make it heat sealable. It also improves its air and moisture barrier properties [8, 9].

Application: Laminated paper is used for packaging of dried foods like soups, herbs, dried fruits, spices, etc. (Fig. 2.2).

2.3 Plastic Films and Laminates

In food packaging, plastics play an important role because of several benefits like high environmental stability, easy processing, lightweight, water resistance, high strength, very good appeal, easy recyclability and many more. Plastics are thus widely used for packaging materials as follows:

- Plastics can be reshaped by heating (they have good flow and molding ability).
- Plastics are mostly chemically inert but permeable.
- Due to low cost and versatile properties, they fulfill the market demands.
- Plastics are lighter than most of other material.

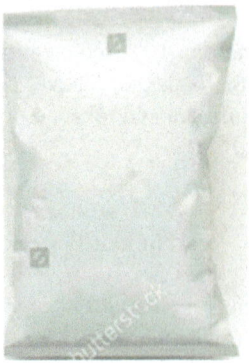

Fig. 2.2 Paper laminates

2.3 Plastic Films and Laminates

- These are transparent, available in different colors, heat sealable, heat resistance and have good barrier properties.

With 34% of all plastics being consumed for packaging, this application represents the largest segment. The two main forms of packaging products are as follows:

- Flexible packaging, comprising films, made by both cast and film blown processes.
- Rigid packaging, comprising bottles and other containers, made by thermoforming and blow molding processes.

Some of the very important examples of flexible food packaging are as follows:

Liquid food (oil, ghee, water, lassi, dahi, milk, etc.).
Semi-liquid food (sauce, chutney).
Solid powders (milk solid, coffee, atta, flours, etc.).
Solid grains (daal, cornflakes, beans, breads, roti, paratha, etc.) and namkeen (mixtures, popcorns, chips, biscuits, dry fruits, etc.).
Flat bottom pouches for drinks (juices, tea, coffee, etc.) (Fig. 2.3).

2.3.1 Materials

Proper selection of materials used in flexible packaging and processes such as film blowing, coextrusion of several layers of different polymer films to provide tailor made packaging solutions, as described later in this section.

The main plastics that are used in flexible packaging are [10, 11]:

- Polyethylene (LLDPE, LDPE and HDPE)
- Polypropylene (PP, BOPP)
- Polyesters (PET)
- Nylon (PA6, PA66 and PA11)
- Ethylene vinyl alcohol (EVOH)
- Fluoropolymers (PTFE)
- Cellulose-based materials
- Polyvinyl acetate (PVA).

The chemical structures of the plastics mentioned above are given in Table 2.2.

2.3.2 Polyethylene

Polyethylene (PE) is currently the most common packaging plastic in use. Polyethylene was commercialized in the 1950s, and since then, it has acquired its dominant position as a packaging material for food and beverages. Its popularity in food packaging can be attributed to its relatively low cost, versatile properties and the ease of manufacturing and processing. Different types of polyethylene with appropriate

Fig. 2.3 Types of plastic films for packaging of food items

physical properties are used for different packaging requirements. LDPE and LLDPE are primarily used for films, while HDPE, which is more rigid, is used for containers. In some of the film applications originally dominated by LDPE polymers, the newer LLDPE polymers are nowadays preferred because of better tensile strength, elongation at break and puncture resistance. Various packaging applications of PE include bags for garments and grocery, trash bags, packaging for different food categories (like bakery, meat, poultry and dairy products and frozen food packaging), packaging of pharmaceutical and cosmetic products and packaging for pesticides, insecticides and fertilizers. Some typical applications are shown in Fig. 2.1.

Properties

- Depending on the type of polyethylene, the density varies from 0.91 g/cc for LLDPE to 0.948 g/cc for HDPE.
- Polyethylene is easy processable with high toughness and flexibility. It has good barrier properties for moisture and water vapor and high chemical resistance.

2.3 Plastic Films and Laminates

Table 2.2 Structure of plastics used in food packaging

S. No.	Name of plastic	Chemical structure
1	Polyethylene	$(C_2H_4)_n$
2	Polypropylene	$(C_3H_6)_n$
3	Polyester (PET)	$(C_{10}H_8O_4)_n$
5	Polyamides (PA66)	$-N(H)-(CH_2)_5-C(=O)-$
8	Ethylene vinyl alcohol	$[-CH_2-CH(OH)-]_x[-CH_2-CH_2-]_y$
9	Fluoropolymers	$-(CF_2-CF_2)_n-$
10	Cellulose	(glucose polymer structure)
11	Polyvinyl alcohol	$-[CH_2-CH(OH)]_n-$

- PE is free of odor and toxicity and has excellent electrical insulation properties with easy heat sealability.
- The PE shows low barrier properties toward gases, oils and fats.
- The PE has high flammability and possesses lower thermal stability compared to some of the plastics used in packaging.
- It is a semicrystalline polymer.
- Higher crystallinity in it gives high stiffness, hardness, tensile strength, opacity, barrier properties and heat and chemical resistance (Fig. 2.4).

Polyethylene is the product of polymerization of ethylene which is obtained as a petrochemical by-product by thermal cracking of ethane and propane. Depending on the conditions of polymerization process, different grades of PE can be obtained.

Fig. 2.4 Packaging products made of polyethylene

2.3.3 Low-Density Polyethylene (LDPE) [12]

LDPE is so named because it contains a substantial amount of branches which hinder the process of crystallization, resulting in relatively low densities (0.9–0.94 g/cc). The high pressure polymerization process employed for the manufacture of LDPE leads to the development of both short chain and long chain branches. LDPE is preferred in film applications due to its superior properties (like toughness, flexibility and relative transparency) which makes it a popular contender for use in applications where heat sealing is necessary. LDPE also may be used to manufacture some products like flexible lids and bottles. This includes linear low-density polyethylene (LLDPE).

Properties

- It has excellent resistance to chemicals like acids, bases and vegetable oils.
- It has high toughness, flexibility and relative transparency with good heat sealing capacity.

Applications PE film finds applications in

- Bags for bread, frozen foods, fresh produce and household garbage
- Shrink wrap and stretch film
- Coatings for paper milk cartons and hot and cold beverage cups

2.3.4 High-Density Polyethylene (HDPE)

It primarily consists of unbranched molecules. Lack of branches facilitates the packing of chains into crystal structure, resulting in a high degree of crystallinity and hence higher densities (0.940.97 g/cc) along with better mechanical properties. HDPE is used to make different types of bottles/containers. Bottles without any colorants are translucent, have excellent barrier properties and stiffness and are good candidates for packaging products with a shorter shelf life such as milk. Since HDPE has good resistance to chemicals, it is used for packaging household and industrial chemicals such as detergents and bleach. Pigmented HDPE bottles have better stress crack resistance than unpigmented HDPE.

Properties

- Excellent solvent resistance
- Compared to other grades of PE, higher tensile strength obtained.
- Relatively stiff material with useful temperature capabilities.

Applications

- It is used in pouches for liquid foods like milk, water, juice and for cosmetics, shampoo, dish and laundry detergents and household cleaners.
- HDPE is used in cereal box liners.

2.3.5 *Linear Low-Density Polyethylene (LLDPE)*

These consist of molecules with linear polyethylene backbones with short alkyl groups attached at regular intervals. In other words, they are PE molecules having short chain branches, but no long chain branches. Density of these resins lies in the range 0.915–0.935 g/cc. These resins can be regarded as a compromise between HDPE and LDPE.

2.3.6 *Polypropylene (PP)*

Polypropylene is a common plastic used in packaging applications. Homopolymer of polypropylene is produced by catalytic addition polymerization reaction. Different grades of PP are used in food packaging like CPP, BOPP, etc., which mainly differ from each other in the degree of crystallinity and stretching. In the case of copolymer of PP, ethylene is used as a comonomer. An organo-metallic catalyst attaches to propylene and works as a functional group. It reacts with the unsaturated bond of propylene to form a long chain polymer. PP is strong and has excellent chemical resistance with high melting point making it good for hot-fill liquids. PP is found in both flexible and rigid packaging, fibers and large molded parts for automotive and consumer products.

Properties

- PP has excellent optical clarity both in biaxially oriented films (BOPP) and stretch blow-molded containers.
- It has low moisture vapor transmission rate toward acids, alkalis and most solvents. PP is a cost-effective plastic, having density between LDPE and HDPE. It possesses high thermal resistance due to the methyl group in the main chain. For the same reason. It has high resistance to chemicals but has poor oxidative stabilities. Addition of antioxidants is therefore necessary for stability in applications involving oxidative environments.
- Polypropylene is a non-toxic, easily processable polymer having good dielectric and insulation properties, excellent clarity and good mechanical properties.
- It is a lightweight material with high-dimensional stability which makes it suitable for replacement of metallic parts in automobiles.
- Ethylene propylene elastomers, produced by copolymerizing propylene with ethylene, show improved resistance to heat and oxidation, as well as favorable tensile and tear properties.

Applications

- Containers for yogurt, margarine, takeout meals and deli foods
- Polypropylene is a versatile plastic with applications ranging from packaging to automobile and textile industry. It possesses properties similar to those of PE and hence competes with PE in several product applications.
- PP is used in packaging applications as an alternative to polyethylene. Due to higher stability of PE, it is generally preferred over PP in applications where the final product has applications in oxidative environment.

2.3.7 Polyethylene Terephthalate (PET)

PET is a linear thermoplastic aromatic polymer, which is commercialized for packaging of carbonated soft drinks due to its excellent gas barrier properties that helps to retain CO_2. Raw materials used for production of PET are ethylene glycol and terephthalic acid. Production of PET involves two steps: in the first step trans-esterification or esterification of dimethyl terephthalate or terephthalic acid, respectively, and in the second step polycondensation of resulting oligomers gives PET. PET is a clear and tough polymer with good gas and moisture barrier properties. It is extensively used as films in food packaging industry.

Properties
PET is important because of their

- Capability for hot filling
- Optical clarity and smooth surfaces for oriented films and bottles

2.3 Plastic Films and Laminates

- Good barrier properties gases like oxygen, water and carbon dioxide
- High impact capability and shatter resistance
- Excellent solvent resistance
- The PET exists in either amorphous or crystalline form. This broadens the range of applicability to a wide variety of packaging applications.
- PET is linear chain thermoplastic polymer due to which it can be easily converted from amorphous phase to crystalline phase by annealing or stretching above glass transition temperature.
- Amorphous PET is highly transparent but is vulnerable to thermal degradation.
- Crystalline PET is high in strength, rigid, dimensionally stable, water resistance and thermally stable.
- Crystalline PET possesses good resistance to chemicals but not at good as PE or PP.
- The PET has a high glass transition temperature and melting point and is 100% recyclable.
- The PET possesses superior gas barrier properties making it an ideal choice in several food packaging applications.

Applications

- PET films are mostly used as one of the important substrate for the manufacturing of multi-layered laminates for food packaging application.
- Ovenable film and microwavable food trays.
- The PET films are used for food and beverage pakaging.

2.3.8 Nylon (PA6, PA66, PA11)

Among the family of polyamides, only a few polyamides are suitable for packaging applications. The best choice of polymer depends on the benchmark required as well as the economical constraints. The two most widely used polyamides are PA66 and PA6. They are often extruded to make fibers (textile industry) or films (packaging) or injected article.

Polyamide 11/Nylon 11 is a bio-based engineering plastic that is prepared from renewable resources like castor plants and is produced by polymerization of 11-amino undecanoic acid.

Polyamide 11 is the first biosourced polymers with a melting point of 190 °C.

Properties
PA6 and PA 66:

- Possesses high strength and stiffness at high temperature
- Even at low temperature, it has high impact strength
- It has very good flow property for easy processing

- Possesses good abrasion and wear resistance
- It has superior resistance to fuel and oil
- Has good resistance to fatigue
- PA 6 has excellent surface appearance and better processability than PA66 (due to its very low viscosity)
- Good electrical insulating properties
- High water absorption and water equilibrium content limits the usage
- Low-dimensional stability
- Attacked by strong mineral acids and absorbs polar solvents
- Proper drying before processing is needed.

PA11:

- Has lowest water absorption property among all the available polyamides
- It has outstanding impact strength, even at temperatures well below the freezing point.
- It is resistant to chemicals (like greases, fuels, common solvents and salt solutions)
- It shows outstanding resistance to stress cracking, aging and abrasions.
- Has very low coefficient of friction
- Shows high fatigue resistance under high frequency cyclical loading condition
- Highly resistant to ionization radiation
- Possesses noise and vibration damping properties
- It has ability to accept high loading of fillers
- It has high cost relative to other polyamides
- Lower stiffness and heat resistance than other polyamides
- Poor resistance to boiling water and UV
- Attacked by strong mineral acids and acetic acid and are dissolved by phenols.

Application

PA6 and PA66 offer very high resistance to puncture, barrier resistance to gases like oxygen, carbon dioxide and aromas and high transparency. All these properties make PA6, PA66 ideal for use in food packaging both as mono or multilayer. Other applications include agricultural films, liquids packaging, protective packaging, etc.

2.3.9 Ethylene Vinyl Alcohol (EVOH)

EVOH is one of the most important polymers for applications with high barrier properties for food packaging. Despite the problem of moisture sensitivity, it is a preferred barrier material for food packaging accounting for 70–75% of barrier films used for retortable pouch. Modern food processing technologies, sterilization with microwave and high pressure processing provide conditions that are less stringent than conventional heating.

In fluoropolymers, all the hydrogen atoms are replaced by fluorine. They are known as high-performance polymers. In polychlorotrifluoroethylene (PCTFE), further another hydrogen is replaced by a chlorine atom. PCTFE offers highest water vapor, gas barrier and high resistance to most chemicals at low temperatures. In many applications, aluminum foil is replaced by it. The PCTFE film although costly but widely in use as a thermoformable blister packaging material coated with PVC for pharma application. This material provides us a wonderful combination of transparency, heat sealability along with the ability to be laminated, thermoformed, metallized and sterilized. Polytetrafluoroethylene (PTFE) is a well-known high melting point, inert polymer used in the form of tape and coatings on packaging machines to reduce adhesion.

2.3.10 Cellulose-Based Materials [13]

Regenerated cellulose film (RCF) is one of the first packaging films prepared from pure cellulose fiber derived from wood. The extracted cellulose is dissolved and then regenerated by extrusion followed by acid treatment. Since CRF or cellophane is not processed in a molten phase or softened by heat, it is not called a thermoplastic material. The naturally derived, high molecular weight cellulose can be made flexible by plasticizing with humectants (glycol-type compounds). The degree of flexibility of cellulose ranges from fairly rigid level to the most flexible, which is known as twist wrap.

Properties
RCF shows inferior barrier properties to water vapor. This property is very important for the products that need to lose moisture, for example, pastries, cakes, biscuit and other flour confections, to achieve the correct texture when packed. Plastic films having high moisture barrier keep the relative humidity (RH) too high inside the package resulting in mold growth. Under dry condition, RCF provides good oxygen barrier to a product. Properties like heat sealability and barrier to water vapor and common gases may further be improved with suitable coating on the film, like nitrocellulose (MS type) or PVdC (MX type) coating. The barrier performances may range depending on the choice and method of coating. The coated films may be colored and metallized. RCF is printable. Paper, aluminum foil, PET, metallized PET and PE lamination with RCF will help to achieve specific levels of performance.

Applications
- Cellophane is used in food packaging as biodegradable wrapper.
- Cellulose acetate derived from cellulose, has high transparency and gloss. It is printable. Laminate of cellophane finds application in confectionery cartons.

2.3.10.1 Polyvinyl Alcohol (PVA)

Polyvinyl alcohol (PVA) is a synthetic polymer that is soluble in water. It is effective in film forming and emulsifying. It also has an adhesive quality. It is without any odor and is non-toxic, is resistant to grease, oils and solvents. It is ductile but strong, flexible and functions as a high oxygen and aroma barrier. PVA also finds application as a coating agent for food supplements and does not pose any health risks as it is not poisonous. Major industrial application of PVA includes food packaging, which accounts for 31.4% of the global share in 2016. Thin and water-resistant PVA film can also handle moisture formation from foodstuff. Its cross-linking density and resistance to moisture are added benefits to its usability in this area. PVA has high tensile strength, flexibility, water solubility, resistance to organic solvent and mild solubility in ethyl alcohol. It has a melting point at 185 °C. PVA packaging film is biodegradable.

2.3.11 Manufacturing Process

2.3.11.1 Extrusion Film Processing [14] and Laminates

Polymer films can be produced through several different techniques such as film casting on a smooth surface, calendaring, film extrusion, etc. Although each one of these techniques hold their own advantages but extrusion film processing can be considered as the most widely accepted and applied technique as it enables us to produce quite high quality extremely uniform films. A large portion of thermoplastics consumed today is extruded in the form of films which are mainly used in packaging sectors with food packaging being the most important. Remaining extruded films are mainly used for agricultural and construction applications. Extrusion process involves feeding and subsequent melting of raw plastic in the extruder. The resulting polymer melt coming out of the barrel can either be casted in the form of a film or be extruded as a tube of molten polymer which can then be inflated to several times its initial diameter to form a film bubble (Fig. 2.5).

Material Selection

A particular grade of a polymer can be classified as film extrusion grade on the basis of following factors.

Density—Polymers chosen for film extrusion process must have an appropriate combination of amorphous and crystalline regions as a completely amorphous polymer would have rubbery nature and pure physical properties, whereas a completely crystalline polymer would be very hard and brittle therefore not suitable for film forming. In case of polyethylene, higher density is related to higher

2.3 Plastic Films and Laminates

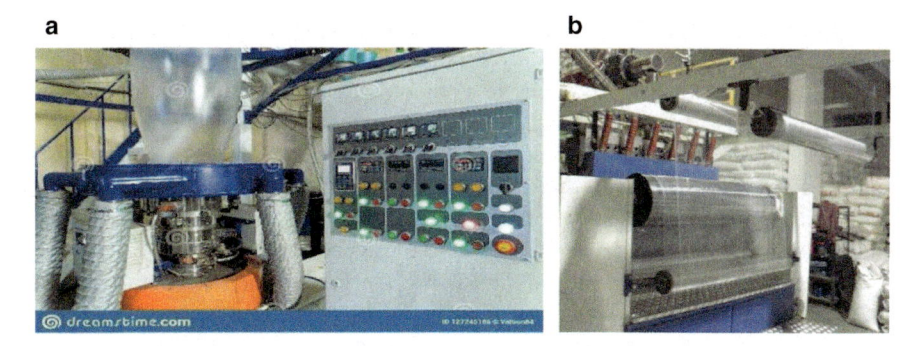

Fig. 2.5 a Extrusion film casting, b blown film extrusion

crystallinity whereas in polypropylenes, due to the different chemical structure it can attain high crystallinity while being relatively less dense. Therefore, as density defines crystallinity in different ratios for different polymers, the appropriate density range in order to classify it as film extrusion grade varies. LLDPE density ranges from 0.9 to 0.939 g/cc, whereas for HDPE used in film extrusion density lies in range of 0.941 to 0.965 g/cc. Density of polypropylene grades that can be cast into films lies in range 0.890 to 0.905 g/cc.

Molecular Weight—MW of a polymer is also one of the major determining factors as an increase in molecular of a polymer can lead to an increase in toughness, tensile strength and most importantly environmental stress cracking resistance which is very important in case of films as polymer films are often subjected to stresses in presence of liquids such as solvents and detergents. Extremely high molecular weights also tend to make a polymer unsuitable for film formation as it will affect its flow behavior.

Molecular Weight Distribution—MWD of a polymer refers to presence of different chain lengths of a polymer. A wide MWD means a combination of varying chain length is present in a polymer, whereas narrow MWD refers to presence of similar chain lengths along the polymer. A narrow MWD is preferred for film extrusion process as it provides higher stress cracking resistance and better optical properties.

Melt Viscosity—It is often expressed in terms of melt flow index (MFI) values (gms per 10 min) measured under standard conditions (2.16 kg/5 kg/20 kg at 190 °C/230 °C). It can also be considered as the most important property as it relates to the processing of a polymer and determines if the polymer can be processed into a film. It defines the flow behavior of a polymer. Polymers with very high MFI values tend to flow very easily and form thin films but will be vulnerable to breaks in flow and therefore non-uniform films, whereas polymers with very low MFI tend to be extremely difficult to process inside the barrel of an extruder as they are quite resistant to flow. Homopolymer with MFI 2–3 is considered suitable for extrusion blow molding, while homopolymers with MFI around 7 are found suitable for cast films.

Extensional Viscosity—Extensional viscosity which is often referred using melt strength is a property specific to only some of the polymer processing techniques that involves stretching of a polymer to a notable extent. Extrusion film processing being one of such techniques depends largely on extendibility of a particular grade. Extrusion blown film requires inflation of a tube of molten polymer to several times its initial diameter, and even during extrusion casting, polymers are subjected to being extended from more than one dimension, therefore making this property a vital requirement, while selecting a suitable grade, as polymers with low extensional viscosity tends to crack during the film formation forces. Melt strength which is the measure of extensional viscosity of a polymer is defined in terms of maximum pulling force which a strand of semi-molten polymer can withstand prior to cracking.

Processing Parameters

There are several processing parameters controlling the output product of as mentioned process, and these properties are mentioned below.

Temperature—The polymers classified as film grade are mostly semicrystalline in nature requiring temperatures reaching 20–50 °C above the melting point of the resin toward end of the barrel. Temperature profiles can be optimized according to the properties of polymer resin being extruded, but on general basis, a difference of 10–11 °C is maintained between different barrel regions. It is generally suggested to choose highest possible extrusion temperature up to the point which still allows for adequate cooling and film wind up.

Screw Speed—Screw speed directly related to the output of the process, the faster the screw runs the more will be the production rate. Increase in screw speed generates heat, therefore requiring less amount of heat to be provided externally. Although extremely high screw speeds can cause temperature and pressure fluctuations which can compromise the quality of the film as it can lead to non-uniformity in the flow causing specks to appear in the film.

Die Width—This is a factor specific to extrusion film casting, for films ranging from 25–75 μm in thickness, the die opening is around 0.5 mm. The more the width of the die the more will be the extrudate stretched thin. As the polymer extrudate passes on to chilled rollers in case of extrusion casting, the films tend to shrink at the edges which can cause beading because of which the edges would be required to be trimmed, therefore causing polymer loss. This can be avoided with a narrower die opening and some other factors such as distance between die and chilled rollers.

Cooling—A controlled cooling is required in film processing processes as non-uniform cooling as indicated by uneven frost line can lead to warping and wrinkling of the polymer films and will also cause non-uniformity in film thickness. Frost line in case of blown films should be as high as possible to obtain high quality films but too high frost line will cause the film to stick while rolling up.

Blow-Up Ratio—It is a parameter specific to blown film extrusion, and it is mainly defined as the ratio of final bubble diameter to that of the diameter of the die. The higher the blow-up ratio the more is the stretching of the polymeric material which can be used to control the thickness of the film. Extremely high blow-up ratios

2.3 Plastic Films and Laminates

are normally avoided as they can cause the polymer films to crack or can cause non-uniformity in film thickness. The values of blow-up ratio for polymer such as LDPE and LLDPE lie in the range of 2–3, whereas in case of HDPE, it can go to relatively much higher values 6–8.

2.3.11.2 Quality Control

Quality of the produced polymer films can be controlled through several processing parameters as discussed above. Some of the prominent properties are mentioned below.

Gloss—It is defined as the percentage of light incident at a particular angle reflecting back at the same angle. It is mainly dependent on processing temperature as increase in processing temperature improves the above-mentioned properties. In case of blown films, rise in frost line also improves gloss.

Strength—Strength of a film can vary along the machine direction and cross direction. In case of cast films, the film gets puller through the roller in machine direction, thereby aligning the chains in MD giving high strength to films in that specific direction. The stretching and thereby alignment can be controlled and optimized by varying the rotating speed of chill rollers. In case of blown films, the blow-up ratio can also be inferred to the measure of film stretching in cross-direction, and therefore, by controlling the blow-up ratio, one can control the strength ratio of films in machine and cross-direction, giving an overall toughness to the film.

Toughness—It is defined as resistance of a film to split under stress. It can be improved by maintaining a proper balance between strengths along machine and cross-direction. Polymers with low MFI and high molecular weight tend to form relatively tough films.

Slip—Packaging often requires films to have lower coefficient of friction which is completely opposite to obtaining high gloss as lower processing temperatures and lower frost line in case of blown films decrease the coefficient of friction of the film but at the same time compromising the gloss property.

Stiffness—Films especially the ones used in packaging are required to be flexible. Cast films are generally more flexible than blown films. In terms of the polymeric material, high density polymer is found to form relatively stiff films.

Heat Sealability—It is one of the foremost requirements in packaging films. Polymers such as polyethylene provide excellent heat sealability. Although during processing, extremely high temperatures can sometimes cause oxidations of films, thus making it less heat sealable.

Gauge Uniformity—It is the measure of difference in film thickness along its center and circumference. A difference of >10% should be avoided to obtain better physical properties. This can lead to poor bubble stability and wrinkling of the film as well as tear in the film. This is mainly caused and can be controlled by avoiding non-uniform heating of polymer melt or also overheating as it may cause the melt coming out of the die to be too fluid in nature. Poor mixing of polymeric materials also leads to this issue.

Clarity—Clarity can be improved by increasing the processing temperature and at the same time cooling rate. Although since cool air is used for the cooling purpose in extrusion blow molding, speed of the air flow must not be very high as it may cause the bubble to chatter and therefore compromising the clarity.

Barrier Properties—They are normally intrinsic trait of the polymer material being used, although in terms of processing factors non-uniform thickness will affect the barrier properties of the film. In case of very thin films also, chances of imperfections such as "fisheye" increase which may also reduce the barrier properties to a good extent.

Laminates and Coextrusions: [15]

Flexible packaging uses flexible film that is generally made by the process of blow film process. The thickness of these films can be adjusted as per requirement (Fig. 2.6).

Flexible packaging uses films either as a single component monolayer film or as a combination of more than one plastic in more than one layer. Combination of plastics can be achieved by two methods: lamination and coextrusion.

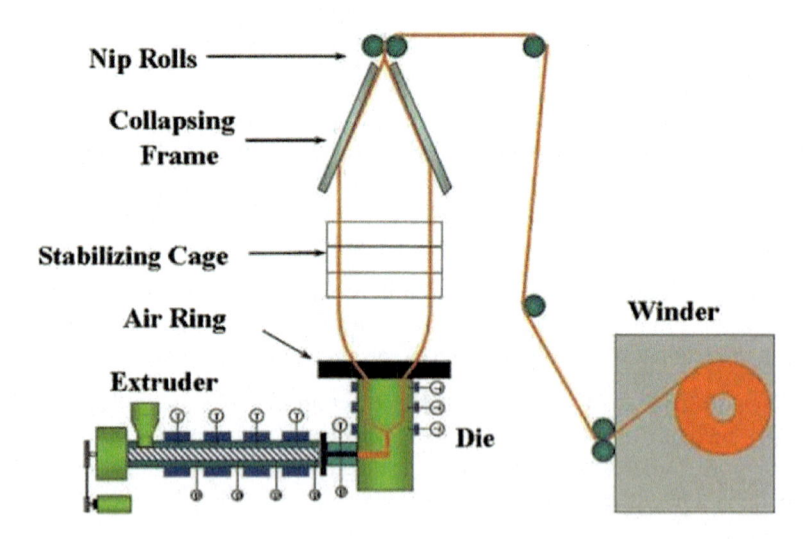

Fig. 2.6 Blow film process

2.3 Plastic Films and Laminates

Lamination process includes bonding together two or more plastics or bonding plastic to another material such as paper or aluminum (as discussed in the section on metal). Bonding of two surfaces can be commonly achieved by use of liquid- (water, solvent) or solids-based adhesives. First the adhesives are applied on one surface, and then, both the films are passed together through high pressure rollers to bond them together. For thermoplastic polymers, lasers can be used to laminate instead of adhesives. Lamination process makes reverse printing possible, in which the printing is buried between layers and is not subject to abraded or affected by heat sealing process.

Film lamination by adhesive

Plastic film lamination combines two or more films with an adhesive. Whereas in coextrusion, two or more layers of molten plastics are combined during the manufacturing process. Lamination is possible using molten polymer as the molten layers act as adhesive. Adhesive less laminate formation, e.g., laser lamination of thermoplastics is also a good choice. The choice of making a laminate or a coextrusion depends on several factors like product need, type of pack (as per shelf life and barrier properties), handling at every stage and the run length.

Coextruded films can only be surface printed unlike laminated films, where one of the films should be reverse printed so that the print can be sandwiched during lamination. This process makes the laminate rub resistance, gloss and clarity. Any alternative approach to provide the required protection by the product, (either by a single plastic material or by a form of coextrusion) several factors must be considered like thickness of the film, barrier property, stiffness, sealing ability and handling difficulties. Sealing time may be varied to overcome poor heat transfer and heat retention. Once the sealing is done, it should be allowed to reopen the seals.

Stiffer materials (which are also a thicker material) can withstand and display better than package made from a thinner material. Due to the different level of stress in each film web, lamination will undoubtedly face curling. In applications where the film undergoes through processes like cutting and then pushed or pulled, the cut edge may require to be smooth and flat to give trouble-free feeding through the packaging machine.

A wide range of adhesives are available for lamination. Water-based adhesives like PVA and other which remain flexible and have a long shelf life may not be effective in adhering polymers with inert surfaces and high moisture barrier like: polyolefins. A long drying period is required with these adhesives before the proper use of these laminates. Water dispersion occurs with substrates like paper and paperboard. Cross-linking adhesives like polyurathane can be used for high barrier plastics. Some important applications of these adhesives are that the film and the gravure roller adhered under pressure (with a coating weight of 1–3 g/m^2). Care should be taken to avoid the formation of carbon dioxide bubbles which may damage the visual properties of the laminate. In few cases, even the adhesive may react with the coating of film to produce discoloration. Scientists are trying to move away from organic solvent-based adhesive systems to the use of water-based and 100% solids systems

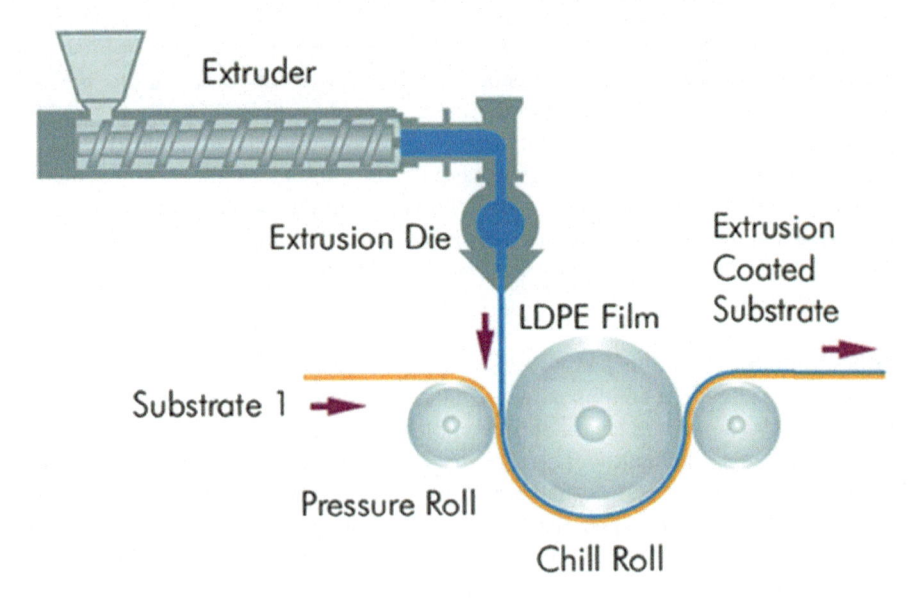

Fig. 2.7 Lamination process

to reduce solvent emissions. 100% solid adhesives cross-link as a result of applied heat, UV or EB radiation after application. Tacky hot melt adhesives with a wax and EVA content and the use of PE in extrusion laminating would also be considered 100% solids adhesives.

Extrusion lamination—In this process, one surface of the film is coated with molten PE and then combined with another film by pressing, while the PE is molten. A small amount of PE (typically 7–10 g/m^2) is used as adhesive. The laminate can be made stiffer by increasing the thickness of the total structure. Application of primer on the surfaces of the films is a need to achieve the PE strength.

In thermal lamination, two layers each having good heat sealing properties are adhered by pressing the films using heated nip roller system. Application of adhesives increases the final weight of the laminate as compared to that of the original components. This lamination process of the films depends on the sealing points, as under stress the films shrink on heating at high temperature. It causes creasing or stretch under tension. As the films approach the elastic limit, curl may be formed because of slightly shrinkage of one layer (Fig. 2.7).

In **coextrusion process**, two or more layers of plastics in molten condition are combined during the film manufacturing. This process is more rapid compared to lamination but requires materials with certain thermal characteristics that allow coextrusion process. Recycling is difficult for both coextruded and laminated products as they combine multiple materials. Combining materials leads to the additive advantage of properties from each individual material and often reduce the total amount of packaging material required (Fig. 2.8).

2.3 Plastic Films and Laminates

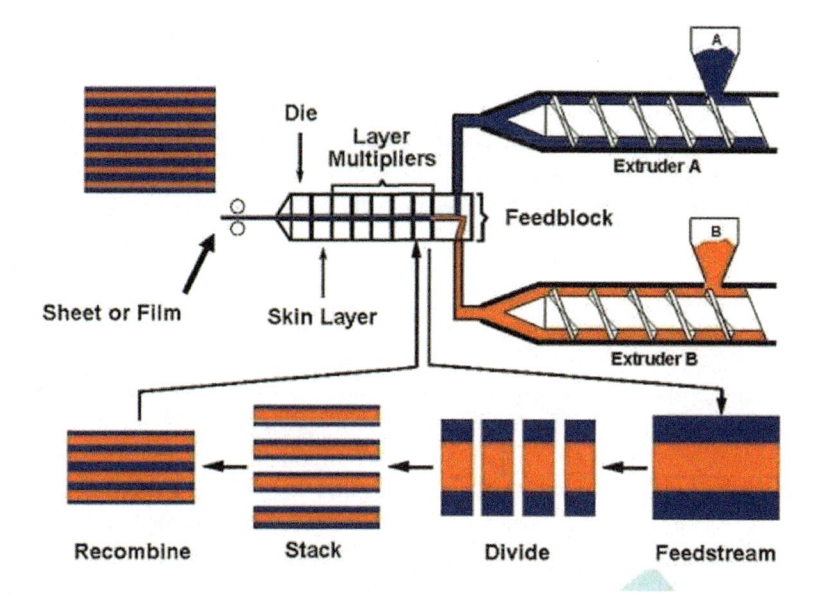

Fig. 2.8 Coextrution process

Flexible packaging may be considered as the most convenient and economical way to package food items, beverages and other products. In other words, flexible packaging with plastic ensures a long shelf life of the products and in efficient distribution with minimum losses.

2.4 Aluminum Foil [16]

Aluminum foil is an aluminum alloy having 99.0% aluminum, and rest is manganese, copper or magnesium. The thickness of aluminum foil is in between 0.004 and 0.24 mm.

2.4.1 Foil is Made by Two Methods

i Rolling method
ii Converting method.

2.4.1.1 Rolling

In this process, aluminum sheet is pulled between two rotating rollers under pressure. At every rolling step, the thickness of the sheet is reduced by 50%. The foil surface which touches the roller has bright finish, while other side has matte finish. The rolled foil is annealed to make it soft, and it is done in furnace.

2.4.1.2 Converting

In this process, foil is either converted into laminate or printed. Foil can be laminated with variety of plastic films or paper. The adhesive used for lamination consists of water-based, solvent-based or of different kind of waxes.

Foil can be printed by any standard method of printing like flexography or gravure.

Aluminum foil is available in a wide range of thicknesses, with thinner foils used to wrap food and thicker foils used for trays. It is available in a range of thicknesses from 7–20 μm when it is used as wrapping for foods and 50–100 μm when it is used as trays for street foods.

2.4.2 Properties

Like all aluminum packaging, foil provides an excellent barrier to moisture, air, odors, light and microorganisms. It is inert to acidic foods and does not require lacquer or other protection. It is resistant to ultraviolet light and afford excellent protection to chocolate, butter, cheese, etc. Although aluminum is easily recyclable, foils cannot be made from recycled aluminum without pinhole formation in the thin sheets.

The number of pinholes in foil can be checked by the following method:

1. Cut a length of 30 cm of foil from the width of a roll, and hold it up against the sun or a bright light.
2. Check which part of the foil seems to have the most pinholes and then carefully cut a 10 × 20 cm sample.
3. Count the number of pinholes in the sample, and multiply the result by 50 to find the number of pinholes per m^2 (Fig. 2.9).

Packing made from aluminum foil is most widely used in the food industry.

- Aluminum foil is non-toxic, so it does not damage the foods wrapped in it, but instead protects them. Aluminum foil is used in food containers, bins, bottle caps, soft packages for liquids or bulk food stuffs, trays, laminates and many other types of containers.
- Foil protects foods against sunlight exposure which can cause them to go bad.

Fig. 2.9 Aluminum foil application

Foil does not melt from high temperature, nor does it lose its shape or impart any bad smell to the food wrapped in it. Foil with water and greaseproof paper on one side is used in the packaging of butter, cheese, ice cream and other dairy products.

- Aluminum foil having 0.009 mm thickness laminated with paper with gold color is used for decorative purposes. These laminates are also used by label manufacturers.

2.5 Plastic Woven Sack

Over a period of time, plastic woven sack has become very popular as one of the important packaging media for the bulk transportation of food products like food grain, wheat, sugar, pulses, wheat flour, gram flour (besan), etc. However, it is also observed that the woven fabrics are also used for the manufacturing of bag with as consumer package with extrusion lamination by means of polyethylene from inside to improve upon barrier properties against the ingress of moisture content during storage and transportation.

These bags are normally manufactured with a capacity of 5 and 10 kg, mainly for the packaging of rice, wheat flour (atta) for domestic and export market. In order to provide the carrying facility, two number of handles made from same fabrics are also provided at the opening of the bag. In the recent days, the laminated printed bags are also made to enhance a particular brands which are highly accepted by the modern consumers for shopping purpose. These bags are preferred by the consumer for easy handling, good mechanical strength and safety. Moreover, the used bags are normally reused for carrying any kind of dry products. Hence, the plastic woven bags are also considered for multiple use bag [17–22].

2.5.1 Materials

Thermoplastic materials like high-density polyethylene (HDPE) and polypropylene (PP) are mainly used for the manufacturing of woven fabrics which are fabricated into different sizes of bags or sacks with varying capacity depending upon the application. The popularity of this packaging medium is mainly due to certain important factors like easy availability of polymeric resin, light in weight as compared to jute and cotton fabric, very good chemical resistance and lower conversion cost. It is also observed that different countries are using different types of polymeric materials either HDPE or PP which is completely dependent on availability and supply of these materials and the price in those country.

Plastic woven sacks are normally manufactured for the bulk package of having capacity of 20, 50 kg. In addition, flexible Intermediate bulk containers (FIBC) or jumbo bags are also manufactured for higher capacity of package ranging from 500 to 3000 kg which are mostly used for dry food products like guar gum, soybean, tea powder, etc., for domestic and export market. India is exporting huge quantity of empty FIBC to Europe, USA for the packaging of dangerous Chemicals and the exporter do need to maintain the quality of FIBC as per International Maritime dangerous Goods (IMDG) codes as the export consignment of FIBC's are mostly transported by sea. The plastic woven fabrics are made from HDPE or PP yarn by means of circular weaving machine [23, 24].

2.5.2 Properties

Properties of high-density polyethylene (HDPE) fabrics are as follows.

- Light in weight
- Density of HDPE ranges from 0.94 to 0.965 g/cc.
- Good corrosion resistance
- Good mechanical properties
- Good barrier properties
- Water repellent
- Offers very good chemical resistance
- Practically, the fabrics are free from any kind of odor and taste.
- Amenable to print on the surface
- Suitable for the use of this fabrics in contact with food products
- Does not allow for bacterial growth
- Fairly good resistance to UV radiation.

Properties of polypropylene (PP) fabrics are as follows.

- Light in weight
- Density of polymer is 0.91 g/cc. Good stiffness but tendency to fibrillate.
- Good corrosion resistance

2.5 Plastic Woven Sack

- Good mechanical properties
- Water repellent
- Offers very good chemical resistance
- Practically, the fabrics are free from any kind of odor and taste.
- Amenable to print on the surface
- Suitable for the use of this fabrics in contact with food products
- Does not allow for bacterial growth
- However, resistance to UV radiation is not as good as HDPE manufacturing process.

2.5.3 Manufacturing

As per the demand of customer, one has to choose those PP woven sacks which are easily printed and manufactured. However, the artwork and number of color in printing are decided by the customer. The end users adopt different kinds of color combinations and designs in the printing of these sacks to convey the massage(s), characteristic(s), quantity- and quality-related details and handling instructions, etc. For some kinds of specific applications like filling of Hydroscopic Materials, e.g., Chemicals, Fertilizers, Food Products, etc. The Process of manufacturing PP woven sacks are given below [22, 25]:

- Manufacturing of tape
- Extrusion of film
- Quenching of film
- Slitting of film
- Orientation of tapes
- Annealing of tapes
- Winding extrusion.

Extrusion
The process of manufacturing PP woven bags involves mixing raw materials starting with PP or HDPE pellets and other additives, and extruding the raw materials into a yarn PP resin is heated with feeler of $CaCo_3$ and pigment, melted and extruded as a flat film. The flat film is then slit into tape yarn by the slitting unit and stretched and annealed. Next, a take-up winder winds the heat-oriented tape yarn onto a bobbin.

Weaving
The process of weaving of yarn is like the weaving process of textile fabrics. Circular weaving machine is used to weave circular fabric from flat tapes. These tapes are commonly termed as Raffia. In fact, Raffia is fibrous material used for tying plants, seems to have come the leaves of a palm tree of the genus raphia. Circular looms are used to convert Raffia Tapes into fabric cloth by weaving which further produce circular cloth of desired width. Yarn is taken from the bobbin of machine in the

Fig. 2.10 Circular loom machine

circular loom's creel stand, and then, it is formed into woven cloth which is tubular in shape. It is an automatic and continuous process. The woven cloth produced is wound on rotating pipes and cut into required dimensions (Fig. 2.10).

Finishing and Stiching
The woven fabric is then cut into desired sizes of making the bag of 5, 10, 25 and 50 kg and the printed. The printed cloth is then stitched manually at different sizes of bag.

However, a valve is also provided in one corner of the cut piece of fabric prior to stitching process.

Few of the important designs of plastic woven fabric-based bags are given in Figs. 2.11 and 2.12:

2.5.4 Applications

- Cereals
- Food grains
- Sugar
- Milk powder
- Plastic granules
- Wheat
- Flour.

2.5 Plastic Woven Sack

Fig. 2.11 Flexible intermediate bulk container

Fig. 2.12 Plastic woven sack

2.6 Jute—As Packaging Medium

Jute is considered to be one of the oldest traditional packaging materials in the Indus Valley Civilization since 3rd millennium BC. In the mid-1500s to early 1600s, villagers, mostly Bengalis, used to make clothes from jute with the help of hand-spinning wheels. In addition, they used to make rope and twine from white jute fiber for their use around the home and made bags for carrying food grains and other agricultural products. In seventeenth century, this fiber became very popular, and number of jute mills were also opened in Europe [19].

However, the trade of jute had really boomed throughout the eighteenth and nineteenth centuries. But in 1970, the introduction of synthetic fibers like nylon and polythene in the market has made a downward market trend of jute. But in the recent days, jute is again considered as an important natural material, and its popularity in terms of its various application has been increased. In fact, nearly 75% of jute goods are used as packaging materials, burlap, gunny fabric, and various bags. Carpet backing makes up about 15%, and the rest are for yarns, cording, felts, padding, twine, ropes, decorative cloth and heavy duty industrial items.

This is a rain-fed agricultural crop with little requirement of fertilizer or pesticides in comparison with cotton. Specifically, 85% of jute production is localized to the Ganges River Delta, which spans throughout Bangladesh and the Bengal region of India. India is the world's largest producers of jute. Jute is a long, soft, shiny bast fiber that can be spun into coarse, strong threads. "Jute" is the name of the plant or fiber used to make burlap, hessian or gunny cloth. It is one of the most affordable natural fiber next to cotton. Jute fiber is also termed as golden fiber for its color and high cash value. The continuous research and development have made it possible to unfold many possibilities of applications of this materials, and generally, the application of these materials was used to be as traditional packaging materials as sack or bags.

These developments are also supported with technical evaluation based on their basic properties, behavioral pattern under varying climatic conditions, conversion possibilities into various forms, performance of such forms for specific end-use applications and subsequently leading to the preparation of standards for the basic materials and code of practice for packaging, industrial and other applications.

2.6.1 Use of Jute Fibers

(i) Jute felt and blends—to make composite boards
(ii) Rope
(iii) Sacks of different capacity
(iv) Bags with handle for consumer package
(v) Twine
(vi) Union fabrics with HDPE/PP
(vii) Laminates with paper, plastics and aluminum foils.

2.6.2 *Manufacturing Process of Jute Fabric [17, 18, 20]*

The following steps are followed for the manufacturing of Jute fabric:

Harvesting

(Mature Jute stacks are harvested by hand)

⬇

Defoliating

(Then they are defoliated)

⬇

Retting

(Retting is a process where the jute stems are bundled together and immersed them into slow running water to remove the non-fibrous materials from the stem and skin of the jute stack).

⬇

Seperating and Combing

(After the jute stalk is retted, it becomes soft and facilitates to separate the long silky fibers easily and comb them into long strings).

⬇

Spinning

(These combined fibers are then spun into yarn. The yarn is then wound into reels).

⬇

Finishing

(The fibers may be subjected to a variety of chemical processes to dye it, provide it with water resistance or make it fire resistance).

⬇
Weaving

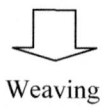

(The finished reels of jute fibers are then shipped out to textile processing units and subjected to weaving to make the fabrics).

2.6.3 Terminology Related to Jute Fabrics [21]

There are number of terminology used in the jute industry. However, the important terminologies related to the weaving of jute fabrics which are used for the fabrication of sacks or bags are given below:

(i) Warp and Weft: These are the two basic components used in weaving to turn thread or yarn into fabric. The lengthwise or longitudinal warp yarns are held stationary in tension on a frame or loom, while the transverse weft (sometimes woof) is drawn through and inserted over and under the warp.

(ii) Ends and Pick—A single thread of the weft crossing the warp is called a pick. Each individual warp thread in a fabric is called a warp end or end.

(iii) A-Twill Cloth—A double warp, 2/1 twill weave sacking jute cloth having 102 ends/dm and 35 picks/dm and weighing 760 g/m.

(iv) A-Twill Bag—A bag made from A-twill cloth, i.e., double warp hemmed twill bag of having standard dimension 112 × 67.5 cm size and weighing 1190 g.

(v) B-Twill Cloth—A double warp 2/1 twill weave sacking jute cloth with 76 ends/dm arid 31 picks/dm and weighing 643 g/m.

(vi) B-Twill Bag—A bag made from B-twill cloth of having dimensions of 122 × 67.5 cm of having masses 1110 g (overhead stitched) and 1115 g (Herakle stitched) for 100 kg, 112 × 67.5 cm of having masses 1120 g (overhead stitched) and 1125 g (Herakle stitched) for 93 kg, 106.5 × 61 cm of having masses 880 g (overhead stitched) and 885 g (Herakle stitched) for 75 kg) and 94 × 57 cm of having masses 730 g (overhead stitched) and 735 g (Herakle stitched) for 50 kg.

(vii) Burlap—A term applied in the American trade to jute hessian.

(viii) Canvas (Jute)—A plain weave cloth made wholly of jute with double warp and single weft interwoven, weighing not less than 407 g/m. The number of warp threads (ends/dm) shall be more than 118, and the number of weft threads (picks/dm) shall not be less than 55.

(ix) Bale (Gunny)—A rectangular- or square-shaped compressed rigid package containing jute fabrics or bags wrapped with bale covering with outer layer stitched and bound by metal hoops.

2.6 Jute—As Packaging Medium

Fig. 2.13 Jute bag for rice

(x) Bleaching—The process of making natural jute fiber or fabric white by chemical means with or without any scouring treatment or removal of natural coloring or extraneous substance.

Lightweight and heavy weight jute bags are commonly known as jute bags, mainly used for packaging of rice, wheat, sugar, coffee bean, cocoa bean, pepper, shelled nuts and other bulky grain products. In general, the capacity of the bag ranges from 50 to 100 kg. Jute bags or sacks are also used in large quantity for packaging of tea and coffee beans.

Few of the important photographs of jute bags for packaging of rice and coffee beans are given in Figs. 2.13 and 2.14

2.6.3.1 Characteristics of Jute Fabrics [19]

- This fabric is also termed as burlap, hessian cloth, gunny cloth
- This purely natural material is made from the bark of the jute plant
- It is composed of cellulose and lignin
- Jute is strong and durable and has ventilating and absorbent properties
- Fabric has got high breathability
- It has also got high moisture-wicking abilities
- Jute fabric has got medium abilities to heat retention
- Low stretchability
- Low prone to pilling/bubbling
- Low thermal conductivity and a moderate moisture regain
- High mechanical strength and thus prevents damage to the packaged product and so the jute bags are still frequently used today due to its excellent qualities

Fig. 2.14 Coffee in jute bag

- Commonly used for making bags, sacks, ropes, textiles fabrics, carpets, rugs, curtains, decorative items, canvas, agricultural erosion prevention, cardigans, sweaters, etc.
- Jute bags are reusable, eco-friendly and biodegradable bags made from the fibers of jute plants. They are commonly used as shopping bags

2.6.4 Benefits of Jute Packaging [17]

- Very good physical and mechanical strength
- Generally light in weight
- Amenable to handle easily
- Facilities high stack due to low slip
- Availability of various grades
- Amenable to various conversion processes and techniques
- It could be bleached, colored and printed.
- It has got good load bearing capacity.
- Memory of regainability of structure and thus facilitates to metallic hooks during handling.
- High resale value

- Environmentally friendly—It is 100% biodegradable and can even be used as compost for the garden. When compared to paper, it takes less time for biodegradation.
- Recyclable—It also uses much less energy to recycle as compared to paper.
- Reusable—It can be reused again for a variety of purposes.
- Sustainable—Jute can be grown throughout the year without the use of pesticide of fertilizer and reaches to maturity within six months. Even the woody part of jute plants known as hurd which could also meet world's wood needs. It has got high potential to have a huge positive impact on the reduction of deforestation.
- Durability—Jute fibers are strong and durable. In fact, they are much tougher and more resilient than paper made from wood pulp and can handle prolonged exposure to water and weather.
- Cost effective and cheaper than plastic and paper bags
- Strong and can carry more weight as compared to promotional carry bags.
- Jute has good insulating and antistatic properties.
- Jute hemp fiber has a reasonable fiber strength, thus, it is good source for pulp for making paper and board, and it also adds to its recycling features.

2.6.5 Limitation of Jute Fabrics [18]

- Being of natural origin the felt, yarn and fabric are sensitive to moisture and microorganisms.
- It is hygroscopic in nature, and thus, it losses the strength with continuous exposure to heat—light and oxygen.

Applications

- Twines and ropes are extensively used for tying, bundling and reinforcement.
- Fabrics are normally used for baling, bundling. Sacks, bags and laminations, etc., for packaging of cereals, pulses, sugar, fertilizers, pesticides and insecticides, chemicals and horticultural produces.
- Jute fabrics can be laminated with barrier films liners like polyolifins, foil or combinations thereof or provided with loose liners of such barriers for the packaging of tea, coffee beans, spices, etc.
- Besides the conventional pillow-tubular type, they could be made as gusseted, rectangular, carry home for better aesthetics and functional as well as reuse shopping bags.
- The fabric is also laminated to kraft paper and polyethylene to get high strength media and converted into bundling and bag applications. Such flexible laminations offer excellent strength and varying combinations for packaging of sensitive commodities or high density products.

- Fabric laminated corrugated boards are used for the packaging of textile garments, suitings and shirtings, spare parts, heavy and light engineering goods, handicrafts, etc.

2.7 Leno Bag

India being the second largest producer of fruit and vegetables in the world requires a lot of packaging related to fresh produce. But the usage of leno bag is still very low in that comparison. Leno bag has a great in India, and the seasons are suitable for dry skin vegetables like potato (third biggest producer), onion, garlic, etc. Leno bags can be made into different forms, like flat yarn leno bags, netted bags, PP leno netted bags, leno mesh bags, leno packaging bags and plastic woven leno bags.

The leno bags are primarily made of the versatile material PP. Leno is a kind of weaving in which adjacent warp tapes are twisted around consecutive weft tapes to form a spiral pair, effectively locking each weft in place. These nags are made primarily with the characteristics of good aeration and strength. Leno bags are suitable for packing and preserving dry vegetables like potato, onion, ginger, garlic, cabbage, etc., and fruits like pineapple, citrus fruits, raw mango, coconut, etc. These have widths between 20 and 72 cm. The length may also be varied as per the requirements. Leno bags on an average weigh 50 g (or less) [26, 27].

2.7.1 Advantages of Leno Bags

- Leno bags have superior aesthetics.
- Good mechanical properties
- Chemical inertness
- Easy to handle and store
- Reusability and recyclability
- Lighter weight as compared to jute bag and thus cost effective
- Moisture and chemicals resistance
- No odor to the packed produce and is food grade material.

2.7.2 The Process of Manufacturing

Leno bag manufacturing process involves the following steps (Fig. 2.15):

The leno bag manufacturing process includes the following steps:

PP granules along with additives are melted and extruded into a film through an extruder where it is heated externally at a temperature of 190–250°C. Next, the molten films are cooled in water and a blade is used to cut the film into filaments.

2.7 Leno Bag

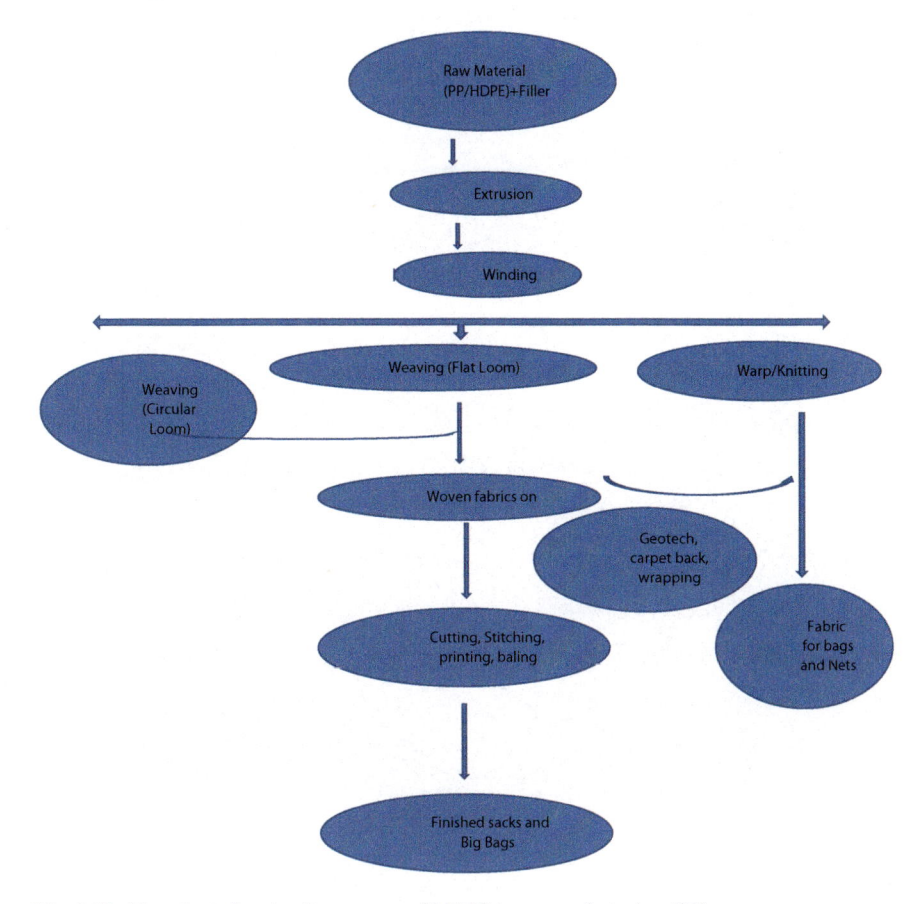

Fig. 2.15 Flowchart showing the process of LENO bag manufacturing [28]

The filaments are then stretched at high magnification at high-temperature oven until a flat yarn appears. The flat filament is passed through the hot roll to heat-set the filament. It subsequently shrinks at a low traction speed and processed by the cold roll at low temperatures. Finally, these flat filaments are wound on to the bobbin to make a spindle.

Weaving A cloth is made by interweaving the flat warp and the flat wept by a weaving machine. The circular loom weaves the tube cloth as there are many spindles on the warp frame of the circular loom.

Laminating: Net bags with trademarks and white films need to undergo this step. the albuginea and the trademark film are laid in the middle of the mesh and laminated.

Sewing leno Bags Stitching strength index of the bags is one of the most critical step in bag making. For laminated cold-cut bags, crimping is generally used because it may tear out the stitches on the cold-cut edges from the warp threads along with the weft threads.

Fig. 2.16 Leno bags

2.7.3 Application

- For the packaging of vegetables like cauliflower and cabbage.
- Extensively used as transport package for potato, onion, etc. (Fig. 2.16).

References

1. Wood B (2012) Packaging outpaces overall economy. Plast Technol. http://www.ptonline.com/columns/packaging-outpaces-overall-economy. Accessed 16 June 2020
2. Packaging market research report. https://www.fortunebusinessinsights.com/industryreports/food-packaging-market-101941#summary. Accessed 16 June 2020.
3. Food Packaging Market Size, Share & Trends Analysis Report by Type (Rigid, Flexible). https://www.grandviewresearch.com/industry-analysis/food-packaging-market/toc. Accessed 16 June 2020
4. Piergiovanni L, Limbo S (2016) Food packaging materials. Springer, Basel

5. Parkar J, Rakesh M (2014) Leaching of elements from packaging material into canned foods marketed in India. Food Control 40:177–184. https://doi.org/10.1016/j.foodcont.2013.11.042
6. Deshwal GK, Panjagari NR (2019) Review on metal packaging: materials, forms, food applications, safety and recyclability. J Food Sci Technol 1–16
7. Deshwal GK, Panjagari NR, Alam T (2019) An overview of paper and paper based food packaging materials: health safety and environmental concerns. J Food Sci Technol 1–13
8. Ojha A, Sharma A, Sihag M, Ojha S (2015) Food packaging—materials and sustainability—a review. Agric Rev 36(3):241–245
9. Robertson GL (2013) Food packaging: principles and practice. CRC Press
10. https://www.tri-cor.com/how-polyethylene-film-is-made/#:~:text=This%20plastic%20film% 20manufacturing%20process,they%20become%20molten%20and%20pliable.&text=The% 20molten%20plastic%20is%20pushed,of%20plastic%20called%20the%20bubble
11. http://www.madehow.com/Volume-2/Plastic-Wrap.html#:~:text=Raw%20Materials,is%20m ade%20directly%20from%20ethylene
12. https://www.google.com/search?q=PROPERTIES+OF+PLASTICS+USED+IN+PLASTIC+ FILM+MANUFACTURING&rlz=1C1CHBF_enIN762IN762&sxsrf=ALeKk02uX3TBCdN uD71PU4XpcN9rnU0xVw%3A1624947268781&ei=RLraYJv1LvyZ4-EPkrm58A4&oq= PROPERTIES+OF+PLASTICS+USED+IN+PLASTIC+FILM+MANUFACTURING&gs_ lcp=Cgdnd3Mtd2l6EAM6BwgjELADECc6BwgAEEcQsAM6BwgjELACECdKBAhBG ABQhBtY7WNg44gBaAFwAngDgAGMB4gB6TySAQ0wLjkuNi40LjAuMi4zmAEAo AEBqgEHZ3dzLXdpesgBCcABAQ&sclient=gws-wiz&ved=0ahUKEwibp--lmLzxAhX8 zDgGHZJcDu4Q4dUDCA4&uact=5
13. http://www.plastemart.com/plastic-technical-articles/cellulose-based-plastics-address-need-for-more-sustainable-raw-materials-from-food-healthcare-coatings-and-construction/2356
14. McKeen LW (2017) Production of films, containers, and membranes. In: Permeability properties of plastics and elastomers, 4th edn.
15. https://www.flexpackmag.com/articles/86903-the-role-of-laminating-in-the-flexible-packag ing-film-process#:~:text=A%3A%20Laminating%20is%20the%20process,using%20adhe sives%20or%20by%20extrusion
16. https://www.aluminum.org/product-markets/foil-packaging
17. https://wikidiff.com/gunny/sack
18. https://saraexim.co/jute-bags-vs-plastic-bags/
19. https://en.wikipedia.org/wiki/Jute
20. https://sewport.com/fabrics-directory/jute-fabric
21. Indian standards IS: 5476 (1986) First Revision. Glossary of terms relating to jute
22. Seminar Paper on Plastic woven Sack. Published by Indian Flexible Intermediate Bulk containers association
23. Ulain Q, Panchal J. Total quality control for manufacturing of plastic woven laminated bags using six sigma. Int J Eng Dev Res 2–4
24. https://www.google.com/search?rlz=1C1RLNS_enIN723IN723&sxsrf=ALeKk03yhoyUxcf9 fY49PmNk4XJJFne8Jw:1597859756170&source=univ&tbm=isch&q=images+of+circular +loom+machine&sa=X&ved=2ahUKEwitaq566frAhXTH7cAHbAqC9sQ7Al6BAgKECg &biw=1350&bih=636#imgrc=SU4hdMOZCACioM
25. Profile of Indian Food Industry and role of Plastics in Conservation of Food Resources, Plastics for Food Packaging (2005) Published by Indian Institute of Packaging
26. http://www.manishflexipack.com/manufacturing-process/index.html
27. http://ittaindia.org/http://www.plastemart.com/goods-plasticsproducts/manufacturers/leno-bags/1301
28. IS:5818 (1988) Indian standard, specifications for lacquers and decorative finishes for food cans. Bureau of Indian Standards, Manak Bhavan, India
29. Coles R, McDowell D, Kirwan MJ (eds) (2003) Food packaging technology, vol 5. CRC press
30. Marsh K, Bugusu B (2007) Food packaging—roles, materials, and environmental issues. J Food Sci 72(3):R39–R55

Chapter 3
Semi-rigid Materials—Manufacturing Processes and Its Application

N. C. Saha, Anup K. Ghosh, Meenakshi Garg, and Susmita Dey Sadhu

3.1 Introduction

As per dictionary meaning, the word "semi-rigid" means the materials which are stiff and moderately rigid but not flexible in nature in terms of its hardness property. These materials are having adequate mechanical strength and other functional properties. These materials can be made either from cellulosic materials like paperboard or the different packaging formats made from paperboard like folding cartons, lined cartons, paper-based laminated composite carton, bag in box, composite containers, etc. Similarly, there are number of packaging formats are also made available from polymeric materials like thermoformed containers, multilayer squeezable tube, etc. These materials are extensively used in packaging application as primary packaging materials. However, paper-based folding board cartons are mainly used as intermediate or secondary packaging material. The details about the manufacturing process, properties and the applications of different semi-rigid packaging materials are discussed [1, 2].

3.2 Paperboard

Paperboard has more weight per unit area and also thicker than paper. It has also multiple layers. It is generally used for making shipping containers like trays, boxes, cartons, etc. It is rarely used as primary packaging material for food and is not used to contact food directly. According to the International Standards Organization (ISO), the material weighing more than 250 gsm/m^2 is known as paperboard. In other words, materials more than 300 μm (micron) will be referred to as paperboard. The properties are dependent on variables, i.e., sources of fibers, extraction process, the machine used and the treatment given for finished products [1, 3] (Fig. 3.1).

Fig. 3.1 Paper board

3.2.1 White Board

Several thin layers of bleached chemical pulp are used to make white board. It is used in the inner side of carton. It can be coated with wax or laminated with polyethylene for better heat sealable strength. This paper is highly recommended as a primary packaging material for food [4].

3.2.2 Solid Board

It is developed from multiple layers of bleached sulfite. It has good strength and durability. After lamination with polyethylene, it is used to make cartons for beverages also like fruit juices, other soft drinks, etc. [5].

3.2 Paperboard

3.2.3 Chipboard

Chipboard cannot be used as a primary packaging material for food. It is developed from recycled paper and often contains impurities of the original paper. Because of this reason, it is unsuitable for direct contact with food, printing, folding, etc. To improve its properties like strength and appearance, it is usually lined with white board. It is the least expensive form of paperboard and is used to make the outer layers of cartons for tea, cereals and other type of foods.

3.2.4 Fiber Board

Fiber board can be solid or corrugated.

Solid type—It has white board at inner side and kraft paper at outer side. These layers provide good protection against impact and compression. Solid fiber board after lamination with metal or plastic can improve barrier properties against air and moisture and is used to pack dry products such as coffee, tea, fruit powder, milk powder.

Corrugated type—Corrugated board is developed from two layers of kraft paper; in between these two layers, there is a central corrugating (or fluting) material. These corrugated-type fiber board is resistance to impact abrasion and crushing damage. Hence, it is widely used for shipping large amount of wholesale food and case packing of retail food products [6, 7].

3.2.5 Ovenable Board

A paperboard can be placed in a convection or microwave oven with food. The base board is SBS. A heat-tolerant polymer such as polyethylene terephthalate (PET) is applied to the surface to provide liquid holdout. It is used for microwave oven for food products [4].

3.3 Folding Carton

Duplex or triplex board is used to make folding cartons. The duplex board of 220 gsm to 300 gsm is used to make folding cartons by use of different processing steps like creasing, scoring, slotting gluing, etc. (Fig. 3.2).

Fig. 3.2 Folding carton

3.3.1 Properties of Folding Carton

Cartons can be easily handled, stored and distributed at retail outlet under refrigeration conditions. These cartons have pilfer proof-closing system. Folding cartons stack the product easily because of their stiffness property. Empty cartons can be collapsed easily and provide excellent space utilization [8].

3.3.2 Application of Folding Carton

These are used as intermediate package for the food products like butter, cheese, sweets, cookies, etc.

3.4 Lined Carton

The term indicates that this package is made of paperboard and then lined internally with appropriate packaging materials. It consists of recycled plies in the core and lined on one or two sides with a higher-quality bleached fiber. It is typically produced on cylinder-type machine, high-quality printing and good folding properties. It is extensively used to make folding cartons for confectionery, bakery items, etc. [3].

3.4 Lined Carton

3.4.1 There Are Different Types of Lined Board

- **Single white (SWL)—lined paperboard**—Made from two different layers of pulp—the top liner of single white lined (SWL) board is 100% new pulp or high-quality recycled pulp. The back is usually gray or brown.
- **Clay-coated-news back (CCNB) lined board**—SWL board with a pale gray back lining. It is also termed as recycled board with white liner.
- **Double white (DWL)—lined paperboard**—A board lined on both faces with a bleached stock.

3.4.2 Properties of Lined Carton

- These are cheaper in cost than metal containers.
- These materials provide excellent barrier properties against moisture, air and light. Hence, shelf life of product is increased.
- These have rectangular shape, hence easy to stack at the outlets.
- These are available in market in different capacity—may be of 100 ml, 200 ml, 1 kg, 2 kg, etc.
- In these cartons, solid, semi-solid or liquid food product can be packed.
- Outer surface can be printed easily.
- These cartons have notch or spout facilities which provide easy opening.
- Lined cartons can tailored according to requirement.

Application of lined cartons
Animal fat, khoa, milk powder, malted milk food, etc. are packed in lined cartons under vacuum (Fig. 3.3).

Fig. 3.3 Lined carton

3.5 Paper-Based Multilayer Composite Carton

It is prepared by combination of two or more materials having different physical or chemical properties which are combined with each other to produce a composite paper-based material with characteristics of different properties from the individual material [7]. The most common materials used in paper-based multilayer composite carton are paper, plastic and aluminum foil. These materials are all good individually also, but the combination is much better.

3.5.1 Advantages

- The paper-based multilayer composite materials are lighter in weight and are less resource-consuming than one single component having the same properties.
- Less amount of waste is produced if we are combining different types of flexible packaging material.
- Multilayer paper-based composite carton can contain, protect, preserve the food better, and it can facilitate handling during transportation and while in use. It also gives information through its label, and it acts as a silent salesman.

In practice, it is difficult to find one single packaging component whose properties meet all the demands of industry and consumers. Hence, our present-day packaging consists of multilayer composite materials in the form of laminates. By combining two or more materials, one can develop required properties into the package which can perform several functions. Each separate material layer can provide one or more properties. The composite of many thin layers of different materials gives a package and the product the best possible protection with a use of least amount of material. Increase in variety and a self-service have created a demand for packaging which are attractive, reusable and provide better protection for the products. This means "customized" multilayer packaging is required for fast development in present market (Fig. 3.4).

3.5.2 Materials Used

Paperboard and paper: Paperboard is the major material and contributes 75% part in composite cartons. It provides smoothness, stability, strength and finishing to the printing surface.

Polyethylene: Polyethylene protects against water vapors and helps the paperboard to stick to the aluminum foil firmly. Composite cartons contain near about 20% thermoplastics.

3.5 Paper-Based Multilayer Composite Carton

Fig. 3.4 Paper-based multilayer composite carton

Aluminum: It protects the carton against moisture, air and light. At ambient temperature, it also preserves the nutritional value and flavors of the food in the package. Composite cartons contain about 5% aluminum.

These materials, plastic, aluminum and paperboard, are layered together using heat and pressure to form six layers which protect the content from light, oxygen, air, dirt and moisture. The product remains fresh without any use of preservative by using aseptic technology.

Applications: Composite cartons are used for packaging of beverages, dairy products, like butter milk, condensed milk, puréed and diced fruit and vegetables, soups, juices sauces, etc. (Fig. 3.5).

1. Exterior PE
2. Paper
3. Interior.

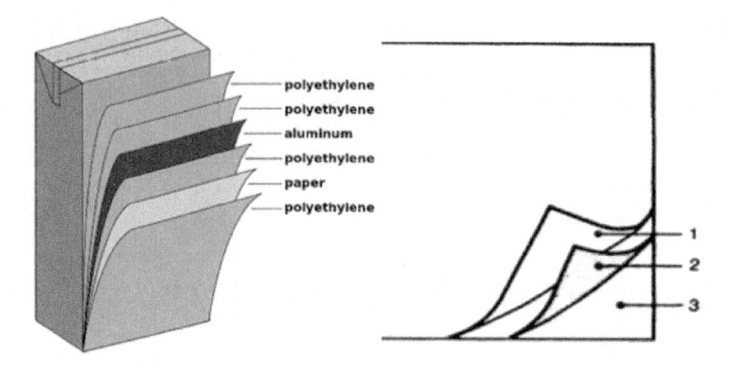

Fig. 3.5 Paper-based three-layer and six-multilayer composite carton

3.5.3 Types of Composite Cartons

Brick-shape cartons: These types of cartons are produced by combining thermoplastic material with paperboard and aluminum foil. This process helps in protecting the product from various types of spoilage. The aluminum foil protects the product from oxygen and light. The inner plastic layer is made from polyethylene which helps in sealing the liquid. The outer paper layer provides stiffness and helps in giving shape to the carton and provides maximum utilization of available storage space and transportation. Good printing quality is possible for display and shelf appeal and also providing nutritional and other type of information regarding the product. The product is free from microorganism because of aseptic packaging. Different type of opening devices like pull-tab opening, custom-designed cap and closures to provide convenient access to the contents can be used [9].

Prism-shape composite cartons: These types of beverage cartons are comfortable to hold, and its shape attracts children. These prism-shape cartons offer effective display and high space efficiency. These are shelf-stable at ambient temperature, and the shelf life of the stored product depends upon the barrier properties of the multilayered composite carton [5, 9].

3.6 Bag in Box

A **bag in box** is used for the storage and transportation of different type of liquid and powder food products. It is made from laminated bag, made from aluminum film or other plastics, kept inside a corrugated fiber board box. The food company which is filling the bag with its product generally removes the tap, fills the bag with liquid or powder and replaces the tap, and then the bag is placed in the box. The bags are available as singles or as web for semi-automatic and fully automatic machine, respectively. In automated filling systems, the bag is separated on line either before automatically filling or after filling [10].

These types of packaging can be made using form seal fill (FSF) technology, where the bags are shaped on-line from roll of film, then the FlexTap is inserted and then product is filled on a rotary head filler (Fig. 3.6).

3.6.1 Advantages

- Bag-in-box packaging is inexpensive.
- It has superior seam strength.
- In bag-in-box fills, foaming or splashing will not take place if filled without air.
- Packaging material is lighter in weight as compared to other plastics.

Fig. 3.6 Bag in box

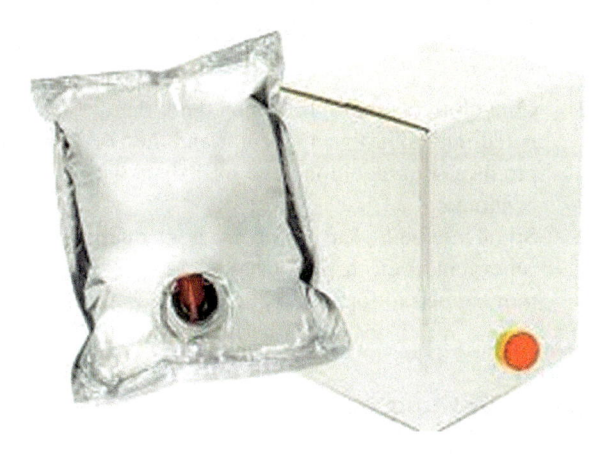

- Bag in box has a low-carbon footprint.
- These bags can be used with manual, semi-automatic and fully automatic filling machines.
- These types of bags minimize shipping and warehouse storage requirements.
- It has environmental benefits. Through bag in box, contents of 1.5–1000 L can be filled. Hence, less packaging and labeling is required.
- Dispenses without surging or glugging.

3.6.2 Applications

These types of containers are used for both dry and liquid products. The food product is kept in flexible bag which is then packed in cardboard carton. It can be used for liquid products like purees, sauces, vinegar, coffee, dairy, juice, lemonade, tea, water, etc. It is also used to pack wine, beer, milk, syrup, etc. Ketchup or mustard in fast food outlets can be served in bag in box. Among dry products, crackers, cookies, biscuits, cereals, etc. are packed in this type of material.

3.7 Composite Containers

A composite can is a rigid structure which is either spirally or convolute bound. Its one or both ends are either openable or permanently fixed. Earlier it was present only in round shape, but later other shapes like oval or rectangular are available in market to influence users with package design. In composite cans, paper is the major layer followed by polypropylene, high-density polyethylene (HDPE) and aluminum [7, 11]. These multilayers may enhance the properties of composite container.

3.7.1 Composite Containers Are Made from Two Methods

1. Convolute method—In this method, the web of papers of width of a can body is fed over adhesive applicator and then on rotating mandrel. This method can produce square, oblong and oval shapes. It is widely used, cheaper, faster and economical.
2. Spiral method—Layers of different qualities of papers are pulled over glue rollers and onto a stationery mandrel at an angle and carried around it by a moving belt to form a continuous tube. It can produce only cylindrical shape.

3.7.2 Advantages

- The right mix of constituent materials often results in composites which are corrosion resistant.
- They can resists damage from weather and harsh chemicals
- Composites are lighter in weight, compared to most woods and metals.
- Composites can be designed to be far stronger than aluminum or steel.
- Composite containers are eco-friendly to the environment because of its disposal property as it can incinerate or crushed for biological removal.
- The stability of the can for a volume up to 1 L is as good are even better than tin or glass container.
- 20% cheaper compared to tin.
- Composites are durable and need little maintenance.

3.7.3 Properties

Composite containers have a wide range of thermal and electrical conductivities. They also have good strength, better modulus, and their density is also low. Composite containers have excellent resistance to fatigue, creep rupture, corrosion and wear. It has also low coefficient of thermal expansion (CTE) [1].

3.7.4 Application

Composite containers are used for keeping refrigerated dough, some types of roasted snacks frozen fruits and vegetables, dried fruits, nuts, chips, powdered foods, dried, chips, spices, confectionary products, etc. Composite cans are used for milk powder packaging also; earlier metal can was used. These containers having square and rectangular cross sections with rounded corners can be used for dry food products

3.7 Composite Containers

Fig. 3.7 Composite container

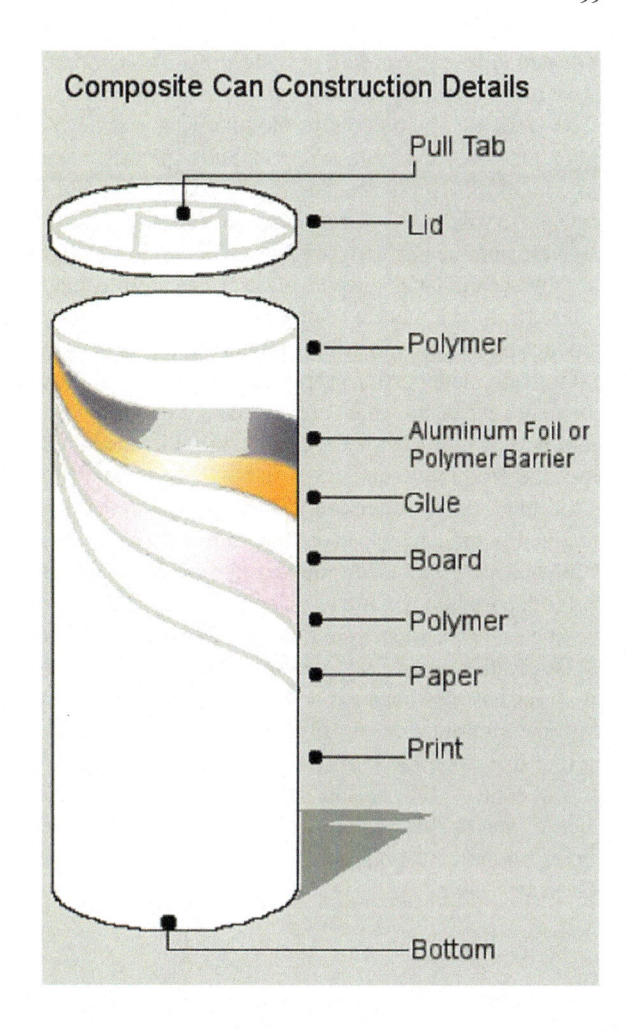

Composite Can Construction Details
- Pull Tab
- Lid
- Polymer
- Aluminum Foil or Polymer Barrier
- Glue
- Board
- Polymer
- Paper
- Print
- Bottom

such as tea, powder, granular mixes, savory snack, etc. It is also used for liquids like non-carbonated drinks [1] (Fig. 3.7).

3.8 Multilayer Squeezable Tube

Multilayer laminated tube is considered to be new-generation packaging format. This is also termed as squeezable **Lami tube** which is cylindrical, hollow piece with a round or oval profile, made up of a combinations of plastics, aluminum and paper [7]. In general, on one end of the tube body, there is a round orifice, which can be closed by different caps and closures. The orifice can be shaped in many different ways: Plastic nozzles in various styles and lengths are most typical. The other end is sealed either by welding or by folding. Typical tube sizes range from 3 to 300 ml. Most tubes are designed to be dispensed with hand pressure.

Historically, metal collapsible tube was first developed in 1841 by an American painter for the packaging of paints. Subsequently, the application of this tube became popular in 1890 for the packaging of personal care like tooth paste. Metal tubes served the industry for many decades, and different types of inner coatings were developed to overcome the problem of reaction of product with the metal of the tube.

However, in 1950, the pure plastic tube, entirely made of from polymeric materials, entered into the market which was enable to provide satisfactory performance of product-package compatibility. Unfortunately, single-layered plastic tube was not having adequate barrier properties, and thus, the application was only restricted to cosmetics products which are not sensitive to the ingress of moisture and gases. The pure plastic tube also had a problem that it would spring back to its original shape after the product is squeezed out.

Around 1970, American Can Co. (Now called as American National Can Co.) invented a tube which had layers on both the surface and contact layers with a sandwiched layer of aluminum foil as barrier and an intermediary layer of paper for collapsibility. This newly developed tube became very popular in USA for the packaging of personal care like tooth paste and also for pharmaceutical products.

But in the twenty-first century, the application of this tube for processed food products has also been experimented by developing innovative multilayer structure with the combination of different types of polymeric materials and aluminum foils to improve upon the barrier properties like moisture, oxygen gases and flavor and thus to enhance the shelf life of processed food products like dairy products, mainly cheese, mayonnaise, spices, mainly ginger paste, onion paste, garlic paste, liquid jiggery, honey, etc. and now, these types of packages are becoming popular in Indian market.

3.8.1 Manufacturing Process

The manufacturing process is consisting of the following steps:

Manufacturing of Web stock: The manufacturing of web stock is the first step of manufacturing process of multilayer tube. It is basically a sandwich structure, consisting of different layer of polymeric materials like low-density polyethylene (LDPE) or high-density polyethylene (HDPE) and aluminum foil. Sometimes, ethyl vinyl alcohol (EVOH) as co-polymer is also used as barrier layer in place of aluminum foil as middle layer in between the layers of polyethylene with the help of tie layer or bond layer. This web stock is manufactured either by lamination process or co-extrusion process for seamless tube.

(i) *Lamination process*: Thermoplastic films (known as substrates) are first manufactured in a Blown film or Cast film process with suitable polymer combination and thickness. Substrates are then taken to an extrusion lamination machine

3.8 Multilayer Squeezable Tube

and laminated along with barrier substrate (like EVOH film) with the help of intermediate tie (maleic anhydride grafted co-polymer) or bond layer or aluminum foil. The thickness of the tube for multilayer web stock depends upon the tube diameter and the quantity to be packed. For laminated tube, wide width web stock is taken to the slitting machine and then slit according to the specific length or diameter of the tube.

(ii) *Co-extrusion process for seamless tube*: Thermoplastic polymers, intermediate tie (maleic anhydride grafted co-polymer) or bond and barrier polymers (if any) are fed to individual Extruder Hoppers, melted suitably and passed through a multilayer co-extruded die to extrude cylindrical tube of suitable diameter. Tubes are cut in-line process according to the tube length.

Printing: Printing of web stock is the second step of manufacturing process of multilayer tube. The plain web is then printed on the surface by means of either letterpress, flexographic, screen or digital printing process in line.

Tube forming: The tube formation is the third step of manufacturing process. The printed web is then subjected to the side seaming machine to form the tube body and then taken to tube heading machine, where shoulder and neck is made either by means of compression or injection molding process. Different types of suitable caps like flip top caps, screw caps, etc. are then fitted on to the tube depending on the market requirement.

3.8.2 Types of Tube

Depending on the structure, the tubes are broadly classified into two categories;

(a) Laminated tube with seam
  (i) Plastic-based laminated tube (PBL)
  (ii) Aluminum-based laminated tube (ABL)
(b) Plastic-based extruded tube without seam.

(a) Laminated tube with seam: This type of tube has got shoulder and a joint at the center, whereas extruded tube does not have at the center. However, the innovative technology has made it possible to make the junction in such a manner that the joint would look very invisible. Out of two types of laminated tube, plastic-based laminated tubes (PBL) do not possess the residual deformation as compared to aluminum-based laminated tube (ABL) which means that they absorb the air while using. PBL tube can be made transparent. In addition, the tubes can also be subjected to hot foil stamping [12, 13].

(b) Plastic-based extruded tube without seam: Extruded tubes have more abilities in the external decoration due to large external surface. Plastic tubes do not deform due to elasticity of the materials. While the plastic tubes are squeezed, the tube absorbs the air and assumes the original form. As compared to plastic-based laminated tube, this kind of tube will have more capacity in terms of volume. Photograph of laminated tube and its parts:

The photographs and parts are given below in Figs. 3.8 and 3.9.

Advantages of multilayer Laminated tube

(i) Easily squeezable and eliminates wastage and permits the consumer to use the entire quantity of the product.
(ii) Cracking does not take place as the tube has excellent flexing properties form the plastic layers, and the tube remains absolutely flat and wrinkle free during use.
(iii) Printing is done prior to tub formation; therefore, the following properties are increased.

- Aesthetic appeal is superior.
- Printed material is absolutely scratch proof due to special printing technology, and print peeling problems are eliminated.
- Background of white color is imparted from the outer polyethylene layer and is not a printed color so there is no possibility of print peel off.

(iv) Multicolor printing including total effects reproducing photographs can be done, since printing is undertaken at the roll stage.

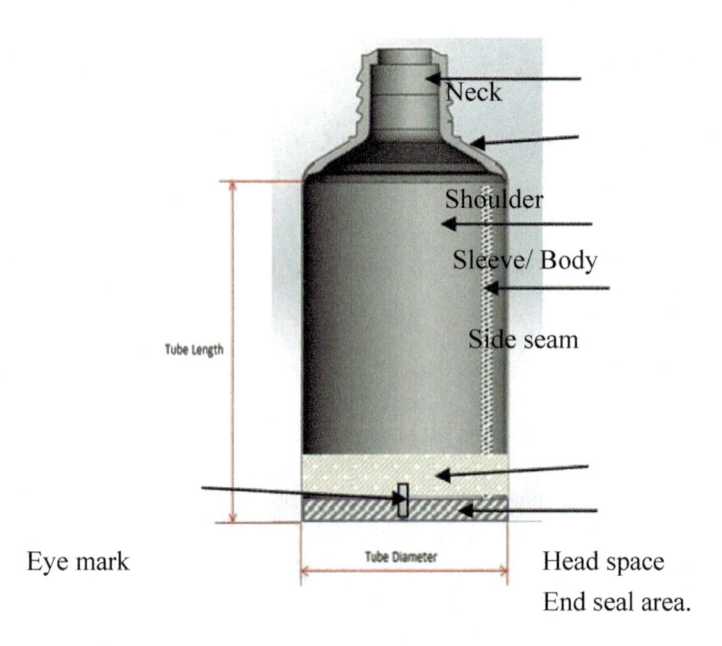

Fig. 3.8 Parts of tube

3.8 Multilayer Squeezable Tube

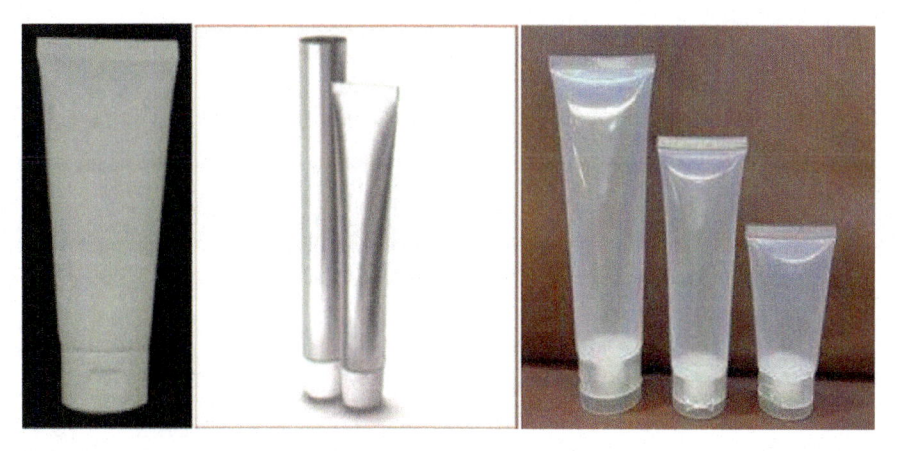

Fig. 3.9 Photographs of different types of tube

(v) Completely eliminates possibility of black aluminum oxide deposition as it has non-metallic plastic neck.
(vi) The production process is completely different, and metal dust is not created during production, thereby eliminating the possibility of metal particles.
(vii) Non-toxic and most hygienic, as lubricants are not used in the manufacturing process.
(viii) Lacquering is not required.
(ix) Ultrasonic or high-frequency sealing provides additional safeguards against leakage.
(x) Can withstand pressure of over 30 psi.
(xi) Since laminated tubes are wrinkle resistant such problems do not arise and in fact, partitions used during transportation in cartons are eliminated, resulting into savings on shipper containers as well as transport costs.
(xii) Temper-proof: Makes adulteration impossible as the tube cannot be re-filled or re-used.
(xiii) Protective- Contents remain well protected and free from contamination.
(xiv) Lightweight- Easy to handle and reduces transport and handling costs.
(xv) Elegant-Attractive and also possesses a bright surface and lend themselves to trouble-free painting and thus enhances shelf life and customer appeal.
(xvi) Excellent aroma barrier: The film features excellent aroma barrier properties; hence, the flavor of the product is retained inside the pack. Also the packed product does not attract flavors from outside.
(xvii) High-flexibility without flex crack.

3.8.3 Applications

(i) Mostly used for cosmetics and personal care products like shaving creams, tooth pastes, hair creams.
(ii) Pharmaceuticals like antiseptic creams, lotion.
(iii) Industrial products like adhesives.
(iv) Processed food products like cheese, mayonnaise, liquid jaggery, honey, ginger pastes, garlic pastes, onion pastes, etc. [12, 13].

3.9 Thermoformed Container

One of the most common applications of thermoforming is packaging, particularly semi-rigid packaging for food and consumer goods. It has changed the way food business used to be done and the way food used to be marketed. For example, earlier, ice creams used to be served in paper cups; with the invention of thermoforming, the ice creams now look more colorful, lucrative and yummy with variety or flavor. Today, in airports, highways or any large food chain outlets, the food is well packed, fresh, hygienic and ready to be consumed on the go (Fig. 3.10).

Another important sector where thermoforming has played the role of a game changer is the packaging of fresh fruits and vegetables. Cut vegetables are packed in thermoformed containers of polystyrene. Being transparent the freshness and the sizes of the cut vegetables are clear to the buyer. Due to the visibility, it is more appealing and convincing. Nowadays, the festival packs of the sweets and namkeen during festive season also have benefitted from the clear, transparent and glossy look of the thermoformed packs.

Some other important examples of thermoformed plastic containers in the field of food packaging are the plastic containers made up of various plastics which not only help to store food but are also microwaveable. Thus, reheating and cooking made easy.

Fig. 3.10 Thermoformed containers for storing food

3.10 Thermoforming Process

Thermoforming is a technique specific to thermoplastics as it involves softening a plastic sheet and then forming it into a mold cavity which would not be possible with thermosets because of their high resistance to plastic deformation, and they also do not soften when heated [8]. Forming of the plastic sheet can be done through applying vacuum, pressure or mechanical force. Application of thermoforming lies largely in food packaging sector for manufacturing ice cream and margarine tubs, snack tubs sandwich packs, meat trays microwave containers, etc. It is also used for sectors such as pharmaceuticals and electronics for products such as skin and blister packs, small tools, etc. Thermoforming provides the advantage of very high production rates with good dimensional accuracy, it also enables us to make very intricate parts that might have been difficult with some other techniques [1]. The only few disadvantage in this case lies in it being expensive due to expensive molds and large material loss in sprues and runners and the limitation in the size of the products that can be made [8] (Fig. 3.11).

3.10.1 Advantages of Thermoforming

1. Low tooling cost relative to injection molding
2. Prototype speed (prototype tools are typically less involved than with other methods)
3. Fabricated parts through thermoforming include those with very high area-to-thickness ratios
4. Extremely large parts can be fabricate.

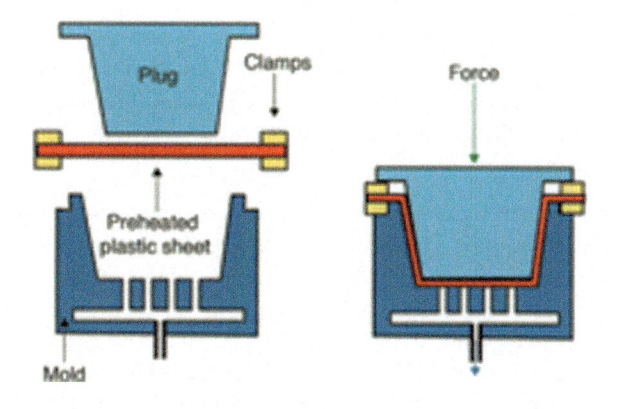

Fig. 3.11 Schematic representation of thermoforming process

3.10.2 Disadvantages of Thermoforming

1. Compared to injection molded parts high cost
2. Due to stretching of the film non-uniform gauge length.

Thermoforming is generally used for food packaging but has many applications in starting from plastic toys to aircraft application to cafeteria trays. Thin-gauge (less than 0.060 inches) sheets are commonly used for rigid or disposable packaging, while thick-gauge (greater than 0.120 inches) sheets are typically used for permanent surfaces for automobiles, shower enclosures and electronic equipment [14]. A variety of thermoplastic materials can be used in this process, including the following:

- Acrylic polymer (PMMA)
- Cellulose acetate
- Low-density polyethylene (LDPE)
- High-density polyethylene (HDPE)
- Polypropylene (PP)
- Polystyrene (PS).

3.10.3 Material Selection

Following are the factors that must be looked upon before characterizing a particular grade as thermoforming grade.

Glass transition temperature—The difference between glass transition temperature and the melting point is a very important criterion. The polymer remains as a soft material above its glass transition temperature up to its melting point; this region is defined as the forming range of the material. Therefore, larger the difference between the melting point and T_g the more optimum the material will be for thermoforming.

Crystallinity—Amorphous materials are termed as easy to transform as they demonstrate isotropic flow and provide better dimensional stability, i.e., they do not warp. They also display higher impact strength and are thus suitable for structural application.

Melt strength—Melt strength is an another important criterion to be looked upon, it is the property holding together the polymer sheet above its softening point, in case of low melt strength materials, polymer tends to flow as it softens which can deform the plastic sheet during thermoforming. Although melt strength is a function of temperature and extrusion rate but materials such as ABS that provides high melt strength in the low extrusion temperature region approaching the thermoforming region provides good sagging resistance during the process.

Molecular weight and MWD—Molecular weight of a polymer is directly proportional to its entanglement density it is observed that on increasing the molecular weight by about four times entanglement density also increases in a similar fashion.

3.10.4 Process Factors

There are several different process parameters which affect the properties of the thermoformed product. These are discussed as below.

Temperature distribution—In case of thermoforming a temperature distribution along the length of the heating elements rather than a fixed uniform temperature is something that should be extensively avoided as this can affect the temperature of the sheet before forming which in turn lead to non-uniformity in both product dimensions as well as properties. Normally, it is hard to thermoform a sheet that has a temperature distribution across it of 30 °C or greater.

Plug assist—The most important process parameter in case of thermoforming is the type of forming process. It is observed that use of plug assist increased the minimum thickness value in case of thermoformed products of conical, hemispherical and cylindrical shapes. Also using plug assist resulted in uniform thickness distribution than negative (female mold) forming.

Air pressure—Providing external air pressure apart from the applied vacuum can really add to the properties of the formed product with very low extra tooling cost. Additional pressure helps form parts with finer details and sharpness. This additional pressure even allows to make textured surfaces and undercuts to allow additional components to be molded in and that too at a high production rate. Even after providing so many advantages also, the tooling costs compared to processes like injection molding is about 20–30% less [1, 14, 15].

3.10.5 Quality Control

There are a few major issues to be dealt with in case of thermoforming to obtain a quality product.

Warpage—Improper temperature control in case of thermoforming can cause unequal heating and cooling of the surface of the plastic sheet; this in turn leads to warpage in the formed product. Therefore, an uniform temperature distribution is the utmost requirement for thermoforming.

Dimensional inconsistency—This is another issue that can occur due to variations in temperature across the sheet surface, due to temperature inconsistency shrinkage occurs through the molded form but in varying ratios thereby causing inconsistent dimensions.

Thickness inconsistency—Vacuum applied during thermoforming must be properly vented as inconsistency in venting of vacuum can add to the inconsistency in thickness of the formed part which will compromise the quality of the product. This also affects the aesthetics of the formed product.

Texture and gloss—Tooling material selection and surface finish of the tool has a drastic effect on the texture and gloss of formed product. Tool surface porosity can cause unwanted dimpling which in turn will lead to gloss reduction. Therefore, it goes without saying that selection of tooling material is a major factor in obtaining a high-quality product.

Chill marks—Wavy undulations called "chill marks" often appear on the surface of the product due to trapped air and unequal cooling of the material. This is generally caused by non-uniform temperature in tool design and improper vacuum venting. This therefore adds to the importance of maintaining uniform temperature and proper vacuum venting.

Thermoforming is the process of heating a thermoplastic polymeric sheet to its softening temperature and stretching it over/into a single-sided mold and holding it while it cools and solidifies into the desired shape. The thermoplastic sheet is first clamped into a holding device and then heated in an oven with the help of either convection or radiant heat until it softens. The sheet then is kept horizontally over a mold and pressed into or stretched over the mold using vacuum pressure, air pressure or mechanical force. The softened sheet takes the shape of the mold and is held in place till it cools down. The excess material is then trimmed away, and the formed part is released. Excess material after collection can be reground, mixed with unused plastic and reformed into thermoplastic sheets.

The thermoforming process involves following steps:

1. Design: This is the first step where the design of the packaging material is done with the help of software.
2. Material selection: The next important step is to select the material depending on the customer requirement. This includes the size, shape, texture, color, finish, etc.
3. Tooling: This is the step which involves the manufacturing of the mold. Generally, the molds are made up of aluminum.
4. Thermoforming: Thermoforming is done by heating the plastic sheet to a pliable temperature, then giving it shape of the mold using vacuum pressure, air pressure or a combination of both. Once the material has taken the shape of the mold, it is cooled and removed from the mold, retaining its final shape [16].
5. Trimming: Once thermoforming of the plastic is done, some of the original article needs to be trimmed away.
6. Secondary operations: Secondary operation involves the other finishing operations after the product is given shape.

3.10 Thermoforming Process

Fig. 3.12 Vacuum forming process [Copyright@2008CustomPartNet]

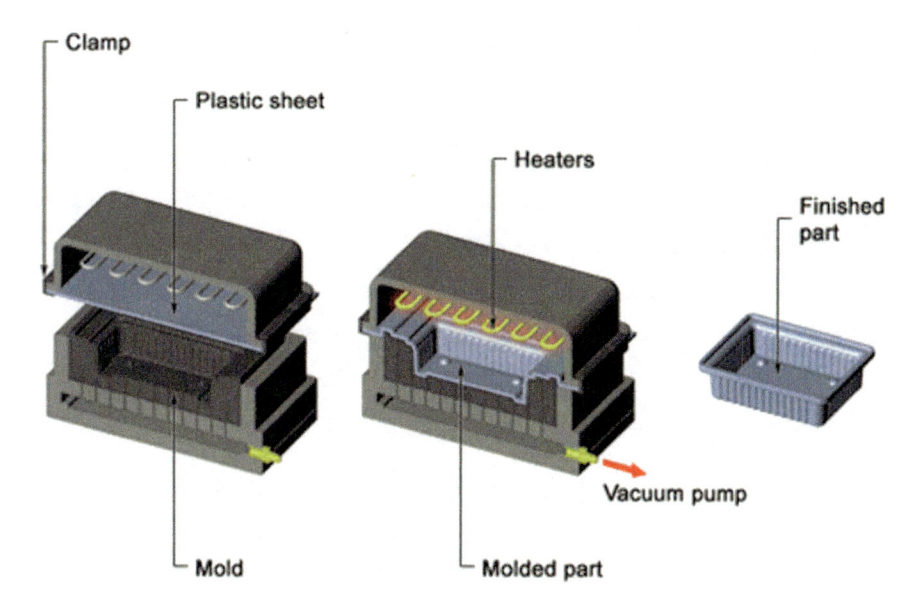

As mentioned above, there are different methods of shaping the thermoplastic sheet in the mold. These types of thermoforming include the following:

Vacuum forming—In this method, vacuum (typically 14 psi) is created between the mold cavity and the thermoplastic sheet which forces the sheet to conform to the mold to form the part (Fig. 3.12).

Pressure forming—In this method, air pressure (typically 50 psi, but up to 100 psi) is applied on the back side of the sheet in addition to the vacuum underneath the sheet, to help force it onto the mold. This additional force allows the formation of thicker sheets with finer details, textures, undercuts and corners (Fig. 3.13).

Mechanical forming—In this method, mechanical force is used on the thermoplastic sheet to give it the shape of the mold by direct contact. A core plug is used to push the sheet into the mold cavity and force it to the desired shape [15, 16] (Figs. 3.14 and 3.15).

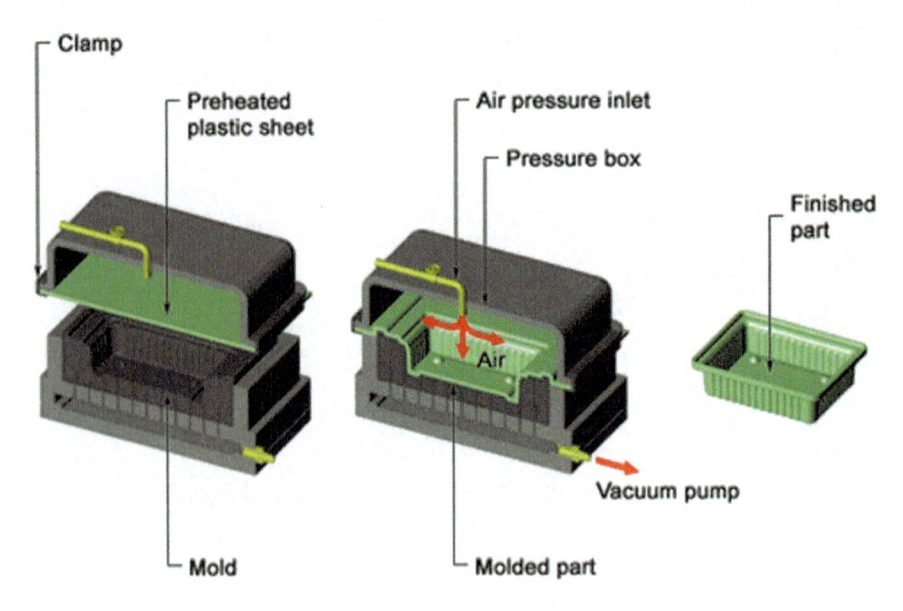

Fig. 3.13 Pressure forming process [Copyright@2008CustomPartNet]

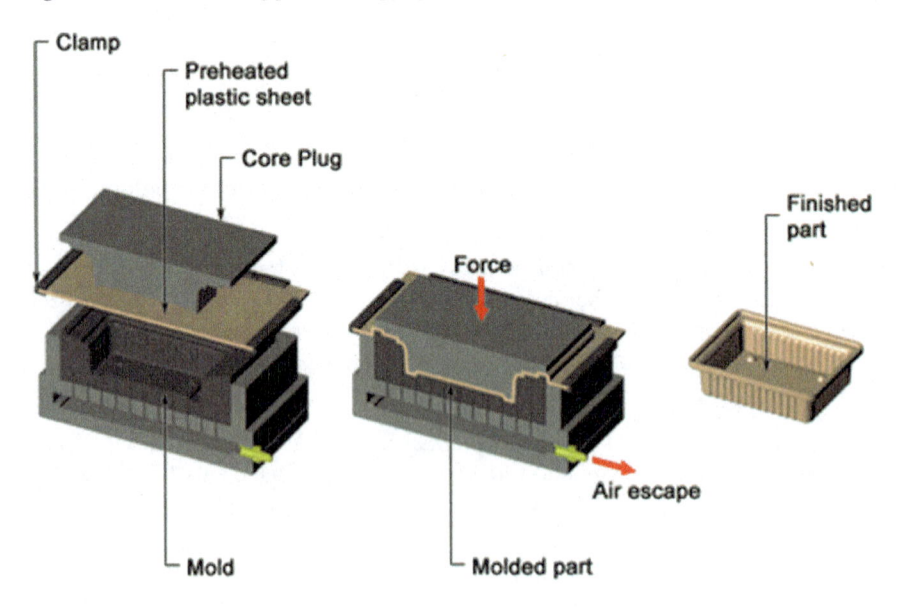

Fig. 3.14 Mechanical forming process [Copyright@2008CustomPartNet]

Fig. 3.15 Thermoformed transparent trays

References

1. Marsh K, Bugusu B (2007) Food packaging—roles, materials, and environmental issues. J Food Sci 72(3):R39–R55
2. Fundamentals of Packaging Technology, Published by Institute of packaging Professionals, USA
3. Deshwal GK, Panjagari NR, Alam T (2019) An overview of paper and paper based food packaging materials: health safety and environmental concerns. J Food Sci Technol 1–13
4. Kirwan MJ (2003) Paper and paperboard packaging. Food Packag Technol 241
5. Twede D, Selke SE, Kamdem DP, Shires D (2014) Cartons, crates and corrugated board: handbook of paper and wood packaging technology. DEStech Publications, Inc.
6. Piergiovanni L, Limbo S (2016) Food packaging materials. Springer, Basel
7. Coles R, McDowell D, Kirwan MJ (eds) (2003) Food packaging technology, vol 5. CRC Press
8. Robertson GL (2013) Food packaging: principles and practice. CRC Press
9. Anukiruthika T, Sethupathy P, Wilson A, Kashampur K, Moses JA, Anandharamakrishnan C (2020) Multilayer packaging: advances in preparation techniques and emerging food applications. Compr Rev Food Sci Food Saf 19(3):1156–1186
10. Lolis A, Badeka AV, Kontominas MG (2019) Effect of bag-in-box packaging material on quality characteristics of extra virgin olive oil stored under household and abuse temperature conditions. Food Packag Shelf Life 21:100368
11. Cirillo G, Kozlowski MA, Spizzirri UG (eds) (2018) Composites materials for food packaging. Wiley, New York
12. http://www.stepinhost.com/digvijaygroup.in/laminated-collapsible-tubes.html
13. https://packagingsouthasia.com/packaging-production/how-the-multilayered-laminated-collapsible-tube-was-invented/
14. McCool R, Martin PJ (2011) Thermoforming process simulation for the manufacture of deep-draw plastic food packaging. Proc Inst Mech Eng Part E: J Process Mech Eng 225(4):269–279
15. https://www.sciencedirect.com/topics/engineering/thermoformed-product
16. https://www.sciencedirect.com/topics/materials-science/thermoforming-process

Chapter 4
Rigid Packaging Materials-Manufacturing Processes and Its Application

N. C. Saha, Anup K. Ghosh, Meenakshi Garg, and Susmita Dey Sadhu

4.1 Introduction

Rigid packaging materials are normally stiff and rigid in nature. The materials are having the features like heavy weight, more hardness and stronger materials as compared flexible packaging materials. There are different forms of **rigid** packaging materials which are made of either paper, wood, plastic, metal and glass. The paper-based materials like corrugated fiber board box and plastic-based forms like plastic containers and plastic crates are covered under rigid packaging. In addition, metal containers with high mechanical strength with opaque in nature are considered as most important rigid packaging materials for food packaging. Another important form of rigid packaging material is glass container, available in natural as well as colored, transparent which are having extensive application in packaging of processed food products [1].

4.2 Metal Containers

Metal is used in various applications from rack system to fruit cans in packaging. For food packaging, elements are used in very pure form. Hence, production of food packaging material starts from ore for mining and then refined for extracting the metal [2]. Main properties of metals which make them suitable for food packaging are

1. High density, compact structure and poor diffusibility of light, gas and vapor make them suitable for food packaging.
2. Metals have unrivalled toughness against stress and strain.
3. Metal packages can achieve a long shelf life because of high thermal conductivity and can be molded in required shape easily.

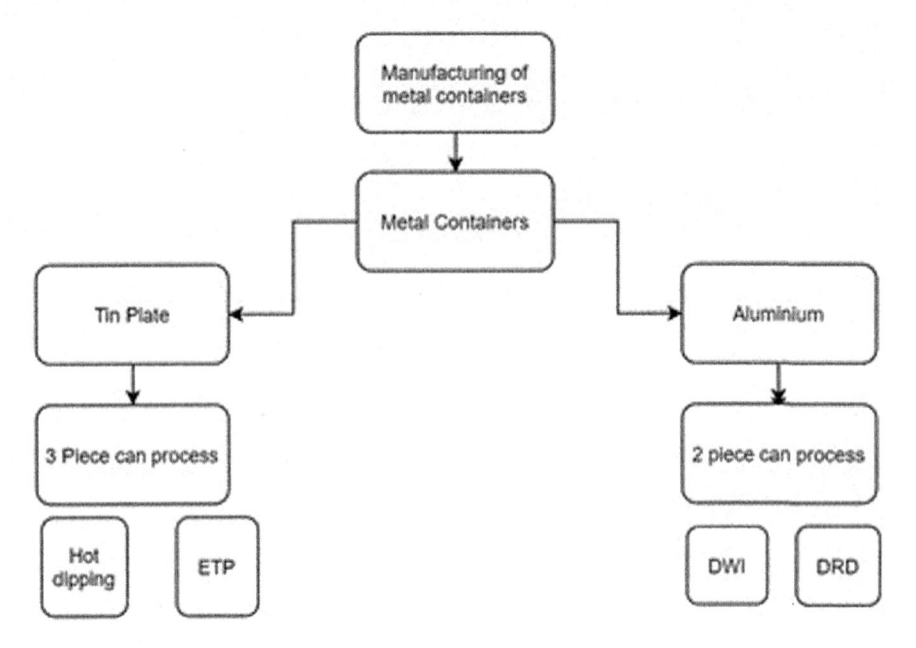

Fig. 4.1 Types of metals used in containers

4. Magnetic property and high density make easy collection of waste, and metals can be thermally recycled without any changes in original performance.

Generally, four types of metals are used in food packaging: aluminum, tin, steel and chromium [3].

Primary material for metal packaging and for production of food cans is Tin plate and aluminum as shown in Fig. 4.1.

4.2.1 Tinplate

Tinplate is made from steel. Iron alloys are generally called steel. In food packaging, carbon steels are majorly used because of good binding property of carbon in alloy. They contain various minerals like manganese—1.6%, silicon—0.6%, copper—0.6%, phosphorous—0.4% and sulfur—0.05% and carbon should be less than 1%. Minimum thicknesses in use are 0.18 mm for single reduced (SR) and 0.12 mm for double reduced (DR) plate. Steels are hard, cheap, strong, durable and easy to shape materials. But they are not inert for food contact. In fact, coated steels only are used in food packaging. It protects from rust, steel is protected by applying layer of chromium or tin. Protective coating on iron should be given to protect it from corrosion and oxidation. Electropolishing process is used to make Grade 304 stain less steel corrosion resistance from chemicals and making steel smooth and shiny. After

4.2 Metal Containers

this process, it is easy to clean also. Grade 304 steel can be degraded by corrosives salt. Second most common type of steel is Grade 316 stainless steel. Its mechanical properties are similar to 304 stainless steel, but Grade 316 stainless has better resistance to salts like chlorides because it has higher amount of molybdenum content. Tinplate is made from steel and tin by electrolytic coating of steel with tin or by dipping of steel plate in hot molten tin viscous liquid to avoid corrosion [3, 4].

Electrolytic chromium-coated steel (ECCS)—In this, instead of tin, chromium is used to provide protection from corrosion. Hence, it is called electrolytic chromium-coated steel (ECCS) or tin free steel (TFS). ECCS has good heat resistance and is cost effective but is less resistant to corrosion as compare to tinplate.

4.2.2 Coating/Lacquering

Cans are coated with an organic layer to protect the integrity of the can from the food effect, and it also prevents chemical reactions between the metal of can and the food. Can coatings should tolerate the production process and sterilization processes, it should be universally acceptable for all types of food and beverage, prevent chemical migration into food that effects human health, adhere to the can, protect the metal of the cans, maintain its organoleptic properties and preserve the food for several years [5, 6].

Two methods which are generally used for the application of protective coatings in metal cans are roller coating and spraying.

1. **Roller coating process**: For external coating of cylindrical can, roller coating is used.
2. **Spray coating process**: For inside surface where physical contact is difficult, spray coating is used.

Lacquers and decorative finishes are usually categorized into three types,

Category A: It is used on internal surface of food can.

Category B: It is used on external surface of food can.

Category C: It is for decoration purpose.

The category A type lacquers fare used for class I processed foods and for class II non-processed foods. Class I materials can be divided into three types depending up on the end use.

Type 1—It is an acid-resistant lacquer which is suitable for colored fruits like purple grapes.

Type 2—It is a sulfur-resistant lacquer which is suitable for peas, meat and fish products.

Type 3—It is a sulfur impermeable lacquer suitable for meat products. It requires long sterilization at high temperature. Examples of lacquers which are used in food

Table 4.1 Types of coatings used in metal containers

Polymer	Property	Application
Vinyl compounds	Resistant to acid and alkaline products, have good adhesion and flexibility, but are not suitable for high-temperature processes such as retorting	Used for canned fruit juices, beers, wines and carbonated beverages. These are also used for exterior coatings
Epoxy phenolic compounds	These are acid resistant, suitable for acidic products and have good properties as a base coat under vinyl and acrylic enamels. These are flexible and have good heat resistance	Used for different steel cans like for beer, meat, fish, soft drinks, fruits, vegetables, etc.
Epoxy amine lacquers	Excellent heat resistance, adhesion and abrasion resistance. These provide flexibility and produce no off flavor	Used for dairy products, beer and soft drinks, and fish
Phenolic lacquers	Resistant to acid and sulfide compounds and are inexpensive	These are used for acid fruits, meats, soup, fish and vegetables
Polybutadiene lacquers	Have good adhesion, high heat and chemical resistance	These are used for soups, soft drink, beer and vegetables (if zinc oxide is included to the coating)
Acrylic lacquers	These are expensive coating materials. They are heat resistant and provide an excellent white coat	These are used for both internal and external coating of fruits and vegetables

are oleoresins, vinyl, phenolic and epoxy lacquers. These are the permitted lacquer materials.

Category B and C are a type of secondary lacquers which does not come in contact with food. These are used for improving the life of can and for beauty purpose.

Different types of polymers are used for coatings which are applied in a liquid state and dried after that by solvent removal, heat-induced polymerization or by oxidation method [4, 6]. Different types of materials used in coating of cans are given in Table 4.1.

4.2.3 Manufacturing of Three-Piece Cans

Three-piece can is hermetically sealed and consist of a body and two end pieces. First steel is rolled into about 1.8 mm thick rectangular strip. Then, coating is applied according to requirements of the food, more acidic food requires thick coating on

4.2 Metal Containers

the inner side of the strip. Extra coating may be applied to prevent interaction with foods to improve the surface brightness and to resist corrosion. During processing and storage, cans for food packaging are subjected to external and internal pressure [4, 7]. To increase its strength, the can body may be rippled or beaded. Properties: The three piece metal cans are rigid, can be produced in various shapes, their material utilization rate is very high, size and shape can be changed easily, mature production process, and can help in many kinds of packaging products. Applications: Three piece metal cans can be used for packaging of food products, medicines and beverages. Manufacturing method is shown in Fig. 4.2.

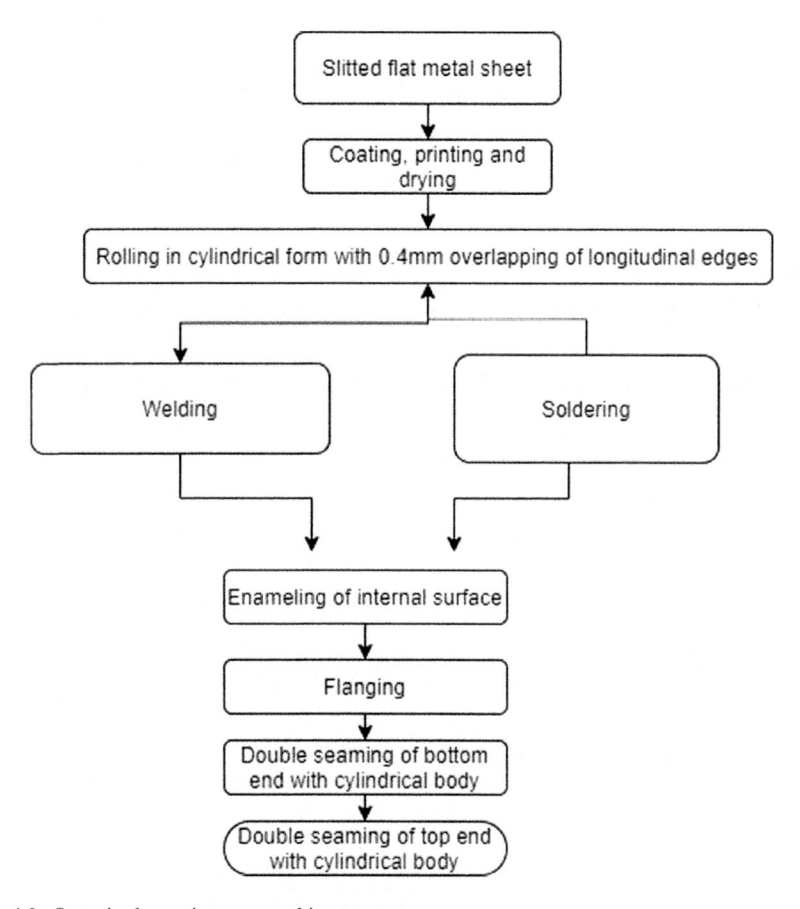

Fig. 4.2 Steps in three-piece can-making process

4.2.4 Aluminum

It is the third element in the Earth's crust. The production of aluminum from bauxite ore is very expensive process [1]. Pure aluminum is produced by electrolytic process. Aluminum is malleable and light in weight, density around 2.7 g/cm^3. Extraordinary malleability property makes it suitable to cast into any form, extrude it and roll it in to variety of shapes. Aluminum strength increases at below zero degree temperature, and it does not become fragile at ultra-freezing temperature. Application of pasteurization and sterilization treatment on aluminum packages is very effective because of high thermal conductivity.

Aluminum is used for production of beverage cans, and steel is used mainly for food cans. Aluminum prepared in thin sheets have a thickness less than 0.2 mm; thinner gauges have thickness near to 6 μm. Standard household foil is 0.016 mm thick, and heavy duty household foil is 0.024 mm thick. Aluminum cans have more recycled content and higher recycling rate than other competing package types, and these are prepared by two-piece can manufacturing process.

4.2.5 Two-Piece Can Manufacturing Process

A side seam is not present in two-piece can. It has body and a cover for the top. There are two main methods of producing two-piece cans.

1. **Draw and Iron (DI)**—In aluminum cans, used for beverage, a circular shape blank is cut and drawn to a shallow cup. The cup is first redrawn, and after that it is passed through a series of ironing dies that would extend and reduce the thickness of the walls. Hence, the base of the can ends is thicker than the walls. This improves the stacking of the cans and lowers the overall cost by reducing metal consumption.
2. **Draw and Redraw (DRD)**—To form a shallow shape, a metal blank is punched in to die (drawn). Redraw process is used to reduce the diameter of the cup formed in the first step. Shallow draw or very short cans do not require second draw. TFS cans are formed by DRD process. First the body is shaped, then can is trimmed, beaded, and the end of can is double seamed on the body of can after filling [3] (Figs. 4.3 and 4.4).

4.2.6 Properties of Metals Which Make Them Suitable for Food Packaging

- Metals are superior packaging material if used with proper coating. It is compact and has high density and malleable. It has high thermal conductivity, and five fundamental characteristics of metals make them particularly feasible for packaging

4.2 Metal Containers

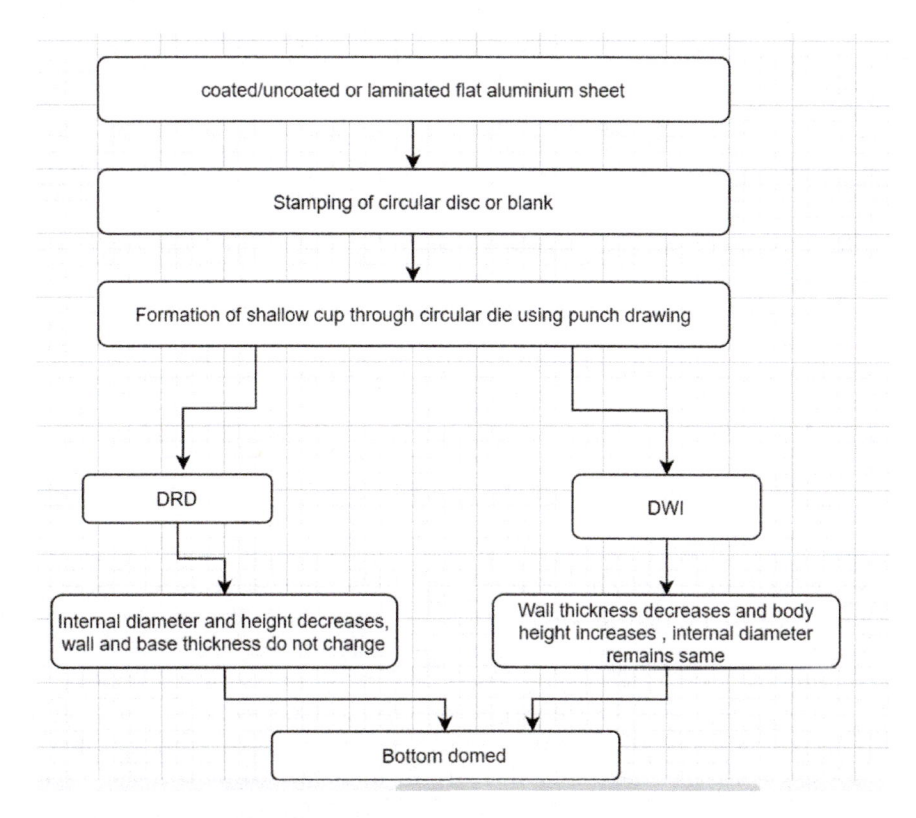

Fig. 4.3 Two-piece can manufacturing method

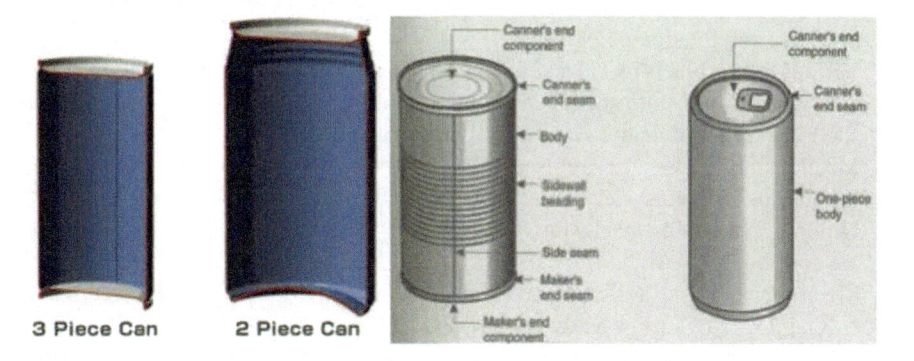

Fig. 4.4 Three-piece tin can and two-piece aluminium can

- Metals show magnetic behavior and have high density values. They provide an opportunity of thermal recycling without any loss of the original behavior, beside this it can be recollected and recycled.
- It shows absolute barrier properties against light, vapor or gas because of compact molecular structure and the high density.
- It is resistant against stress and strain because of its toughness property.
- The malleability property of metal helps to mold it in any shape. This property is because of the metallic bonding among atoms.
- It has high thermal conductivity which makes thermal treatments like pasteurization, sterilization, etc., possible on closed metal can to improve the shelf lives of food.

4.2.7 Applications

- Metals and its alloys are used as a primary or food contact materials. It is used mainly in processing equipment, containers and household utensils. It is also used in foils for wrapping food. It provides barrier between the food and the exterior material like process equipment, storage containers, spray dryers, boilers, kitchen knives, pipes, pots, pans, cutlery, etc.
- Tin containers are used for extending the shelf life of Indian sweets, confections like khoa, rasogolla, gulabjamun, rasomalai, paneer, etc. Animal fat is packaged in tin containers for improving their shelf life. Lacquered tin containers with capacities varying between 1 and 15 L are widely used for packaging of fats and oil.
- Aluminum cans with easy opening ends provided barrier properties against light and air, good for storage of beer.
- Various fish products like mackerel in brine, fish curry, tuna in oil, prawns in brine, etc., were canned in tin-free steel which is coated by polymer which extends the shelf life for more than 2 years at ambient temperature.
- Metal drums, pails, etc., are used for storage of food components at large scale for bulk or for wholesale.
- Different types of metal closures like twist-off, press-twist, pry off and deep-press are commonly used.
- Metal tubes, open trays caps and closures for bottle tops and lids are also in use specially for yoghurt or butter containers.
- Foil or extruded containers are usually made from pure or commercially pure aluminum (Type 1100 and 1050).
- Carbonated beverage ends are prepared from hardest grade (5182) alloy containing 4–5% magnesium and 0.35% manganese.

4.3 Glass Container

History of glass is extremely old. The first glass objects for food holding was used around 3000 BC back. Glass is an amorphous and inorganic product which has been prepared from fusion after that cooled till it become rigid without crystallizing. It can be recycled 100% without loss in quality or purity. Eighty percent of the glass which is recovered is converted into new glass products. A glass container can reach a market store from a recycling bin in one month only. Glass is non-porous and impermeable in nature. It does not interact with food products to affect their color and flavor. Glass has an almost zero rate of chemical interactions, ensuring that the products is safe inside the glass bottle and aroma, flavor and strength of product remain unaltered. Consumers enjoys foods or beverages when packaged in glass. The three main parts of a glass container are the bottom, the finish and the body. These are formed in molds. *Bottom* of the container is made in the bottom plate part of the glass container mold. *Finish* holds the cap or closure, required for opening in the container. *Body* is made in the body mold. It is the largest part of the container and present in between the finish and the bottom. Composition and role of ingredients in glass making are given in Tables 4.2 and 4.3, respectively [1].

Table 4.2 Chemical composition of glass

	Soda lime glass	Borosilicate glass	Amber
Silica	71–73	65–85	72.6
Boron oxide	–	8–15	–
Sodium oxide	9–15	3–9	12.8
Potassium oxide	0–1.5	0–2	1.01
Calcium oxide	7–14	–	11.1
Magnesium oxide	0–6	0–1	0.23
Aluminum oxide	0–2	1–5	1.81
Mixture of iron oxide and titanium oxide	0–0.6	–	0.34

Table 4.3 Role of ingredients used in glass-making process

Ingredients	Function
Silica sand and boron oxide	Former
Cullet	Former, fluxes and energy saving
Sodium and potassium carbonate	Fluxes
Calcium, magnesium and barium carbonate	Stabilizer
Sodium sulfate	Fining agent
Metal oxides	Colorant and bleaching

Glass is strong against pressure, non-permeable, transparent inert to a wide variety of food and very rigid and have excellent barrier properties. Glass is very heavy and fragile which causes safety concerns like presence of chipped glass in food products.

4.3.1 Manufacturing of Glass

The raw material like cullet, broken or recycled glass is introduced in to the glass melting furnace at 1500 °C. In the furnace, the solid material is converted to liquid state, homogenized, and then refined to remove bubbles. After that "gob," a lump of molten glass is transferred to the glass forming process (Fig. 4.5). In food packaging, various methods are used for glass making: the blow-and-blow process, narrow-neck-press-and-blow process and wide-mouth-press-and blow process [8, 9].

Blow-and-blow process The shape of the parison is formed when compressed air blows the gob into the blank mold of the forming machine, and then, final shape is formed by transferring the completed parison into mold where air blows and form required shape.

Narrow-neck-press-and-blow process Narrow mouth is called when less than 38 mm of finished diameter is used and a small metal plunger is used to make the parison shape, and the compressed air blows the container into its final shape. Beer or beverage bottles are formed by the narrow-neck-press-and-blow process.

Wide-mouth-press-and-blow process Instead of using air, a metal plunger is used to form the gob into the parison shape, and the compressed air blows the container into its final shape. Over 38 mm wide mouth is used as final diameter (Fig. 4.6).

After making the finished container, it is transferred to a lehr, large oven for the annealing process. Annealing is used to reheat and slowly cool the container to relieve the residual thermal stress. For strengthening and lubricating the surface, coatings are applied to the glass container [4, 7].

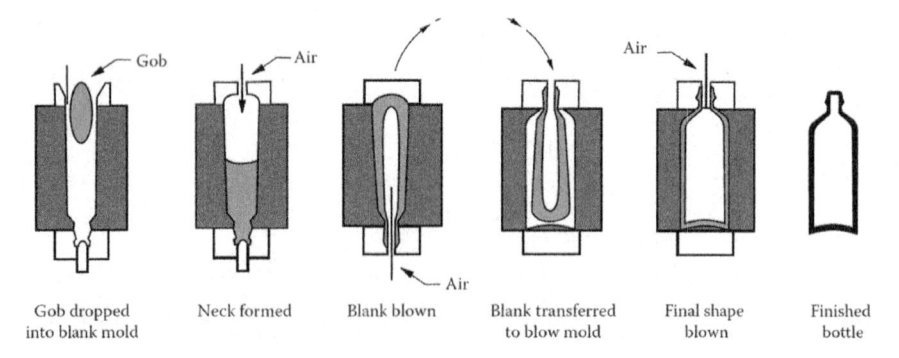

Fig. 4.5 Blow-and-blow process

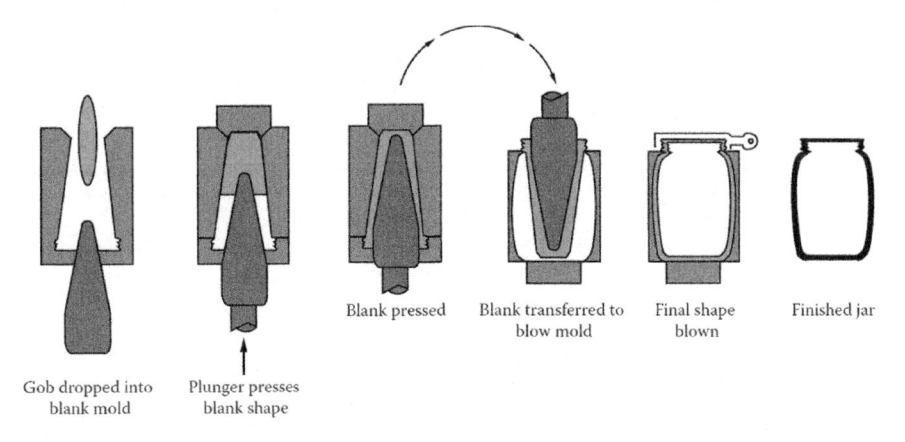

Fig. 4.6 Wide-mouth-press-and-blow process

4.3.2 Types of Coatings on Glass

1. Hot-end coatings—Hot-end coatings are applied to the glass bottles just before entering the annealing oven while still hot. Coating material consist of tin chloride, which later form tin oxide or organotin. These compounds leave rough surface when applied in vapor form which creates a good adhesive surface for the cold-end coatings. Hot-end coating fills minor cracks, provides hardness and compresses the glass surface.
2. Cold-end coating—These type of coatings are applied after cooling the glass containers. These are used to increase lubricity and to reduce the scratching of surfaces. Examples of these coatings are waxes, polyvinyl alcohol, polyethylene, silicone, etc. Cold-end coatings lubricate the surface and make it slippery. It also checks the compatibility of the cold-end treatment with adhesives which are used in labelling (Figs. 4.7 and 4.8).

4.3.3 Properties of Glass

1. Optical properties—Glass is transparent in visible and microwave wavelengths and has low ultra violet transmission coefficients. It is useful for protection of food. It is because of amorphous structure and chemical nature of ingredients. Alkaline oxides improve UV barrier properties.
2. Thermal properties—Glass is amorphous in nature and does not show sharp melting point. Higher concentration of alkali in glass decreases glass transition temperature and melting point. Boron and aluminum oxide increases the thermal strength of glass. Tensile stresses develop on the surface and compressive stresses produce in the inner structure when glass is cooled quickly. Compressive forces are set up on the surface when glass matrices are heated quickly

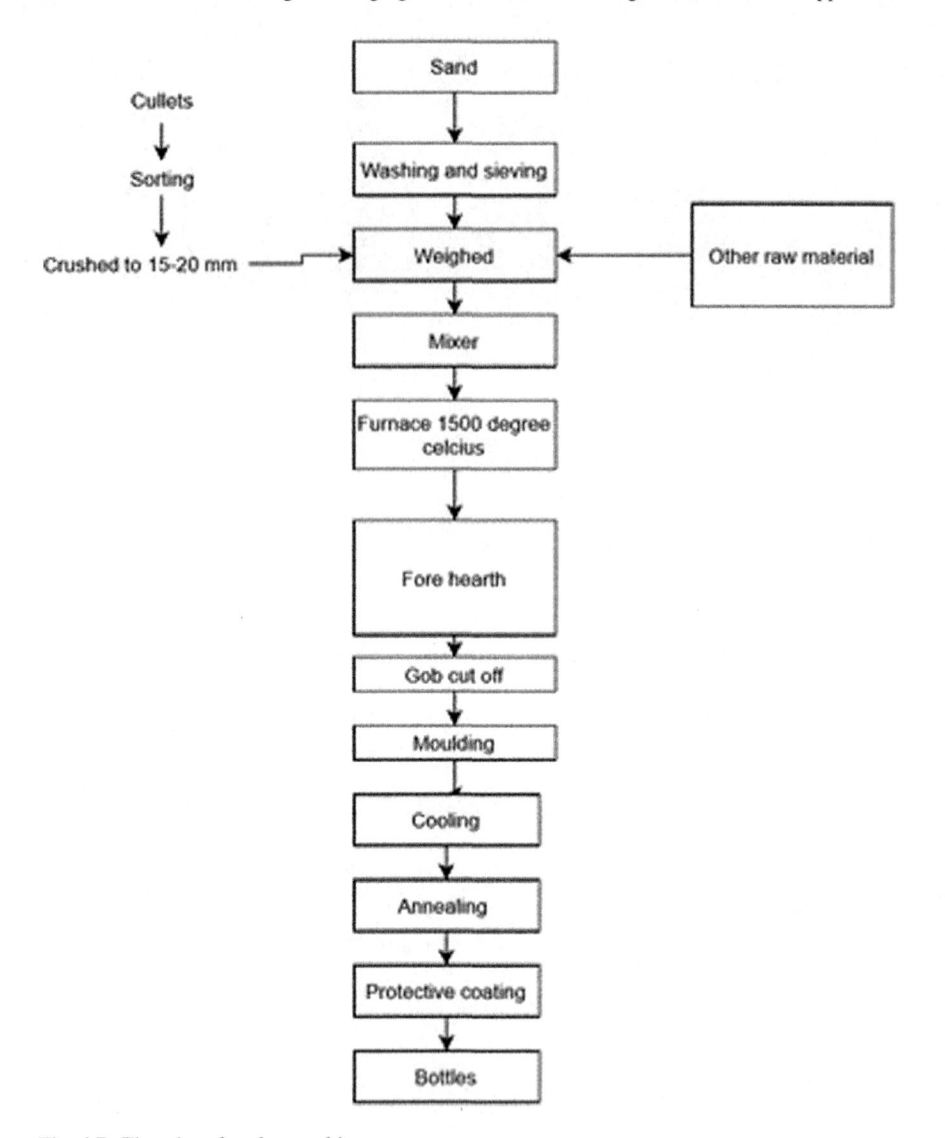

Fig. 4.7 Flowchart for glass-making process

and tensile forces develop on the opposite. These stresses disappear when the temperature equilibrium is slowly achieved [8, 9].

3. Mechanical properties—Glass containers are fragile objects. Glass bottles are tested for vertical load strength, internal pressure resistance, and resistance to impact to check superficial or internal defects (cracks, flaws).

Fig. 4.8 Glass bottle

4.3.4 Applications

Commonly used glass containers are bottles, jars, tumblers and jugs, carboys, vials and ampoules.

1. Bottles are made in various shapes and sizes from glass and have narrow or broad neck. This helps in easy pouring of contents, juices and drinks like flavored/sterilized flavored milk, milk beverages, etc., which are often packaged in these type of bottles. In dairy industry, heat-resistant glass bottles are used. These bottles can tolerate pasteurization and sterilization temperature.
2. Jars are mostly wide mouthed bottles having no neck. They are used for liquid, viscous, solid and semisolid products like fruit pieces, sauces and tomato pastes. They require large size closures.
3. Tumblers are almost like jars used for such products as jams and jellies.
4. Jugs are large sized bottles with handles and used for packaging wine.
5. Carboys are also made from large globular wicker-covered glass bottles for carrying or storing acids or other corrosive liquids.
6. Vials and ampoules which are small, thin-walled glass tubes used in the food industry for small quantities of very expensive ingredients such as flavors and colors.

7. Food products like baby foods, malted milk, flavored milk, beer, soft drinks, meat/fish products, fruits and vegetable products, etc., are packed in glass bottles to improve their shelf life.

4.4 Plastics Containers

Rigid plastic packaging accounts for more than 30% of the total rigid packaging in the food packaging sector in Asia. There are several advantages of plastic rigid packaging over glass or metal packaging. Largely for cost reasons rigid plastic bottles, jars, tubes, cups and trays are increasingly replacing glass and tin cans for food packaging. A wide range of plastics and copolymers are used to make rigid plastic containers [10]. To be used as food containers, plastic have a lot of advantages like:

– Low cost
– Lightweight
– High impact strength
– Available both as clear and colored material
– Squeezability.

  Generally, plastic packaging materials cannot be used at high temperatures. Thus, hot filling and heat processing with plastics are less common. This also limits the types of food that can be packaged:
– Foods which are naturally stable for the estimated shelf life and which can be filled in cold condition (such as dried goods, some pickles, cooking oils, fats, yoghurt, fruit juices containing preservative, beers, vinegar and honey).
– Jams and pasteurized pickles (provided that the product is cooled to below about 60 °C before filling)

The common uses for rigid plastic containers are:

Container	Application
Plastic bottles	Non-alcoholic beverages, cooking oils, ketchups, Sauces
Plastic jars	Honey, spreads, peanut butter, dry foods
Trays and tubs	Butter, fats, spreads, ice cream, jams, condiments
Cups	Drinks, yoghurt tubes honey, spreads

4.4.1 Production of Rigid Plastic Containers

Plastic containers can be made by various methods like:

  4.3.1.1. Injection molding. In this method polymer grains are heated by a screw inside a molding machine and then injected under high pressure into a cool mold. This method is primarily used for wide necked containers and lids [4] (Fig. 4.9).

Fig. 4.9 Blow-molded container

4.3.1.2. Blow molding is a similar method in which plastic containers are made in a similar method to glass bottle making and is performed in one or two stages. Blow molding are of different types:

Injection blow molding: In this method, a polymer is injected and molded around a blowing stick, while molten is transferred to the blowing mold. It is then blown into shape by compressed air.

Extrusion blow molding: In this method, a continuous tube of softened polymer is trapped between the two halves of a mold. The extruded is then inflated by compressed air into the mold.

Stretch blow molding: In this process, by either injection or extrusion molding process, a particular shape is prepared. This articles is then reheated, which causes the plastic molecules to "line up." This gives a crystal clear container of higher strength with better barrier properties to gases and moisture (Fig. 4.10).

4.4.1.1 Injection Molding

Injection molding is a manufacturing process, where molded plastic products are obtained by pushing the molten thermoplastic into a mold and then by cooling and solidifying it. It involves large volume production with complicated shape and good finish within a limited time. The machine has the advantages of high repeatability and low scrap production using this method. The main downside of the process is high initial investment and cost of new mold designing [4].

Material Selection

Polymers come in large number of grades depending upon their properties such as crystallinity, molecular weight, molecular weight distribution, etc. What makes a particular grade injection moldable is an important question that needs to be answered

Fig. 4.10 Injection-molded container

before selecting an appropriate grade for a particular product. Following are the factors that determine the moldablity of a polymeric material.

- **Crystallinity**—Crystallinity plays an important role in selecting an injection-moldable grade as amorphous polymers tend to show minimal shrinkage than compared to crystalline or semicrystalline polymers. They also provides improved transparency and are suitable for high tolerance applications. Amorphous polymers are brittle in nature and has very low chemical resistance which is not the case with crystalline polymers. Therefore, grades with appropriate crystallinity must be chosen depending upon the intended end-use application.
- **Molecular Weight**—Although higher molecular weight gives better strength and other physical properties to polymer due to increased entanglements, but it also increases the viscosity and thereby resistance to flow of the polymer. Since injection molding requires polymer to flow at a high rate inside the mold and therefore in order to completely fill the mold, polymer grades with relatively less molecular weight as compared to those used in extrusion molding are preferred.
- **Molecular Weight Distribution**—MWD of a polymer can be either narrow or broad, narrow MWD although provides good low temperature impact resistance and low warpage, whereas broad MWD offers good processability, and therefore, an optimized MWD is required to attain a balance between required properties and ease of processability.
- **Melt Flow Index**—It is a measure of melt viscosity of polymers. It is defined as the amount of polymer flowing through a capillary under standard conditions during a time interval of 10 min. For injection molding process, homopolymers and impact copolymers of MFI in the range of > 3 are normally preferred, whereas for random polymers, MFI values of 10 or more are preferred.

4.4 Plastics Containers

Processing Parameters

There are several processing parameters controlling the output product of as mentioned process, and these properties are mentioned below.

- **Melt/Nozzle Temperature**—Melt temperature has generally an effect on the viscosity of the polymer melt, and viscosity tends to increase with decreasing melt temperature. Melt temperature also affects the molecular weight of polymer in final molded product, and it is observed that with increasing melt temperature, molecular weight decreases causing a decrease in impact resistance as well as higher energy consumption.
- **Mold Temperature**—Mold temperature has different effects on amorphous and crystalline polymers. In case of amorphous polymers, higher mold temperatures result in less molded-in stress and therefore gives better impact resistance, stress crack resistance, and fatigue performance to the product. In case of semicrystalline polymers on the other hand, mold temperature defines the crystallinity of the formed product.
- **Injection Pressure**—Injection pressure in the molding process needs to be sufficiently high so as to fill the mold before gate "freezes off," but at the same time, extremely high injection pressure can cause premature opening of the mold that can damage the product.
- **Rate of Injection**—Rate of injection can be directly related to the screw speed or the production rate, so the injection rate needs to be as high as possible up to the limits that the machine and mold can withstand.
- **Holding Pressure**—Holding pressure is mainly required to fill the last 5% of the mold and thin wall sections of the mod. Holding pressure needs to be a bit lower than the injection pressure so as to prevent premature opening of the mold and to allow for stress relaxation, but it should be sufficiently high to completely fill the mold.
- **Holding Time**—It is mainly required to allow the venting of gases and to relieve stress. It also allows for the polymer chains to align themselves and achieve highest possible crystallinity. Therefore, holding time needs to be as less as possible but sufficient for all these functions to complete while maintaining maximum possible production rate.
- **Pressure During Plasticization**—Plasticization is generally the first stage in injection molding process. In this stage, screw motion generates controlled low pressure that is generally around 50–300 psi in order to melt the plastic granules or chunks. The pressure needs to be just sufficient to melt the polymers so as to attain minimal energy consumption.
- **Mold Clamping Force**—Clamping force needs to be sufficient enough to hold the mold shut tightly during the filling process and it should there be a bit higher than the injection pressure (Fig. 4.11).

Fig. 4.11 Holding pressure of plastics

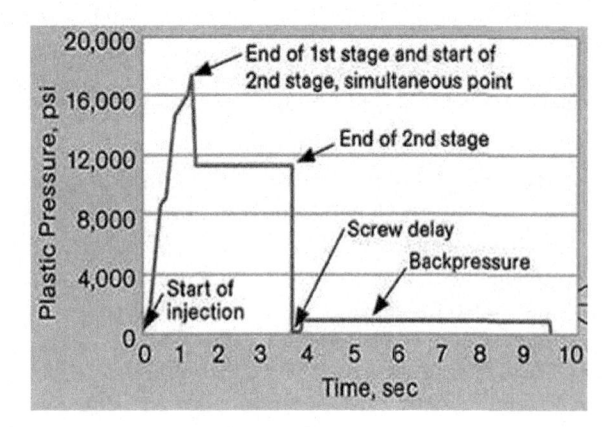

Quality Control

Quality of the injection molded products can be improved by focusing and taking care of the following issues.

- **Warpage**—It refers to the distortion of injection molded product during the cooling cycle. This is mainly caused by large difference in shrinkage ratio of different polymeric materials used in the product. It can be avoided through longer cooling cycle and modified mold design such as increased number of gates and their appropriate location along with appropriate design of undercut, rib and boss.
- **Burnt Deposit**—These as the name suggests are spots on the formed product where the plastic literally gets burned. These are the result of higher processing temperature or longer processing time along with increased speed and lack of proper ventilation. This can be tackled by controlling the as mentioned parameters.
- **Sink Marks**—Sink marks refer to small craters formed in the thicker regions of molded products due to shrinkage in inner regions. Lack of pressure keeping and pressure keeping time is the main cause of sink marks with fast sealing of gates also contributing to it. In order to control this, higher pressures and pressure retention time must be applied. Dimensions of gate, sprue, runner and nozzle must also be increased along with shorter gate land.
- **Burr**—Burr refers to materials used to form the product sticking to the sides in plate like structures, and this can occur due to very high pressures or temperatures or could also be the result of too much plasticization or lack of clamping. Mold wearing with time could also lead to this. Therefore, in order to obtain a high-quality product temperatures, pressure and plasticization must be optimized. Mold should be tightly clamped and should be properly maintained.
- **Incorrect Dimension**—Deviations from intended dimensions of the products may occur as a result of lack of filling or overfilling of the mold due to very high or very low injection pressures and pressure keeping time along with resin and barrel temperatures. This creates the need to optimize the resin and mold temperatures

4.4 Plastics Containers

and also injection pressure and pressure keeping time. Mold designs should also be changed to accommodate any expected overfilling or lack of filling.

- **Weld Marks**—Weld marks often appear on an injection molded product due to low resin temperature and injection pressure and speed. It can also occur due to inappropriate gate location and lack of proper ventilation. This can therefore be easily avoided by increasing the injection pressure/pressure keeping time or speed or rather cylinder temperature. Proper venting space must also be provided along with appropriate gate location.

An Injection Molding Machine Consists of the Following Parts

(1) Hopper
(2) Screw motion
(3) Heaters
(4) Nozzle
(5) Extraction pin
(6) Split molds
(7) Clamping unit
(8) Injection unit
(9) Drive unit
(10) Hydraulic unit.

The diagram of an injection molding machine and the function of each part of an injection molding machine is given below (Fig. 4.12):

Hopper: The plastic granules and other additives are added here.

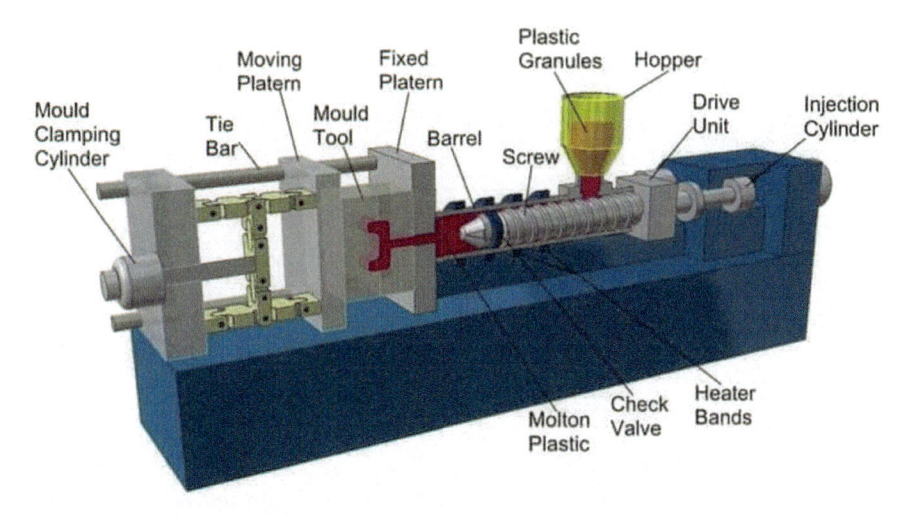

Fig. 4.12 Injection molding machine

1. Screw motion or Archimedean screw: Plastic and the additives are pushed forward here.
2. Heaters: Heaters are used to melt the polymer and to prepare the product with a good finish.
3. Nozzle: The material melts at such a high temperature that it can flow into the mold cavity.
4. Extraction pin: Extraction pin helps like split molds, and the formed article is removed from it for further processing.
5. Split mold or cooling channels: Cooling channels cool the product.
6. Clamping unit: It is used to clamp the tool.
7. Injection unit: Injection unit injects polymer into the mold (Plastics).
8. Drive unit: The driving unit is used to ram the mold in the cavity.
9. Hydraulic system: It helps in ramming the mold by the press (Fig. 4.13).

The working of an injection molding machine is explained below step by step:

- Plastic granules are fed into the hopper from where it comes into the system.
- Then, the material passes through the Archimedean screw whose work is to rotate and push the plastic material in forward direction.
- Next, the plastic comes to the hot zone where it gets heated.
- At the nozzle, the plastic flows into the mold cavity at a very high temperature
- Now, the cooling of the material is done.
- The extraction pin helps in removal of the mold cavity parts and the product.

These are different types of injection molding machines.

1. Metal injection molding
2. Die casting
3. Thin-wall injection molding

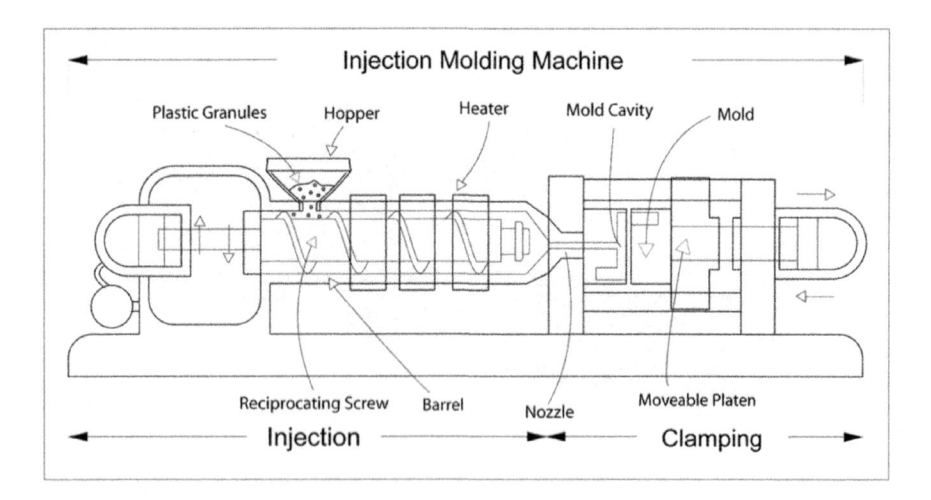

Fig. 4.13 Schematic presentation of working of injection molding

4. Reaction injection molding.

Among the above-mentioned injection molding techniques, primarily thin-wall injection molding machine is only useful for food packaging industry.

4.4.2 Thin-Wall Injection Molding

This type of plastic molding process can be applied to produce thin-walled plastic parts, so that the cost of mold process is much less than any other molding process. The main application of this molding process is mainly in the field of food packaging, medical, computer hosing making, etc., industries.

4.4.2.1 The Advantages of an Injection Molding Process Are

- Fast production
- Less labor expenditure
- High volume production
- Easy manufacturing of the small product parts
- Leaves very less scrap during the production of a part
- Can create holes in the product
- The color of the product can be controlled
- Good finish
- Proper dimension.

4.4.2.2 The Disadvantages of the Injection Molding Process Are

- High tooling and machinery cost
- High cost of molds
- The prototype design has to be created before beginning the process.

4.4.3 Blow Molding

Blow molding is a process specific to making hollow objects. It involves blowing air inside heated plastic to inflate it into a mold cavity to create hollow objects. On basis of stage prior to the molding process, this process is generally classified as either extrusion blow molding or injection stretch blow molding. Extrusion blow molding as the namesake suggests extrudes plastic in the form of tube of molten polymer which can then be used to make different hollow objects such as bottles and biohazard containers. It is especially known to make blown films which is mainly used in packaging applications. In case of injection blow molding, a parison is formed

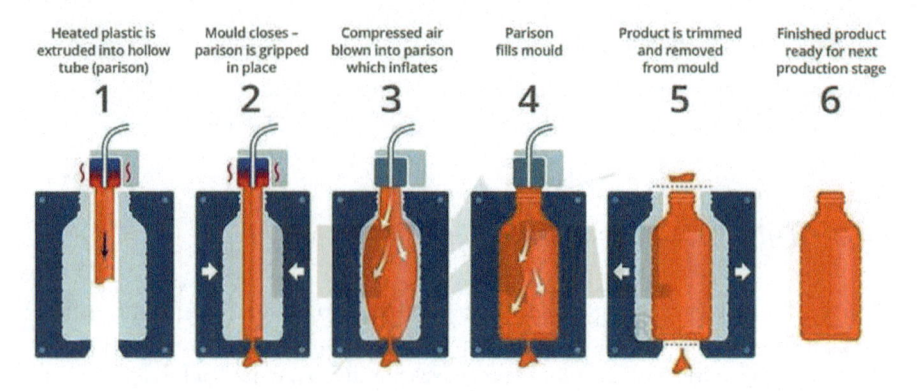

Fig. 4.14 Blow molding process [https://www.petallmfg.com/blog/wp-content/uploads/2017/07/extrusion-blow-molding-process.jpg]

using injection molding around a core rod that is then used to blow air from the inside to inflate the parison into filling the mold. Blow molding allows for high production rate at low tooling and die cost, and it enables us to make complex hollow shapes which might not be very efficient with many other processes (Fig. 4.14).

4.4.3.1 Material Selection

In the process of blow molding, the material selection is a critical step to design the end product. Before selecting the material, following aspects need to be considered:

1. Properties of the plastic resin
2. The cost of the material
3. The processing characteristics
4. Usability of finished parts.

Some of the least expensive materials are also the easiest to process. Polyethylene (PE) and polypropylene (PP) are the most popular blow molding resins. PE is currently less expensive, but PP tends to be stiffer which sometimes offsets the cost difference.

Blow molding involves both flowing and subsequent stretching of polymers, and therefore, before classifying any polymer grade as blow molding grade, one needs to consider the following properties.

- **Crystallinity**—Polymers are normally semicrystalline with varying degree of crystallinity, and the crystalline regions of the polymer makes the polymer harder and more thermally stable but at the same time imparts brittleness to the polymer which should be avoided for blow molding process as the polymer in this process is subjected to stretching. Amorphous region on the other hand imparts elasticity and impact resistance there an optimized value must be chosen depending upon the end-use application.

- **Molecular Weight**—Although higher molecular weight gives better strength and other physical properties to polymer due to increased entanglements, it is also often directly proportional to viscosity. With increase in viscosity, polymers will be resistant to flow and would require high energy input to make them occupy the mold uniformly. It is thus required to select a polymer generally with average molecular weight in range of 80,000–1,20,000 for blow molding operations.
- **Melt Flow Index**—It is a measure of melt viscosity of polymers. It is defined as the amount of polymer flowing through a capillary under standard conditions during a time interval of 10 min. For extrusion blow molding, homopolymers and random copolymers of polyolefins for MFI range 2–3 are often preferred.
- **Melt Strength**—Melt strength refers to the ability of semi-molten polymer to resist a break in flow or cracking under stretching force. It is the most crucial parameter while selecting a grade for blow molding. As the parison is blown to fill the mold, the material gets stretched in which case lack of extensional viscosity or melt strength can cause the parison to rupture. Melt strength also increases with branching or cross-linking of polymer chains; therefore, such grades are termed appropriate for blow-molding process.
- **Molecular Weight Distribution**—Blow molding operations require flowability as well as searchability of polymers. Higher molecular weight chains provide elasticity, whereas lower molecular weight chains act as plasticizers thus enabling the polymers to flow easily therefore bimodal molecular weight distribution is found to be very effective in blow molding operations. It is also observed in some cases that melt strength initially increases and then later on decreases with polydispersity. Therefore, in case of broad molecular weight, distribution grades with polydispersity values where melt strength reaches its peak must be chosen.
- **Processing Parameters**—There are several processing parameters controlling the output product of as mentioned process, and these properties are mentioned below.
- **Feed Zone Temperature**—This is required to be able to just melt the polymer.

Therefore, it is normally kept 5–10 °C above melting point. Improper feed temperatures can lead to lower output and porosity in the parison.

- **Head Zone/Die Zone Temperature**—Head zone temperature should be the desired melt temperature, whereas die zone temperature should be about 5 °C greater than the melt temperature to offset the effect of cooling air.
- **Mold Temperature**—This should be set at values just above the dew point of ambient air in order to maximize cooling and minimizing surface defects or blemishes that can be caused by condensation. A cold mold will give the minimum cycle time, whereas a warm mold will give the best surface appearance.
- **Blowing and Cooling Times**—Cooling cycle time is decided on basis of design parameters and required wall thickness, and it is generally 25–120 s but can go to higher values in case of large complex part designs. The coolant flow through the molds should also be fully turbulent.
- **Quality Control**—Quality of the blow-molded products can be improved by focusing and taking care of the following issues.

- **Parison Deformation**—This can generally occur if the outer wall of the parison is too warm or too cold or it may be the result of variation of wall thickness through cross section. This is also observed in cases of materials with melt viscosity too low or due to too low screw speed or too high processing temperature. Therefore, die must be optimally heated, and a high melt viscosity material must be chosen to avoid this. Melt temperature and screw speed must also be optimized.
- **Appearance of Flow Marks**—This can occur due to presence of contamination or degraded material. Mandrel support being too close to the die can also result in such flow marks. Therefore the die ring should be properly cleaned and melt temperature should not be too high to degrade the material. Head design should also be modified in order to counter such issue.
- **Rough Inner Surface**—Roughness on the inner surface can be the result of melt and die temperature being too low or if the parison is extruded too quickly or because of die chatter. In order to counter this melt and die temperatures must be increased and parison should be extruded at a relatively lower rate. Parison programming needs to be checked to avoid any die chatter.
- **Blown Stripes**—Blown stripes can appear on the parison due to overheating i.e., high melt temperature or longer residence time. This can also be the result of too high shear rate or if the mandrel support is not streamlined. This can be solved with optimized melt temperature and screw speed and repositioning the mandrel support.
- **Parison Folds or Ripples**—This is often observed in case of too thin parison or with high melt temperature; therefore, either the melt temperature needs to be decreased or parison should be given a light pre-blow.
- **Materials Stuck to Edges**—This can be avoided by lower of mold temperature or decreasing the cycle time.
- **Bubbles**—Bubbles are caused because of the entrapped air, and in order to avoid them, screw speed and extrusion pressure must be increased. Proper venting must also be provided to let the trapped air escape.

4.4.3.2 Working of a Blow Molding

Formation of a molten tube can be achieved by blow molding process (referred to as the parison or preform) using thermoplastic material by placing the parison or preform within a mold cavity. The tube is inflated with compressed air, to take the shape of the cavity and cool the parts before removing from the mold. Blow molding can be used to make any hollow thermoplastic part.

1. The first step involves mixing, melting and pushing plastic (by extrusion process) to give it a tube like structure called a parison which will be used to make the end product.
2. A mold is used to provide the desired shape to the part. The mold is divided into two halves that are closed around the molten parison.
3. Air is blown inside the parison to expand the molten plastic against the mold surface.

4. The mold is cooled, so that the plastic is set to the new shape of the mold.
5. The molded plastic part is then removed from the mold separated from.

There are three main types of blow molding [3] which are useful in the food packaging industry (Fig. 4.15).

The main differences among various methods of blow molding are the methods in which the parison is formed. The parison is created:

- Either by extrusion or injection molding
- The size of the parison and
- The method of movement between the parison and blow molds; either stationary, shuttling, linear or rotary.

4.4.4 In Extrusion Blow Molding (EBM)

In this process, plastic granules are melted and extruded into a parison which is captured by closing the parison into a cold metal mold. Through the parison, air is then blown which inflates it into the shape of the hollow container or part. In the next step, the article is cooled sufficiently and is ejected after opening of the mold.

There are two types of extrusion blow molding: continuous and intermittent.

In continuous extrusion blow molding, the parison is extruded continuously, and the individual parts are cut off by a suitable knife.

Intermittent blow molding can be further divided into two processes: straight intermittent blow molding is same as injection molding, whereby the screw turns, then stops and pushes the melt out.

In the accumulator method, an accumulator gathers molten plastic. When one set of mold is cooled and sufficient plastic is accumulated, a rod is used to push the molten plastic and form the parison. In this case, the screw turns either continuously or intermittently.

In Injection Blow Systems (IBS) process, the plastic is injection molded to a core within a cavity where a hollow tube called a preform is made. At the blowing station, the preforms rotate on the core rod to the blow mold where it is to be inflated and cooled. Small bottles, usually of 500 ml or less is typically prepared by this method. The process involves three steps: injection, blowing and ejection, in one integrated machine. Parts with accurate finished dimensions come out (Fig. 4.16).

The process of **injection blow molding** (IBM) involves the production of hollow glass and plastic objects in large quantities. In the IBM process, the polymer is injection molded onto a core pin; which is then rotated to a blow molding station to be inflated and cooled. This is not very commonly used process. It is typically used to make small medical and single serve bottles. The process involves three steps: injection, blowing and ejection [4].

Injection Stretch Blow Molding (ISBM) process is similar to the IBS process described above and in that the preform is injection molded. The molded preform is

➤ Extrusion blow molding ➤

Injection blow molding

➤ Injection stretch blow molding

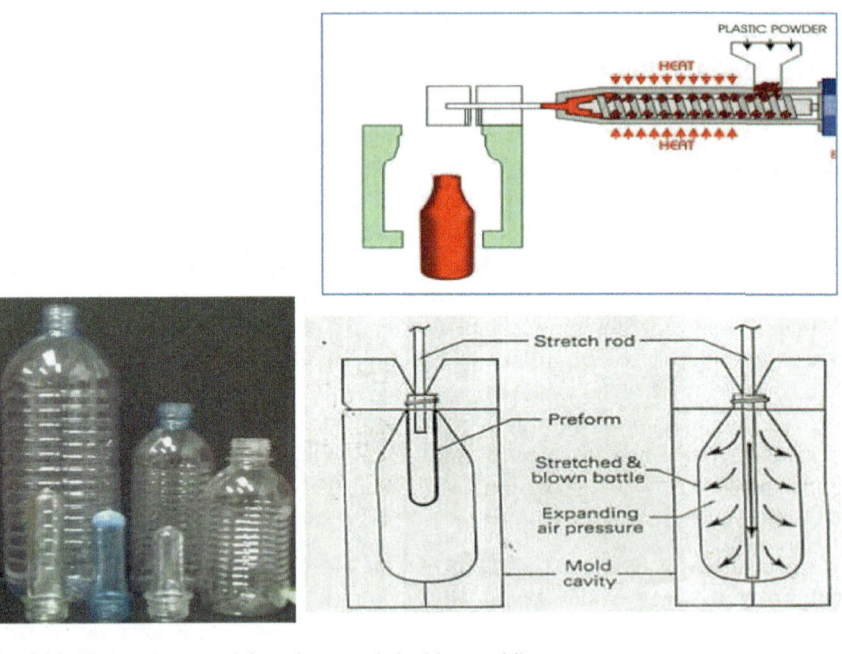

Fig. 4.15 Various commercial products made by blow molding

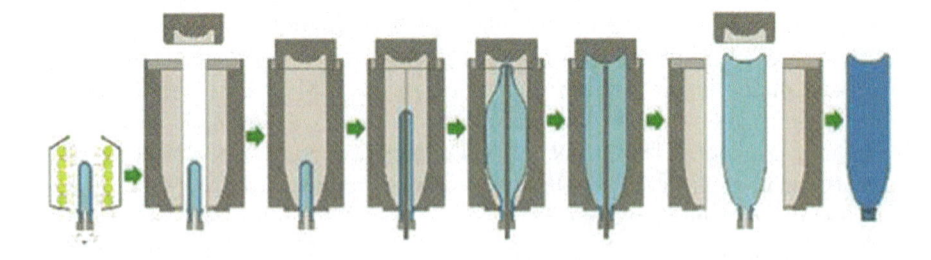

Fig. 4.16 Injection blow molding

then presented to the blow mold in a conditioned state, but before final blowing of the shape, the preform is stretched in length as well as radially. The typical polymers used are PET and PP that has physical characteristics that are enhanced by the stretching part of the process.

Examples of each processes.

Advantages of blow molding

- Affordable tool and die cost
- Production rate is fast
- Complex parts can be molded
- Handles can be incorporated in the design.

Disadvantages of blow molding

- It is strictly limited to hollow parts
- Low strength of the product
- Different materials are used to increase barrier properties making it non-recyclable (multiparison)
- Trimming is important to make wide neck jars spin
- Limited to thermoplastics only.

4.5 Plastic Crates

Crates are considered as one type of rigid plastic material which are mostly used as shipping container. This type of containers are made either of wooden or plastics. The crates are typically used for the bulk packaging, storage and transportation of fresh produces, mainly fruits and vegetables. In addition, specialized types of crates are also designed for the packaging of specific products like milk pouches, soft drink bottles, etc. The dimensions and the weight of crates will vary depending upon the size, weight and characteristics of the item to be packed.

Over the years, the demand of usage of plastic crates have been increased to a great extent due to its most important returnable property. The usages cover a huge range of fresh and processed food products for the carrying and storing of these goods. The use of returnable plastic crates for harvest, packing, transport and storage of fresh produce has repeatedly been shown to reduce damage and post-harvest losses. In the developing countries, there has been a rapid adoption of plastic crates for bulk packaging of selected fresh produce items, and this growth has offered new business opportunities for service providers such as farmers groups or clusters, cooperatives, traders, commercial farmers and foreign agribusiness firms and even the plastic crate manufacturers who are engaged in rental services to farmers. The increased adoption of plastic crates is helpful for the alleviation of human drudgery and to meet the requirement of buyer preferences [11].

4.5.1 Criteria for the Selection of Crates

The following criteria are normally considered while selecting the plastic crates instead of corrugated fiber board boxes:

(1) **Store Space**: The store need to be have more space for storage of plastic crates as compared to corrugated fiber board boxes. The CFB boxes are normally stored in collapsing condition, but the plastic crates are stored by stacking one above another which leads to the requirement of more space.
(2) **Enclosure of Store**: Plastic crates are not hygroscopic in nature unlike corrugated fiber board boxes. These are not affected even if the crates are completely drenched with rain water. Due to this fact, there is no requirement of complete enclosure of the store which leads to less expenses toward infrastructural facilities.
(3) **Handling facility**: The plastic crates can be handled easily as compared to either corrugated fiber board boxes or wooden crates. The plastic crates are light in weight as compared to wooden crates. Moreover, the design itself provides handle slot by which the crates can be easily lifted and handled.

4.5.2 Materials Used for the Manufacturing

Plastic crates are generally manufactured by using two kinds of polymeric materials like high-density polypropylene (HDPE) and polypropylene copolymer (PPCP) [12]. These two polymers offer a number of advantages as given below:

(1) These polymers are dimensionally stable with high resilience factors and also absorb the stresses and impacts which might occur in normal practices either in the manufacturing unit or during handing, storage and transportation.
(2) HDPE/PPCP are chemically inert, and thus, the packaging materials do not affect the food stuff, contained in the crates.
(3) The polymeric materials have no taste or smell and also make the compliance to the prescribed Indian standard.
(4) The materials are resistant to acids and alkalis in wide range of concentrations as well as temperature.
(5) It is also not affected by the synthetic detergent which are commonly used for washing purposes.
(6) However, these polymers might get affected while exposed to highly oxidizing agents such as nitric acid, Oleum or halogens. In addition, these materials can also get dissolved to aromatic and halogenated hydrocarbons.

4.5.3 Manufacturing Process

The plastic crates are manufactured by means of injection molding process. In this process, the molten plastic is injected into the selective mold /die at high pressure to ensure that mold is properly filled. In general, molds are made from metal either steel or aluminum and then machined very precisely to form the features of various parts of the crates which is mainly due to the extremely high force that is exerted on the molds during injection molding process. Though, the injection molding process seems to be very simple but considering the design of crates, manufacturing process in terms of mold design and injection process become complicated in order to have proper finish of every features of the crates as per the specified design [11, 12].

4.5.4 Type of Plastic Crates

The Plastic crates are designed into various types by considering the important factors like easy bulk handling, protection of the produce during handling, facilitate cooling and preventing condensation to avoid decay during storage. Accordingly, the important designs of plastic crates are as follows.

4.5.4.1 Stacking Crates

This type of crates are made in square shape with slightly rounded corners. Due to this, this type of crates can be stacked easily one above another within a limited space. Moreover, the design itself provide a slot for easy handling. However, depending upon the size of the crate, the loss of loading space compared with loose break bulk is between 20 and 30%.

Although stacking crates have a rigid design, some space during the return trip can be gained by putting one crate inside two others, and in order to overlap crates during stacking, some space gaps in the rim of the crate are needed (Fig. 4.17).

4.5.4.2 Stack-Nest Crates

This type of crate has got two swing bars to facilitate easy handling. But the design itself has got vertically tapered shape which reduces the inside volume of stack-nest crate as compared to a crate with a squared design. But this type of crate of having specific size and capacity would provide advantages about the requirement of lesser space for stacking as compared to stacking crates due to its swing bars in the design. If the bars are swung out, the crates can be nested. A swing bar (9 mm) is swung from the outside or from the side over the top of the crate and forming a support for the following crate. The stack-nest crate is slightly weaker than the stack crate,

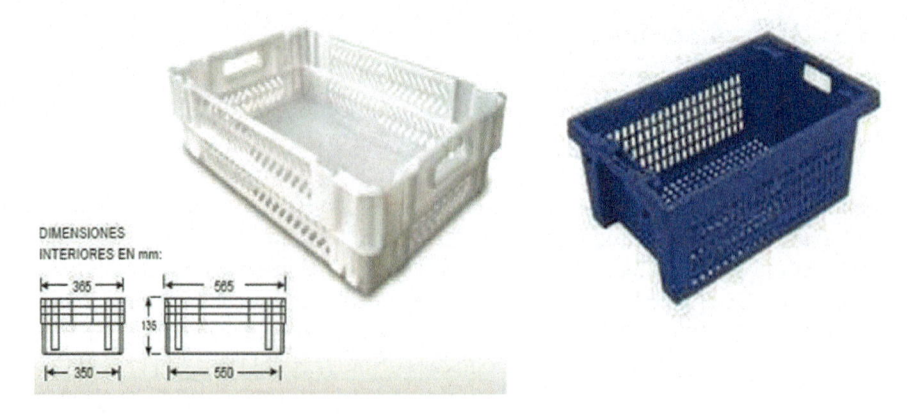

Fig. 4.17 Stacking crates

Fig. 4.18 Stack-nest crate

because the bar is not resting on the corners which is the strongest part of the crate but on the long side of the crate. The swing bar is placed 1–2 cm under the rim of the crate which facilitate easy stacking. The crate on top should be placed within this rim. The swing bar can also be fitted on the bottom side of the crate to make another design (Fig. 4.18).

4.5.4.3 Stack-Nest Crate with Cover

The design of plastic crate is little different where instead of the swing bar, the crate is closed by two cover parts on top of which the next crate can be stacked. The advantage of this design is that this cover can be sealed to prevent the pilferage of produce. However, the crates with cover will be more expensive as compared to the crate without a cover (Fig. 4.19).

4.5 Plastic Crates

Fig. 4.19 Stack-nest crate with cover

Fig. 4.20 180° stack-nest crate

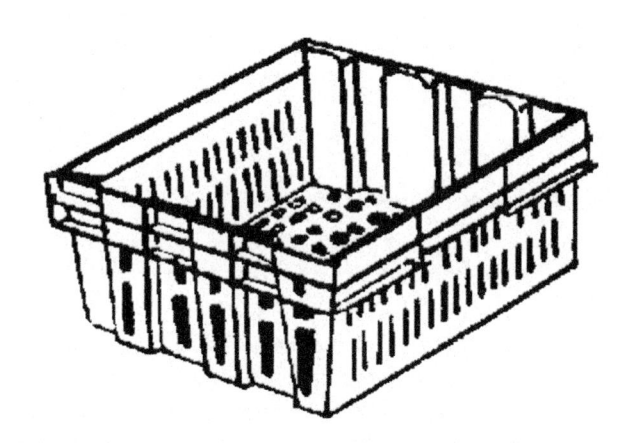

4.5.4.4 180° Stack-Nest Crate

This kind of crate has got a special design where supports are given at several places inside the crate, so that it would like a box and can stacked in a 180° turned position be nested. These supports require extra space, and dirt will assemble in the corners created by the supports (Fig. 4.20).

4.5.4.5 Collapsible Crates

A collapsible crate consists of a base with sides attached to it by plastic or metal hinges. This is not only an innovative attractive design in order to provide savings

Fig. 4.21 Collapsible crates

of space when folded. Moreover, this particular design of crates complies with environmental regulation due to its reusability. Still, it is observed that this kind of crates is not having demand to the market which might probably due to high purchase cost. The interior of the collapsible crates should be smooth to prevent damage to any fresh produce packed inside [13, 14]. If the interior is rough or there is too much venting, a low cost paper or lightweight fiber board liner can be added to the collapsible crates [11, 15] (Fig. 4.21).

4.5.4.6 Advantages of Plastic Crates

As a strong, rigid crate, these plastic crates can be used for many journeys, making the cost per journey relatively low.

- Different sizes and shapes are available to suit different customers' needs. Colors can be used for marketing purposes.
- The containers are easy to clean and to disinfect.
- Plastic crates are strong and weather resistant and, because of their water resistance, the containers can be used in humid areas and during hydro-cooling.
- The crates are also economical by way of saving in weight, no maintenance costs, longer life, less breakage than glass.
- The crates are having high functional property in terms of easy to clean, no special surface protection, easy to stack, less noise and increased safety.
- The crates also enhance sales promotion due to its bright colors, easy to print with names, monograms or advertising slogans.
- The surface of plastic crates is smooth and non-porous and consequently does not harbor bacteria.
- The corners and edges of a correctly designed crate are rounded; crevices in which dirt can accumulate which can be cleaned with steam or hot water and are detergent resistant.

4.5.5 *Disadvantages of Plastic Crates*

- The hard surfaces can damage the produce, and it is advised to use liners.
- The high purchase cost combined with the risk of pilferage could make this type of crate a financial risk.
- Because this crate can be used several times, the extra cost for the return trip should be included in the total running cost.

4.5.6 *Applications*

- Storage and transportation of fresh fruits and vegetables
- Carrying and storage of milk pouches
- Carrying of soft drinks bottle, mineral water, beer, etc.

4.6 Wooden Crates

A wooden crate is considered to be another type of transport containers which has a self-supporting structure, with or without sheathing. The wooden crates are completely different than wooden box. In case of wooden crates, all six of its sides are not completely closed like wooden box. Due to this, crates are having distinct design from wooden boxes. However, in general conversation, the term "crate" is sometimes denoted as wooden box. A wooden crate is typically defined about the placement of battens at edges and corners of the crate [16].

Wood, besides being the most ecological option, is an easily repairable material; if any of the pieces break, it can be easily replaced without having to resort to complete replacement. This type of packaging gives the customer a presentation that refers to the values of tradition and origin, being respectful of the environment. Wood is a natural material which acts as moisture regulator. Wooden crates are able to keep the product fresh for longer duration. The porous structure of wood is a physical inhibitor of bacteria, and thus, the wooden containers are capable for the transportation of goods in hygienic manner for export market.

4.6.1 *Material Used*

There is often a confusion about the terms "timber" and "wood" which are normally interchangeable. The term "wood" is used to refer the substance which makes up the tree. Wood is the tough, fibrous material which makes up the tree under the bark. The primary function of wood is to enable the tree to grow straight and tall by absorbing sunlight for photosynthesis and transfer of water and nutrients for growing tissues and leaves. But the term "timber" might be defined as a wood at any stage after the tree has been felled and retains its natural bodily structure and chemical composition and is suitable for various engineering works. In fact, this is the raw material which is known as timber or processed material. Timber was one of the first structural materials used by man in primitive days, and it continues to be used till today for various purposes [16, 17].

4.6.1.1 Classification of Timbers:

Timbers are broadly classified into two categories:

(a) Coniferous timbers or non-porous timbers or soft wood like blue, pine, chir, spruce, deodar and fir
(b) Non-coniferous timbers or porous or hard wood like teak, mango, rosewood and haldu.

4.6.1.2 Properties of Timbers

The physical and chemical properties of timbers are given below:

- Physical Properties:
- Color—It varies from species to species. Color helps in identification of timbers. e.g., timber of walnut, ebony, rosewood and mahogany are used for decorative items only because of their color.
- Lusture—Certain timber appears to be shiny due to reflection of light from its surface. e.g., Spruce has a characteristics pearly lusture.
- Odor—Different timber has got very characteristics odor due to the presence of chemical deposits and oils. e.g., sandal wood.
- Weight—It depends upon density, moisture content and organic and mineral infiltration deposits in the cell cavities.
- Mechanical Properties:
- Strength as a beam—It indicates the suitability for using as griders, axles, etc. (ii) Shock resistance ability and toughness—It indicates the ability to absorb shocks and blows.
- Elasticity—It is the ability of timber to withstand the compression and resist crushing against the grain or in the direction of grains.
- Shear—It is the failure of timber due to one part sliding over the other which is called shear and the force which causes this failure is called "shearing force."
- Hardness—It is the property of timber to resist penetration by other bodies.
- Seasoning of Timber:

Seasoning is a process of controlled, gradual drying of timber to safe limit of moisture content. Seasoning renders timbers quite stable, thereby ensuring that once any container like wooden crate is made from it, there would be very little or no change in shape and dimensions. Adequate care is required to ensure that there is proper air circulation around each wooden hair. However, the decrease of moisture content over a period of time is a natural process. A well-seasoned wood should not have moisture content more than 18%. During the course of drying process, timber may develop defects, and this is normally avoided by providing adequate protection against unduly rapid drying conditions, care in stacking and top weighting of the stacked timber. Generally, timber gets shrunk during drying [18].

- Objective of Seasoning:
- The main objective of seasoning is to eliminate this shrinkage and the associated drying defects before use.
- To increase the strength and electrical resistance of timber
- Reduction in weight for transport purposes
- Improvement in wood-working qualities including gluing, painting and polishing
- A certain degree of protection from attack by insects and fungi depending upon the moisture content to which timber has been seasoned.

Seasoned timber should be stored under cover protected from rain and other forms of precipitation.

Seasoning is carried out mainly in two ways:

(a) Air seasoning
(b) Kiln seasoning.

- **Air Seasoning**: Sawn material for air seasoning should be stacked under shade, preferably in a shed.
- **Kiln Seasoning**: It is a quick method to attain the desired level of moisture content. For large-scale production of seasoned timber, kilns have become almost an essential adjunct to many timber industries. For special purposes, kiln seasoning is considered to be most indispensable method for seasoning of timber to low moisture contents which normally cannot be attained in humid climatic condition by air seasoning. Kiln Seasoning is also used for sterilization of timber against insects and decay.

4.6.1.3 Characteristics of Good Timber for Manufacturing of Crates

The principal characteristics of timber is mainly concerned to the strength, durability, and finished appearance. However, there are certain important parameters as given below:

- Free from all kind of the defects like objectionable knots, center heart (Pith), insect attack and any kind of decay (rot), warping, splits, etc.
- Presence of narrow annual rings, closer the rings greater is the strength
- Presence of medullary rays in compact forms
- Shock resistance ability
- Retention of shape
- Sufficient hardness
- Light in weight

Fig. 4.22 Vegetable crates

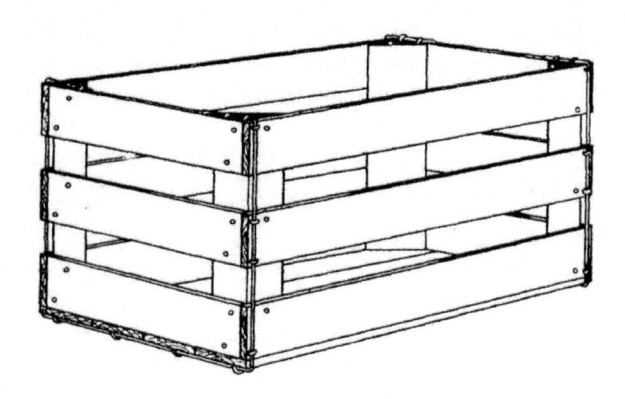

- Good nail holding power
- Easily workable
- Uniform texture
- Economical
- Preferably of whitish in color.

4.6.1.4 Construction of Wooden Crates

The bottom, side and end boards shall be fixed to corner posts as illustrated in Fig. 4.22. The nails used in the manufacturer of wooden crates shall be of clout headed type. If additional battens are provided, bright nails of 25 mm length and 1.8 gauge may be used and clinched along the grain. The ends of the crates should be wired with 3 mm galvanized or rust resistant wire. At each end of the crate, there shall be four staples on each side and at the bottom and four staples at the top (Fig. 4.22).

4.6.2 Design of Wooden Crates

Depending upon the application, the wooden crates are designed into the following types:

(a) Nailed crates
(b) Stitched crates
(c) Wire bound crates
(d) Wooden collapsible crates.

(a) **Nailed crates**
- This crates are mostly used for apple, pear and other fruits.
- Nailed crates are rigid and strong boxes which serve as multitrip containers with a long life time.

- The planks have a thickness of at least 6 mm. Because of the rigidity, the crate is quite heavy and the initial cost is high compared to, for instance, wire bound crates. Spacing between the planks, (bottom, sides) and/or between top side plank and the bottom of the next crate creates is recommended for good ventilation.
- Nailed crates are frequently used for domestic transport e.g., as a field crate or as a crate to transport produce from the producer to the wholesaler or trader.
- An advantage of nailed crates is the possibility to repair the crate.
- Disadvantages of the rigid nailed crate include the high return freight volume. A partial solution to this problem is to put one crate in two other crates which are placed opposite each other; so three empty crates will take up the space of two stacked crates (Fig. 4.23).

(b) **Stitched crates**

- The stitched crates are normally used for tomato.
- Stitched crates are made of thin (3–4 mm) pieces of wood stitched together.
- Corner pieces, mostly triangular, provide the necessary strength to stack crates. This type of crate is mainly used for single journeys (Fig. 4.24).

(c) **Wire bound crates**

- Wire bound crates are used for orange, grape fruits and also for potato.
- As a rigid, cheap crate with a good stacking strength, it is mainly used for single journeys.

Fig. 4.23 Nailed crates

Fig. 4.24 Stitched crates

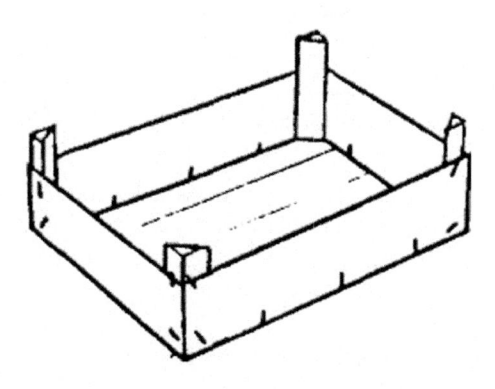

Fig. 4.25 Wire bound crates

Fig. 4.26 Wooden collapsible crates

- Wire bound crates are stitched crates with a wire under the stitches which gives extra strength to the container. The wire also serves as a hinge and as a lock for the lid. These crates provide good ventilation, and fast pre-cooling is possible.
- Wire bound crates are mostly used as returnable crate (Fig. 4.25).

(d) **Wooden collapsible crates**

- This is the recent development of wooden crates.
- It consists of a removable top and bottom part and the crate can be folded using the hinges on the corners.
- The crate performed well in trials and it was possible to be used as returnable crates for many times (Fig. 4.26).

4.6.3 Advantages of Wooden Crates Are

- The crates can be manufactured and repaired locally.
- Wood is relatively resistant to different weather conditions and (sea)water.
- Wooden crates are often used on more than one journey and have a higher efficiency for larger fruits, e.g., watermelons.

4.6 Wooden Crates

- Most crates have good ventilation, and fast pre-cooling is possible.

4.6.4 Disadvantages of Wooden Crates Are

- Untreated wood can easily become contaminated with fungi and bacteria.
- Treatment of wooden crates with paint or other chemicals may cause produce deterioration.
- The material may be too hard or rough for produce like soft fruits, and therefore, liners of a soft material may be needed.
- Disposal of the crates after use
- Manufacturing of wooden crates puts an extra claim on the natural forest resources [18].

4.7 Corrugated Fiber Board Boxes

Corrugated fiber board and boxes are considered to be most dynamic development in packaging which came from simple adaptations. This offers one of the most versatile group of packaging media cutting across all the conceivable product range. The versatility also ensures its amenability to varying distribution practices and climatic conditions. The flexibility projects possibility of designs, sizes and shapes, collapsibility and esthetics. The design features improves productivity, operator and user friendliness.

Corrugated fiber board box is considered to be the most popular packaging media as transport containers. Million tons of agricultural and industrial produces are stored in the factory and transported from the factory to the market in corrugated fiber board boxes. Corrugated board boxes are also used to pack live chickens, dry fish, dry onion and garlics, flowers, fruits, vegetables and a large variety of processed foods in packaged condition.

Corrugated packaging media has got a very interesting historical background. In fact, in 1871, Albert Jones, an American who discovered this idea to provide strength to paper by converting a piece of paper, convoluted it into series of flutings by moistening it and the passing through the two hot corrugating rolls (Fig. 4.27).

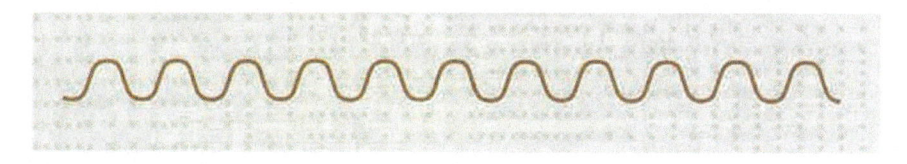

Fig. 4.27 Flute without liner

Fig. 4.28 Flute with single liner

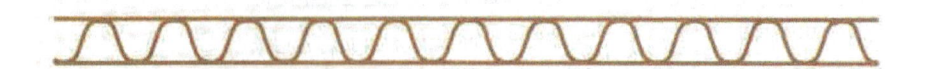

Fig. 4.29 Flute with double liner

Subsequently, in 1874, Oliver Long, another American fellow discovered when a flat sheet of paper was glued to one side of corrugated paper, it retain the shape even if it was stretched and subjected to pressure. This discovery has developed "Single face Corrugated," and thus, the corrugated board industry got the birth (Fig. 4.28).

In 1878, Robert Gair is considered to be the father of the folding carton and was the pioneer in the corrugating industry who discovered the single wall corrugated sheet by pasting another flat paper to the corrugating roll (Fig. 4.29).

In fact, Oliver Long discovered the application of single face corrugating roll as cushioning material for wrapping the fragile items. But Robert Gair had given another dimension of application by pasting single facer to another single facer and then to single liner to make three ply or single wall, five ply or double wall and seven ply or triple wall corrugated board (Fig. 4.30).

Triple wall or seven ply corrugated fiber board generally consists of

1. An outer liner
2. A narrow flute
3. An intermediate liner (test liners)
4. A medium flute
5. An intermediate liner (test liners)
6. A wide or medium flute
7. An inner liner.

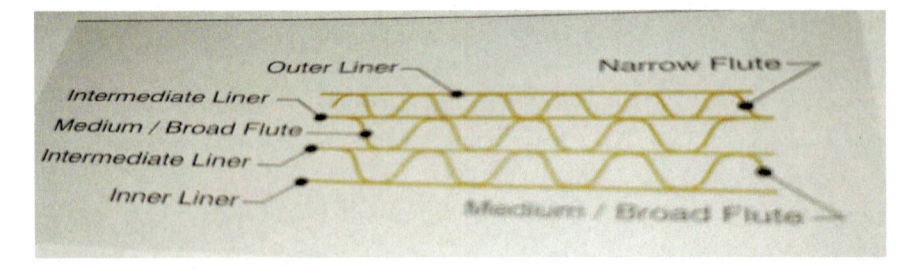

Fig. 4.30 Flute with multiple liner

4.7 Corrugated Fiber Board Boxes

4.7.1 Components of Corrugated Fiber Board

A corrugated board is having the following important components:

1. Kraft liner
2. Test liners
3. Corrugating or fluting medium
4. High performance liners
5. Corrugating adhesive.

1. **Kraft Liners**: The word "kraft" means "strength" in German, and kraft liner is one with high mechanical strength. It is completely unbleached and made of wood fibers by chemical pulping—sulfate process. Kraft liner is to be made from 100% virgin wood fibers. However, in practical, about 25% recycled content is also used. The best kraft liner is made from soft wood, also known as "coniferous" wood, e.g., pine, fir, spruce, hemlock, cedar and redwood. It has long fibers and gives higher strength to paper because of more crisscrossing of fibers. Hard wood, also known as "angiosperms," e.g., beech, birch, maple, oak, gum, eucalyptus, poplar, casurina, etc., has short fibers and gives better surface finish in the paper. The third type of natural source is "grass" which includes "bagasse and bamboo." This has medium length.

Kraft liners can be characterized by its smooth surface.

(a) Relatively brighter luster
(b) The burst factor (BF) will be more than 35
(c) Generally, its RCT (Ring crush Test) is 0.9 times gsm of the paper divided by 100. For example, 200 gsm kraft liner could be $(0.9 \times 200)/100 = 1.8$ KN/m
(d) Dirt content are less and smaller in size
(e) Moisture-absorbing capacity measured as Cobb value is less.

2. **Test Liners**

Test liners is made from recycled fibers. The quality of test liners depends on the quality of recycled fibers and number of times, and these fibers have been recycled earlier. Test liners may be used either as outer or inner or intermediate liners of a corrugated board.

The Test liner is characterized by

(a) Slightly rough and dull surface.
(b) The burst factor is around 25–28.
(c) RCT is almost same as that kraft liner.
(d) Dirt spots are more but smaller in size.
(e) Cobb value is marginally higher, i.e., 40–60 gm/m^2.

In the recent days, test liners are becoming very popular due to the following reasons:

(i) Source reduction: Source reduction is a reduction in dependence on and depletion of virgin raw materials. However, for food packaging purpose where food products is contact with the liner, virgin fibers are recommended.
(ii) It offers RCT almost same as kraft liners.
(iii) The test liners are cheaper as compared to kraft liners.
(iv) It consumes less virgin fibers; hence, it is more eco-friendly.

3. **Corrugating or Fluting Medium**

Fluting medium in corrugated board is like bracings in a truss. It should have just enough strength to connect both the liners without buckling or stretching. Main strength is contributed by liners, due to this, fluting medium or corrugating medium from recycled fibers.

In general, fluting medium provides the cushioning effect to the board. Depending upon the nature of recycled fibers, the fluting medium is also classified into three categories:

Fluting medium (FM)—made from fibers of old boxes

Semi-chemical (SM)—made from fibers of wood chips

Schrenz—made from fibers of general waste materials.

The common grammage (gm/m^2) for fluting medium is 112, 127, and 150. Fluting medium is characterized by

(i) Dull surface with fibrous look
(ii) BF in the range of 18–22
(iii) RCT could be (0.4 × gsm)/100
(iv) Cobb value is more than 100 gm/m^2
(v) Dirt spots are more and prominent.

4. **High Performance Liners**

High performance or high density liners are a lightweight grade of liner board that has very high ring crush rest (RCT) value, and BF is 28–30.

This liners will also have high edge crush test (ECT) value. But these liners do not absorb glue effectively due to compact bonding and lesser porosity.

5. **Corrugating Adhesives**

The main function of this adhesive is to adhere the liner to the medium. It is mostly used as starch adhesive. This adhesive is composed of food grade corn starch, water and additives to control the moisture resistance and fluid properties. It is applied to the surface of the flute tips of the medium, and the peaks in the flutted components structure and those flute tips are then brought into contact with the surface of the liner. A typical process involves using partially cooked as a carrier to transport raw starch to the tips of flutted medium, where it is heated and allowed to cook or gelatinize bonding paper liners together.

However, the poor bonding between the liners are due to the following reasons:

(i) Insufficient cooking of the starch adhesives
(ii) Inadequate amount of adhesives
(iii) Excessive heat during bonding process
(iv) Poorly bound fibers on the paper surface.

The quality of adhesives is also generally assessed by considering important parameters like total solid content (TSS) and viscosity.

4.7.2 Manufacturing Process of Corrugated Fiber Board Box

A corrugated plant can be divided into two major sections:

(i) Corrugating section: In this section, kraft paper is processed into corrugated board.
(ii) Converting section: The corrugated board is converted into boxes.

There are also other supporting sections like utilities, raw material stores, quality assurance and waste management.

The corrugators is also called as "corrugating line" is a long machine which manufactures the blanks of corrugated board from reels to paper. It has two main sections, the "wet end" and "dry end."

- Wet end: The corrugating medium is corrugated by being passed between two corrugating rolls in conjunction with heating, wetting and the application of pressure. Gluing of a liner to the corrugating medium in order to make single faced Web.
- Dry end: Slitting and scoring (or creasing) of the Web and cross-cutting of the resulting out in order to produce blanks of desired size of corrugated board.

The steps followed for the manufacturing of corrugated fiber board box is given below (Fig. 4.31):

Types of Flutes

The flute is the most important component of corrugated board for its functional properties.

Flutes can have different profiles. The flute profile is characterized by: H: the flute height, measured as distance between crest and trough P: the flute pitch, measured as distance between two successive crests. n: the number of flutes per meter.
m: take of factor.

The schematic diagram will indicate about the flute profile (Fig. 4.32).

The following Table 4.4 based on an internationally recognized classification shows the various flute profiles in existence with their geometric characteristics (Fig. 4.33).

Fig. 4.31 Manufacturing of corrugated fiber board box

Wet end ⟹

- Unwinding of the reels of paper
- Preheating of the liner
- Preconditioning of the corrugating medium
- Formation of flutes
- Gluing of liner
- Storage and transfer of single faced web
- Preheating of the single faced webs and of the outer liner
- Glue application to the single faced webs (printed/plain)
- Formation of double wall board.
- Drying of the double wall board.

Dry end

- Slitting and Scoring
- Cross-cutting or Cut-off knife

⟹

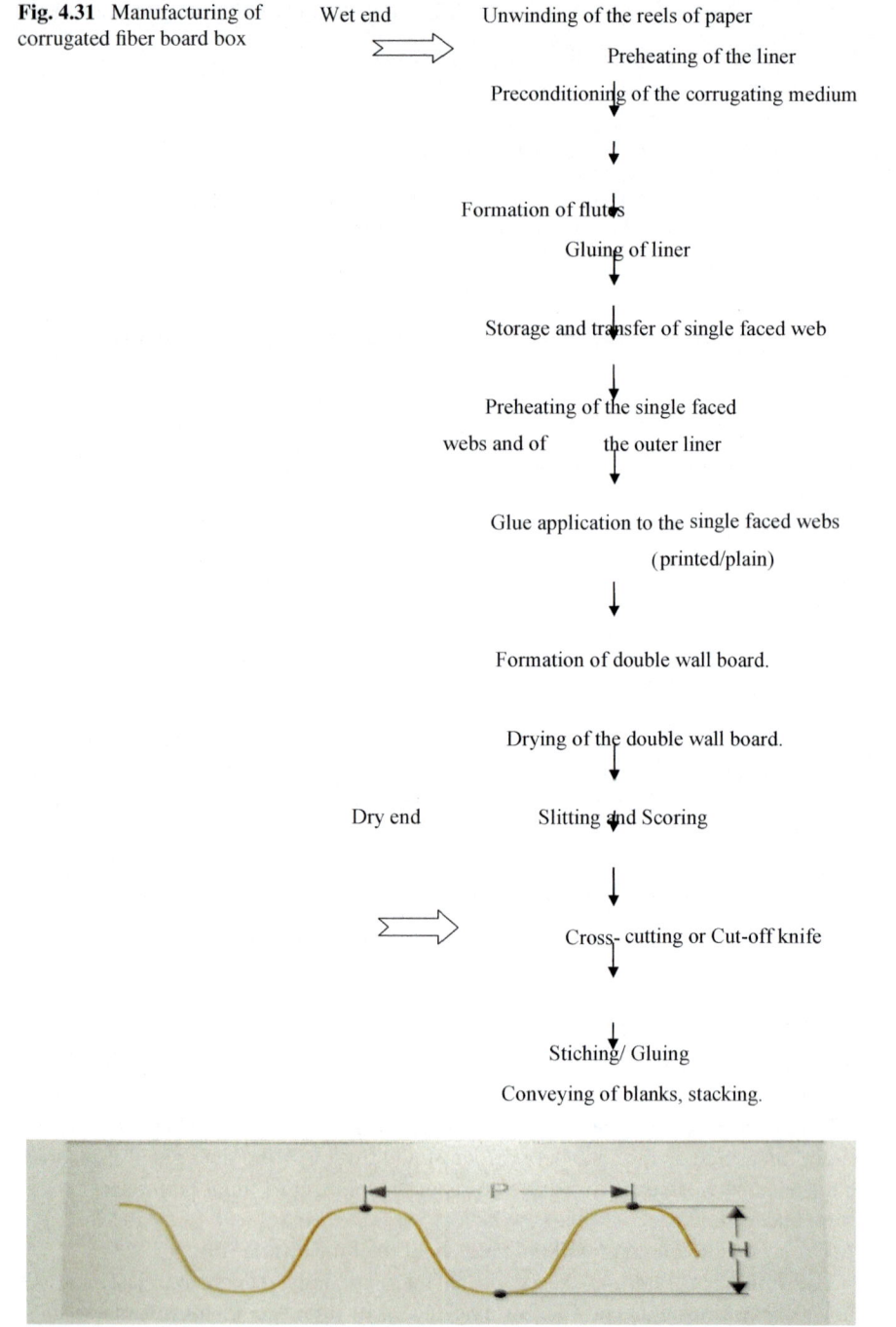

- Stiching/ Gluing
- Conveying of blanks, stacking.

Fig. 4.32 Flute profile

4.7 Corrugated Fiber Board Boxes

Table 4.4 Comparison of different flute profiles

Flute profile	Pitch (mm)	Height (mm)	Flutes per meter	Take-up factor
K (Jumbo)	12.0–21.0	6.5–10.0	68–80	1.60–1.80
A (Broad)	8.0–9.5	4.0–4.8	105–125	1.48–1.53
C (Medium)	6.8–8.0	3.2–4.0	125–147	1.42–1.50
B (Narrow)	5.5–8.5	2.2–3.0	153–181	1.28–1.43
E (Micro)	3.0–3.5	1.0–1.8	285–334	1.22–1.29
F (Small)	2.0–2.5	0.7–1.0	400–509	1.18–1.21
N (Small)	1.5–1.9	0.5–0.7	526–670	1.13–1.18

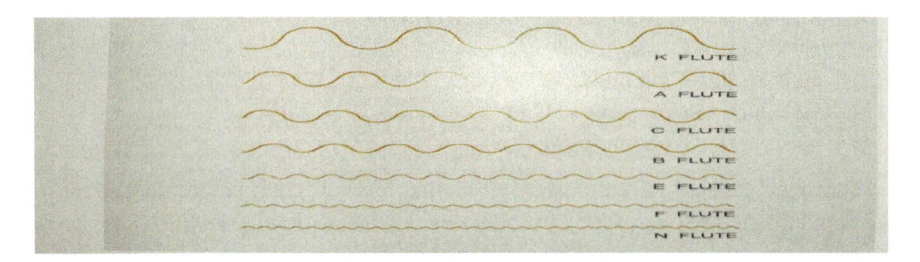

Fig. 4.33 Types of flutes

F and N flutes are recent inclusion to replace solid board. However, the A, B, C and E flutes are commonly used. A is mostly used in five ply with B or C flute on the outer side. C is mostly used in inner face to provide cushioning to the product inside. C flute has higher FCT. B flute is used for three ply as well five ply CFB boxes. E flute has minimum compression strength (ECT) and maximum crushing strength (FCT). It is predominantly used for primary packaging where printing and esthetics are very important [19, 20].

4.7.3 Design of Corrugated Fiber Board Boxes

The corrugated fiber board boxes are manufactured in different designs by considering the load-bearing capacity, space utilization, customer demand and its applications. However, International Standards Organisation (ISO) has adopted the different designs as "International Fibreboard case Code" in order to simply the matter, and these designs are accepted in the international market for exportable consignments.

For export market, slotted type or die type boxes are preferred. Moreover, the boxes are required to be stitched by means of gluing instead of staples to avoid corrosion, while the filled corrugated boxes are exposed to humid atmospheric condition during handling, storage and transportation [21, 22].

The terminology of important designs of slotted boxes and their code numbers are given below:

(1) Regular Slotted Container (RSC)-0201
(2) Overlap Slotted Container (OSC)-0202
(3) Full overlap Slotted Containers (FOL)-0203
(4) Centre Special Slotted Containers (CSSC)-0204
(5) Centre Special Overlap slotted Container (CSO)-0205
(6) Centre Slotted Container (CSC)-0207
(7) Half Slotted Flap Container (HCFS)-0209
(8) Half Slotted Container (HCS)-0214
(9) Full Telescopic design (FTD)-0301
(10) Full Telescopic Half Slotted Box (FTHS)-0320
(11) Half Slotted Container (0422)
(12) Half Slotted Container (0423)
(13) Half Slotted Container (0424)
(14) Half Slotted Container (0425).

The selection of different designs of slotted boxes to be used are dependent on the applications of boxes, type of product to be packed, distribution network, gross weight of the box and the market requirement [23, 24].

The schematic drawings of the blank or magazine of different designs of slotted containers are given below (Fig. 4.34):

In addition, the different fitments made of corrugated fiber board are also coded, and the same is given below (Fig. 4.35):

4.7.4 Advantages of Corrugated Fiber Board Boxes

- Availability of raw materials, i.e., kraft paper and adhesives, stitching pins, etc.
- Technology of box-making process is very simple
- Availability of box making machineries
- Cost effective
- Amenable to make display box
- Tare weight of box is less resulting to reduction of freight cost
- Maximum utilization of space due to collapsing nature of boxes
- Recognized as eco-friendly packaging materials
- Facilitates to have excellent printing on the outer surface of the box
- Easy to handle in the shop floor due to collapsing in nature as compared to wooden box [25].

At the same time, the corrugated fiber board boxes are also having the following limitations:

- Strength properties of boxes are gradually influenced by the environmental condition.

4.7 Corrugated Fiber Board Boxes

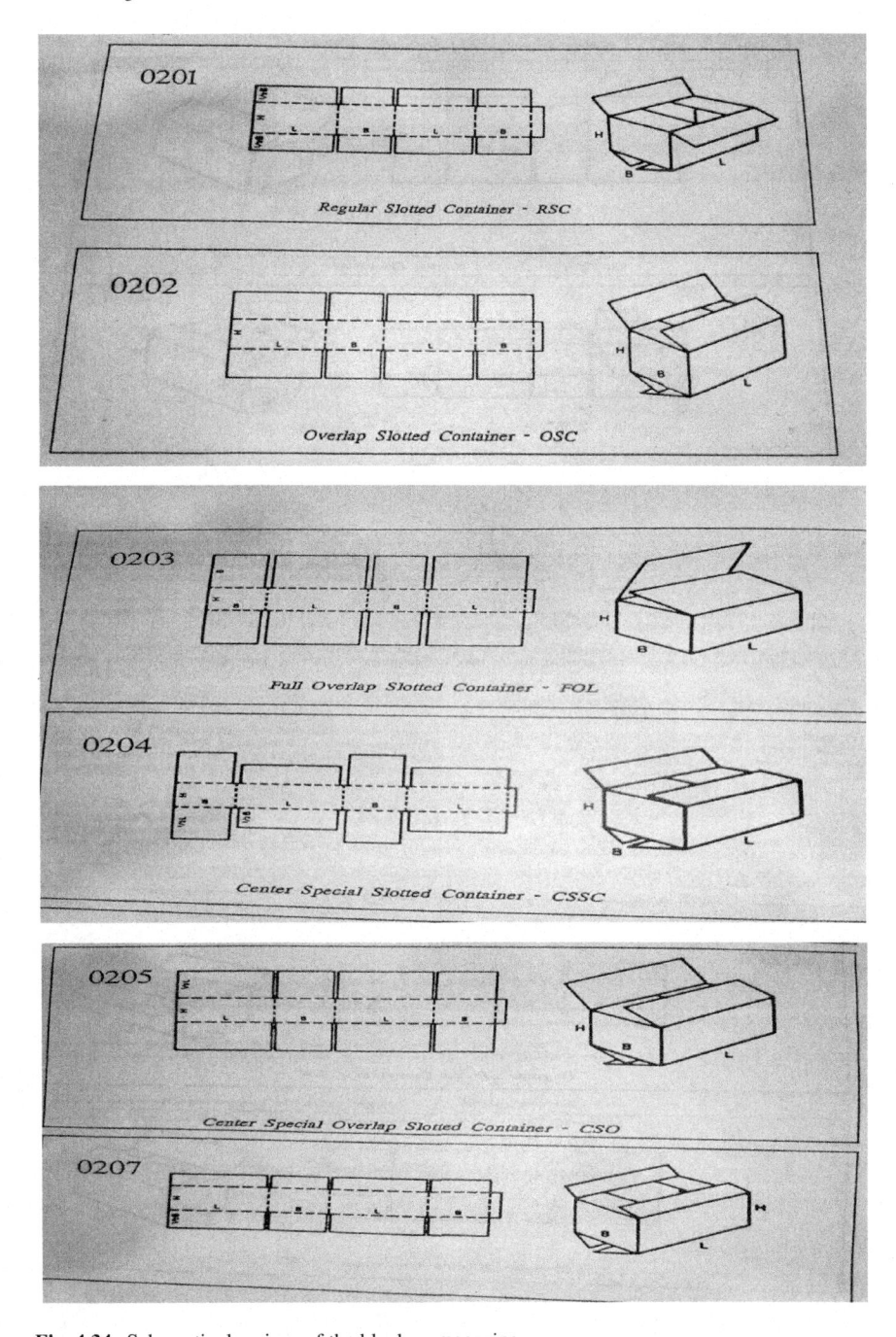

Fig. 4.34 Schematic drawings of the blank or magazine

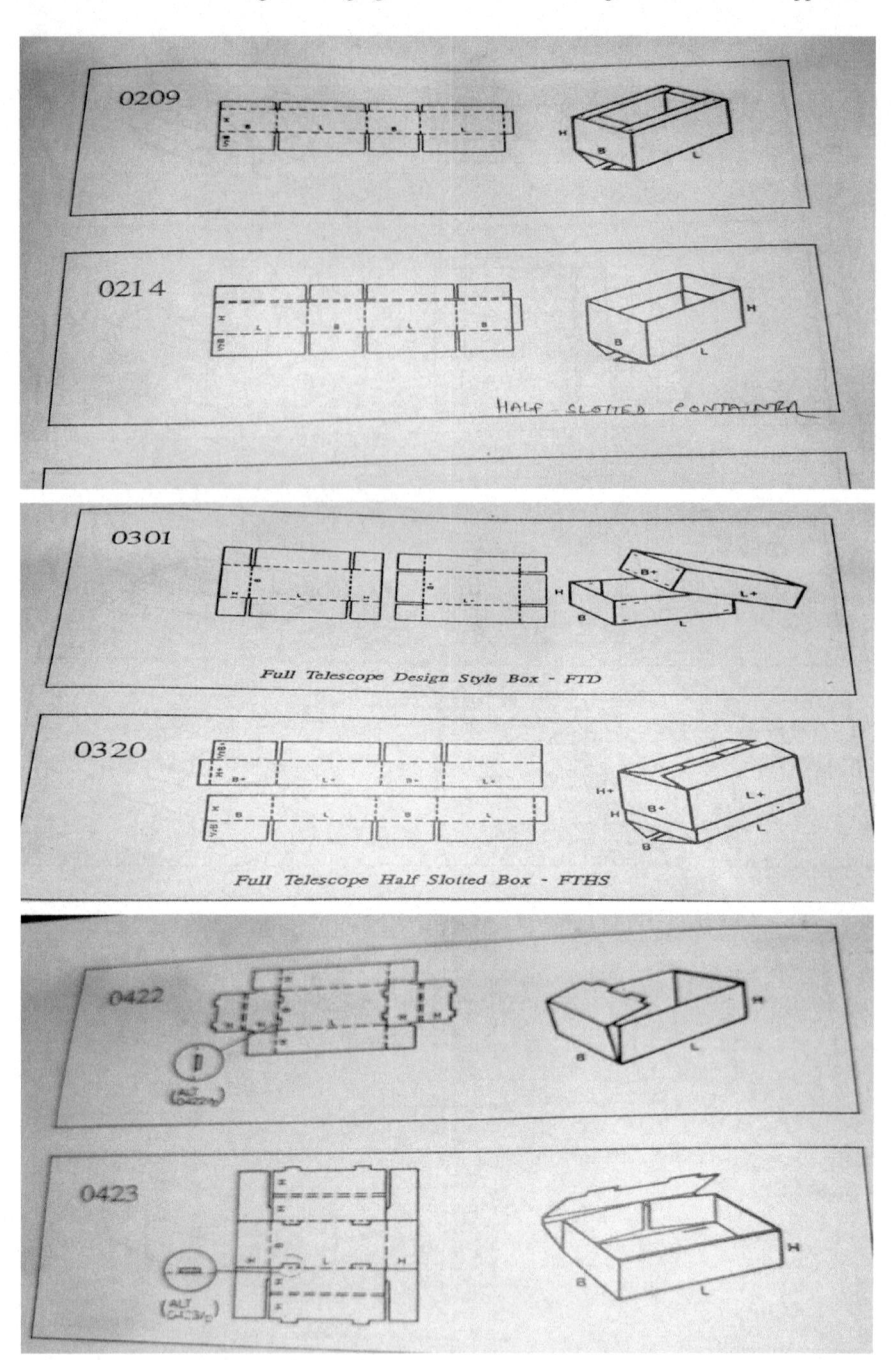

Fig. 4.34 (continued)

4.7 Corrugated Fiber Board Boxes 161

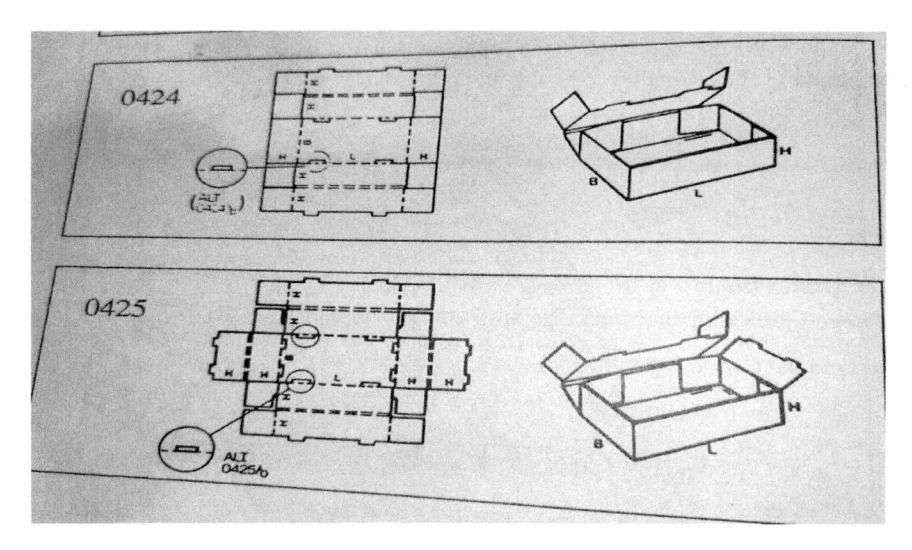

Fig. 4.34 (continued)

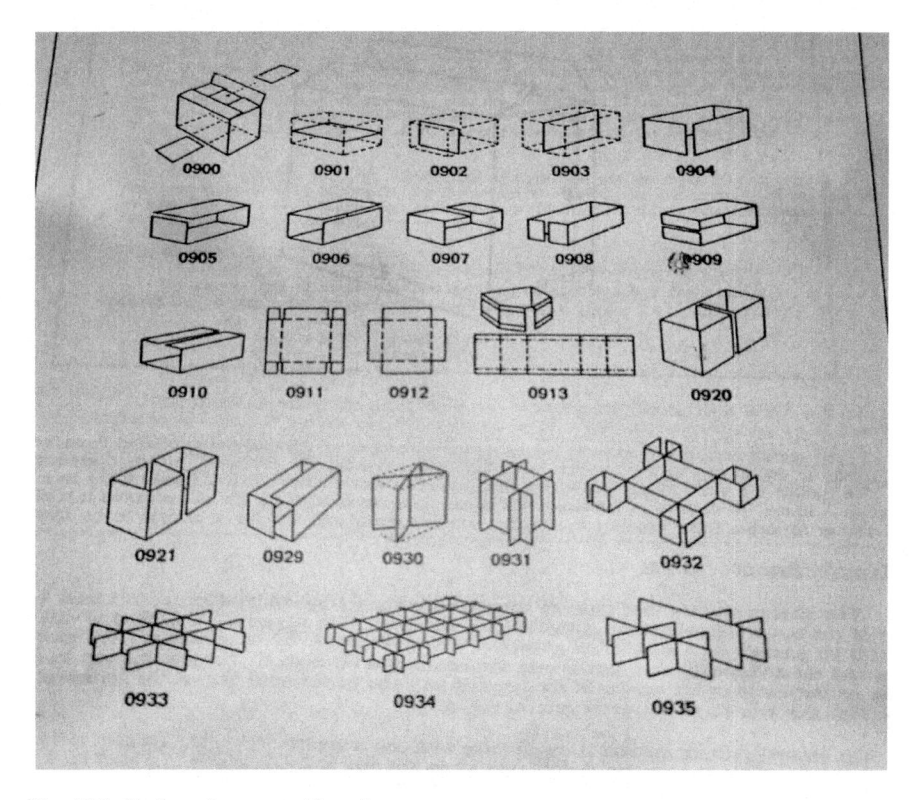

Fig. 4.35 Coding of corrugated board

- Requires special condition for storage to maintain the strength properties.
- Shortage of best quality of kraft papers in India.
- Lack of technology in the converting machineries.

4.7.5 Applications

The Corrugated fiber board boxes are used as Mono Carton, display package and as transport packaging materials. The different applications are given below.

(a) Mono Carton (Primary Package).

- Baby feeding bottle
- Torches
- Electrical items
- Electronic goods
- Alcoholic beverages
- Ready-to-serve beverages, etc.

(b) Display package

- Fresh fruits and vegetables
- Gift items
- Crockery items
- Utensils
- Jewelry items etc.

(c) Transport packages

- Fresh fruits and vegetables
- Processed foods in packaged conditions
- Flowers
- Agricultural commodities
- Industrial goods
- Chemicals
- Pharmaceutical products
- Health and beauty products
- Personal care products
- Household goods like refrigerator, washing machine, etc.

References

1. Piergiovanni L, Limbo S (2016) Food packaging materials. Springer, Basel
2. Parkar J, Rakesh M (2014) Leaching of elements from packaging material into canned foodsmarketed in India. Food Control 40:177–184. https://doi.org/10.1016/j.foodcont.2013.11.042
3. Deshwal GK, Panjagari NR (2019) Review on metal packaging: materials, forms, food applications, safety and recyclability. J Food Sci Technol 1–16

References

4. Robertson GL (2013) Food packaging: principles and practice. CRC Press
5. Ojha A, Sharma A, Sihag M, Ojha S (2015) Food packaging–materials and sustainability—a review. Agric Rev 36(3):241–245
6. IS:5818 (1988) Indian standard, specifications for lacquers and decorative finishes for food cans. Bureau of Indian Standards, Manak Bhavan
7. Coles R, McDowell D, Kirwan MJ (eds) (2003) Food packaging technology, vol 5. CRC Press, Boca Raton
8. ASTM (2010) ASTM C162-05—Standard terminology of glass and glass products, ASTM vol 15.02 Glass; Ceramic Whitewares. ASTM International, West Conshohocken. https://doi.org/10.1520/C0162-05R10
9. Grayhurst P (2012) Glass packaging. In: Packaging technology. Woodhead Publishing, pp 109–121
10. Marsh K, Bugusu B (2007) Food packaging—roles, materials, and environmental issues. J Food Sci 72(3):R39–R55
11. Rapusas RS, Rolle RS (2009) Management of reusable plastic crates in fresh produce supply chains—a technical guide. UNFAO, Regional Office for Asia and the Pacific. RAP Publication 2009/08, 52p
12. Indian Standard IS:15532:2004-Plastics Crates for Fruits and Vegetables, Published by Bureau of Indian Standards
13. http://www.weepallet.com/plastic-pallet-manufacturing-process/
14. Fundamentals of packaging technology. Institute of Packaging Professionals, USA
15. http://www.stepinhost.com/digvijaygroup.in/laminated-collapsible-tubes.html
16. https://www.embalex.com/en/wooden-crates-for-transport-and-protection-of-goods_-theperfect-packaging/news/2421. https://en.wikipedia.org/wiki/Crate
17. Indian standard IS: 7698-1983 (reaffirmed 2003): Specification for returnable wooden crates for vegetables
18. https://advancedcrates.com/crate-manufacturing/
19. http://www.postharvest.org/RPCs%20PEF%202013%20White%20paper%201301%20pdf%20final.pdf
20. Core Emballage Limited (1999) Corrugated box handbook. Core Emballage Limited
21. Wright PG, Mckinlay PR, Show EYN (1968) Corrugated fibreboard boxes. APM Packaging
22. Indian Institute of Packaging (2012) Packaging technology educational, vol 1. Indian Institute of Packaging
23. Indian standard IS: 1141-1993 (reaffirmed 2004), Seasoning of timber-code of practice
24. Indian standard IS: 7698-1993 (reaffirmed 2003), Specification for returnable wooden crates for vegetables
25. Saha NC (2018) Presentation on Transport packages-its design aspects and Technical specifications during 47th Annual Conference of Federation of Corrugated Box Manufacturers of India

Chapter 5
Packaging Techniques for Fresh Fruits and Vegetables

N. C. Saha, Meenakshi Garg, and Susmita Dey Sadhu

5.1 Introduction

All kinds of horticultural produces, mainly fresh fruits and vegetables, are considered to be more nutritious foods as these foods are having high protein content, minerals and vitamins. Fresh produces are having living tissues, high in water content and diverse in terms of morphology, composition and philosophy. The respiration process continued even after harvesting of fresh fruits and vegetables. During this process, oxygen is used, and carbon dioxide is released. The rate of deterioration is proportional to the high respiration rate. Rapid respiration results in fast ripening/aging of the produce. The availability of oxygen gas is restricted as it results into the changes in Chemical reactions, break down of cells, production of small quantities of alcohol, off odour and finally bringing about decay or spoilage of the fresh produces. However, the respiration rate of individual fruits would vary depending upon the type, variety and degree of maturity of fruits and vegetables.

Due to this, fresh produces are very perishable in nature, and these produces have got very limited shelf life ranging from few hours to few weeks at ambient temperature. Packaging plays an important role for the preservation of fresh fruits and vegetables. In fact, packaging is required not only for food preservation and protection but has also assumed a multifunctional role by serving as a symbol of value addition, an assurance of quality and quantity/number, a conveyor of convenience and ultimately a tool for marketing of food products. In the recent days, the importance of packaging has gained enormously. This is mainly due to growing consumer awareness and willingness to pay for value added and hygienic products. In addition, increase in exports market has also influenced to develop innovative packaging technologies to enhance the shelf life of fresh produces. Moreover, increase in environmental concerns has also imposed newer parameters for package performance, giving an impetus to the development of innovative eco-friendly packaging materials [1].

The marketing cycles of fresh produces start from the small farmers–large farmers–commission agent–wholesaler–retailer and finally to the consumers. Damages of fresh produces during the entire distribution channel are considered to be the biggest challenge as the produces are normally undergone to mechanical hazards during handling, storage and transportation. Due to this, packaging plays a crucial role to provide the protection of fresh produces against all kinds of mechanical, chemicals and biological hazards in order to reduce the post-harvest losses. In addition, packaging also reduces wastes, adds value to the products and makes the product qualitatively and quantitatively acceptable by the consumers. In case of horticultural produces, packaging is also required to enhance aesthetic value but also to keep the produce in good condition for longer duration. In short, packaging is required to deliver the fresh produces to the hands of consumer in fresh and sound condition.

In the post-harvest period, biochemical processes, such as respiration, transpiration and ethylene autocatalysis, shorten the shelf life of the fresh produces. Additionally, spoilage and pathogenic microorganisms can proliferate with greater intensity. The speed of these spoilage mechanisms is significantly affected by environmental conditions, such as temperature, relative humidity, oxygen and light. Moreover, the fresh produces get bruised during transportation. The physical injury initiates vigorous biochemical reactions in the damaged cells, and as a result, the bruised product begins to deteriorate rapidly.

Considering these aspects, the package design is always dependent on the requirements of the produce within the framework of the handling and marketing system. The main causes of deterioration in fresh produce are metabolic changes, mechanical injury and attack by pests and diseases. There are a variety of packages developed for packaging of different types of fruits and vegetables. But there is no rigid rule that a particular type of packaging material or style would be used for a particular type of product. This would be completely dependent on many factors like type of fresh produces, storage condition, handling and transportation system and finally the preferences of consumers [2, 3].

Consequently, whole or cut fresh produce is the most challenging category of food which needs to be packaged with desired shelf life for marketing. The role of packaging in fresh produce has been changed over the years. It is no more now, restricted only to provide its basic functions like protection, preservation and presentation rather innovative packaging materials and techniques are now developed to extend the shelf life of fresh produces to meet the market requirement and also satisfy the modern consumers. At the same time, it is also important that the packaging materials should have adequate mechanical strength so that packages would be able to withstand the mechanical hazards like shock, vibration, etc., which would occur during handling, storage and transportation.

The packaging of fresh produces essentially revolves around the following two important aspects:

- Interaction of the environment on the physiology/pathology of fruits and vegetables and product/package

- Influence of the transportation and storage environment on the mechanical strength of the package.

Besides, there are many intrinsic and extrinsic factors that also affect the shelf life of fresh produces which have also spurred the necessity for the development of a large variety of packaging technologies to maintain the quality and nutritional value, as well as safety of the goods with extended shelf life [4].

The innovative packaging technologies would contribute the following:

- Add value to produce
- Strengthen brands
- Monitor produce and environment changes during storage
- Be a tracking tool
- Provide antitheft prevention
- Allow unique identification and communication with consumers
- Increase consumer satisfaction
- Reduce food waste.

Over the years, consumer lifestyles and preferences have also been changed. The consumers demand for qualitative, nutritious and value added food has increased enormously. In additions, the consumers are more concerned for hygiene, safety aspects and the reduction of food wastes. All these aspects have created a great demand for innovative packaging technologies to meet all the requirements.

Considering this, the constant research and development in packaging technology has been able to develop newer packaging materials with adequate functional properties, and these materials have been introduced in different innovative packaging technologies to cater the demand of market requirement. However, a number of innovative packaging materials and technologies like modified atmosphere packaging (MAP), controlled atmosphere packaging (CAP), active packaging (AP), antimicrobial packaging, breathable films and molded pulp tray, etc., are extensively used in recent days, for the enhancement of shelf life of fresh fruits and vegetables.

5.2 Breathable Films

Breathable films make an important category of packaging materials which prevent fresh fruits and vegetable from undergoing degradation. Fresh fruits and vegetables require fresh air to breath and to remain freshness, but water droplets should be barred as it causes damage to the fresh produce. Breathable film prevents water vapor molecules from passing through the film but allows the air to enter. Typically, a breathable film should have optimally dense matrix with enough room for water vapor molecules to penetrate though the film but not enough space to allow liquid water to make its way into or through the film. This can be described as a film with a bunch of small holes or pores that filter out large objects, but allow smaller objects to pass through (Fig. 5.1a, b) [5–16].

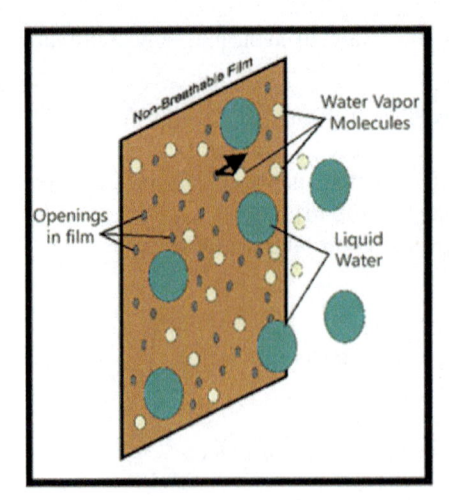

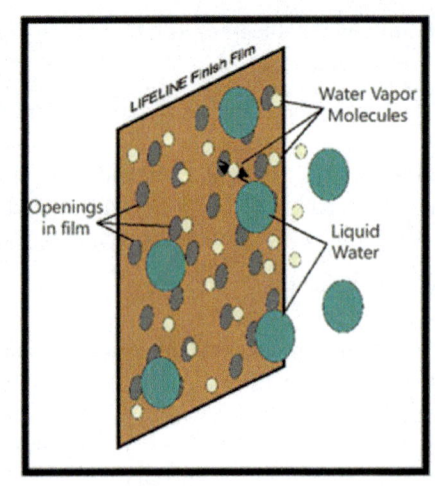

Fig. 5.1 **a** Non-breathable film, vapor barrier, **b** lifeline breathable film

Breathable films are commonly known as semipermeable membranes which are made of polyethylene and used as breathable film in packaging industry. Thus, it can be said that a film with more amorphous nature will be more porous or breathable.

Besides chemical composition, the thickness of the film is another factor that influences vapor permeability. The thicker the film, the more difficult it becomes for water vapor molecules to make their way through the film. That is why one should apply finishes on the films in thin layers.

5.2.1 Manufacturing Process

Typically, breathable films are made by hot stretching of the polymeric film to ensure better orientation and optimum gap between two molecules. The film extrusion can be performed through a cast or blown process. The formulation needs to be modified so as to fulfil the process requirement and to the required film properties. The stretching can be done in line during the film extrusion but can also be performed offline. Often a machine direction orientation (MDO) line is used for creating the microvoids. An alternative mechanism like intermeshing stretching is also used.

In fact, the number of micropores to be made on the film is dependent of the respiration rate of fresh fruits and vegetable. According to the respiration rate, the horticultural produces are classified in to different ways. Dates, dried fruits, nuts, etc., come in very low class and have less than 5 mg CO_2/kg h at 5 °C. Apple, beat, citrus fruits, garlic, grapes, onion and potato have respiration range 5–10 mg CO_2/kg h and counted in low class, whereas apricot, banana cabbage, carrot, fig, cucumber, mango, tomato, peach and plum have 10–20 mg CO_2/kg h and have moderate class. Avocado, blackberry, radish, cauliflower, carrot (with top), beans, sprouts, broccoli, okra asparagus, mushroom, peas, spinach and sweet corn have respiration rate more than 20 mg CO_2/kg h and come in moderate to extremely high class. Potatoes, onion, cabbage, apple and citrus fruits have long storage life due to low respiration rate. Brussels and spinach have short storage life because there respiration rate is high.

There are several patented process of manufacturing microporous breathable films. However, most commonly used materials for breathable films are as follows:

- Polyethylene-based breathable films
- Polypropylene-based breathable films.

Materials for breathable films have typical properties such as high absorbency, permeability and high strength.

5.2.2 Types of Breathable Films

Based on the manufacturing technique, breathable films are available in the following two types:

5.2.2.1 Microporous Breathable Films [5]

One important characteristic of a microporous breathable film is that billions of micropores connected to each other are present in the film. The pore sizes are smaller than a drop of water; however, the pores are much larger than water vapor molecules, so air and gases can readily penetrate the film and pass from one side of the film to the other. These types of films can also be used for applications such as filtration and protective garments. Formation of micropores in plastic films can also be done by increasing stress level high enough and the rate of deformation fast enough. Mostly unfilled plastic films undergo stress cracking and form micropores under stress condition. Under such conditions, the films appear opaque due to light scattering from the small pores that are generated.

5.2.2.2 Monolithic Breathable Films [6, 7]

Another category of breathable multilayer films suitable for a wide variety of uses is monolithic films. These are continuous and free of pores. They find important applications in, e.g., protective apparel, surgical gowns, surgical drapes, aprons, roofing material, house wrap, etc. The mechanism of this type of breathable films lies in their capability of absorbing gases and water molecules on one surface, transferring the molecules through the film and releasing the molecules on the opposite surface. Monolithic film's ability to serve as a liquid barrier is independent of the surface tension of the liquid into which the film is exposed. This is one of the advantages that monolithic films have over microporous films.

The major application areas for breathable films include

- Fresh fruits like apples, mango, chikoo, orange, banana, papaya, kiwi, grapes, etc.
- Fresh vegetables like tomato, lemon, ginger, okra, broccoli, etc.

5.3 Molded Pulp Trays

Molded pulp, also named molded pulp or molded fiber, is a packaging material that is typically made from recycled paperboard and/or newsprint. It is used for protective packaging or for food service trays and beverage carriers. Other typical uses are end caps, trays, plates, bowls and containers. For many applications, molded pulp is less expensive than *expanded polystyrene* (EPS), vacuumed formed PET and PVC, corrugation and foams, since it is produced from recycled materials, and can be recycled again after its useful life cycle. Molded pulp products can be made waterproof with a spray or dip coating of wax. Molded pulp trays as eco-friendly packaging have gained commercial importance in the recent years (Fig. 5.2).

5.3.1 Manufacturing Process [1]

The molded pulp trays are manufactured by considering the following stages:

1. *Mixing*: The raw material is soaked in water and mixed until the desired pulp consistency has been achieved. Additives, such as sizing agents or fillers, are introduced to produce the required functions, commonly utilized also in the papermaking industry.
2. *Forming*: The pulp part is formed to shape by custom designed tools. The tools consist of a metal net supported by a perforated mold. The machine is lowered

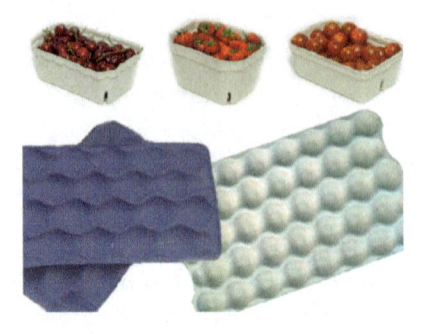

Fig. 5.2 Molded pulp trays for fruits

into the slurry, and the water is pulled through the metal mesh by applying a vacuum. The suction helps to bind the fibers.
3. *Pressing and drying*: The wet part is moved to a heated mold. The part is compressed and dried completely by two matched halves of a mold. The surface becomes relatively smooth, and good dimensional accuracy is achieved. The pressing also improves the mechanical properties of the product, aiding stacking and nesting for more efficient storage and shipping. Pressure is necessary to achieve good bonding between fibers. During this phase, vacuum can be applied when the part is pressed, thus removing excess steam.
4. *Trimming and quality inspection*: Protruding edges are trimmed, and all the scraps or rejected products are returned to the pulp mixture and re-used. For specialty products, the produced parts can undergo some further processes, such as debossing or special treatments. The part is finally inspected for quality.

5.3.2 Properties of Molded Pulp Packaging

The molded pulp trays are having important properties which are given below:

- Excellent buffer and shock resistance
- Delicate quality, no deckle edge or wrinkle appearance
- Cut down the packaging process labor hour
- Flexible of the material with diversity usage
- No wastewater and wastes during manufacturing
- Additives may be added as required. Molded pulp products are the new packaging materials, an excellent alternative to the plastic product
- The mechanical behavior depends on its size, on its particular geometry and on its wall thickness. The lack of control over dimensional tolerances affects the cushioning property of the products and leads to considerable variation in material strength
- Molded pulp products have the potential to be environmentally sustainable
- Molded trays are 100% recyclable
- They are Reusable and biodegradable as the raw materials like waste paper are cellulosic materials. The finished product has a moisture content of about 4–8%, roughly corresponding to the equilibrium moisture content of the paper under the humidity conditions at which it will be stored or used.

5.3.3 Applications [17, 18]

In the recent days, its application is increased due to its eco-friendly characteristics. However, the important applications of these materials are as follows:

- Fresh fruits like apples, custard apples, etc.
- Fruits trays with stretch wrapped package

- Egg trays
- Egg cartons
- As cushioning materials for other application.

5.4 Expanded Polystyrene (EPS) Containers [19]

Expanded polystyrene (EPS) or polystyrene foam appeared in the Indian Market in the late 50s and the application in packaging as molded tray and boxes started much later in the early 80s and has now gained tremendous popularity for the numerous advantages which are offered by this packaging media. In fact, this material has been utilized as a material of choice in food packaging applications, cost-effective and energy-efficient insulation in construction applications as well as cushion transport packaging material for shock-sensitive goods and many more.

Expanded polystyrene is a white foam plastic material produced from solid beads of polystyrene. It is primarily used for packaging, insulation, etc. It is a closed cell, rigid foam material produced from Styrene (which forms the cellular structure) and a blowing agent. Both styrene and pentane (blowing agent) are hydrocarbon compounds and obtained from petroleum and natural gas by products.

5.4.1 Manufacturing Process

The manufacturing process of expanded polystyrene involves three stages as follows:

1. Pre-expansion
2. Maturing/stabilization and
3. Moulding.

Pre-expansion: Polystyrene is produced from the crude oil refinery product styrene. For manufacturing expanded polystyrene, the polystyrene beads are impregnated with the **foaming agent pentane**. Polystyrene granulate is pre-foamed at temperatures above 90 °C.

Maturing/Stabilization: This temperature causes the foaming agent to evaporate and hence inflating the thermoplastic base material to 20–50 times its original size. After this, the beads are stored for 6–12 h allowing them to reach equilibrium.

Moulding: Then, beads are conveyed to the mold to produce forms suited as per application. During final stage, the stabilized beads are molded in either large blocks (block molding process) or designed in custom shapes (shape molding process).The material can be modified by the addition of **additives such as flame retardant** to further enhance fire behavior of EPS.

5.4.2 Properties [20]

- It is very light in weight and thus saves freight cost.
- Very low thermal conductivity due to its closed cell structure consisting of 98% air. This air trapped within the cells is a very poor heat conductor and hence provides the foam with its excellent thermal insulation properties. The thermal conductivity of expanded polystyrene foam of density 20 kg/m^3 is 0.035–0.037 W/(m K) at 10 °C. It therefore ensures survival of perishables for long distances through adverse climatic condition.
- Low moisture absorption. Even when immersed in water, it absorbs only a small amount of water. As the cell walls are waterproof, water can only penetrate the foam through the tiny channels between the fused beads.
- Excellent cushioning affect is provided and therefore protects the contents against all static and dynamic loads during transportation.
- Can be colored, printed and labelled for attractive packaging.
- Its surface is smooth and abrasion resistant and thus does not injure the delicate skin/tissues of fresh produce.
- It is food grade and hygienic.
- One of the serious limitations of polystyrene foam is its rather low maximum operating temperature ~80 °C. Its physical properties do not change within its service temperature range (i.e., up to 167 °F/75 °C) for long-term temperature exposure.
- Its chemical resistance is nearly equivalent to the material upon which it is based—polystyrene.
- Flexible production process makes EPS versatile in mechanical strength which can be adjusted to suit the specific application. EPS with high compressive strength is used for heavy load bearing applications, whereas for void forming, EPS with a lower compressive strength can be used.
- Generally, strength characteristics increase with density; however, the cushioning characteristics of EPS foam packaging are affected by the geometry of the molded part and, to a lesser extent, by bead size and processing conditions, as well as density.
- EPS offers exceptional dimensional stability, remaining virtually unaffected within a wide range of ambient factors. The maximum dimensional change of EPS foam can be expected to be less than 2%.
- The dielectric strength of EPS is approximately 2 kV/mm. Its dielectric constant measured in the frequency range of 100–400 MHz and at gross densities from 20 to 40 kg/m^3 lies between 1.02 and 1.04.
- Molded EPS can be treated with antistatic agents to comply with electronic industry and military packaging specifications.
- It is resistance to chemicals. Water and aqueous solutions of salts and alkalis do not affect expanded polystyrene. However, EPS is readily attacked by organic solvents.

- EPS is also resistant to aging. However, exposure to direct sunshine (ultraviolet radiation) leads to a yellowing of the surface which is accompanied by a slight embrittlement of the upper layer. Yellowing has no significance for the mechanical strength of insulation, because of the low depth of penetration.
- EPS is flammable. Modification with flame retardants significantly minimizes the ignitability of the foam and the spread of flames.

5.4.3 Applications of Expanded Polystyrene [21]

- Expanded polystyrene trays are extensively used for the packaging fresh fruits and vegetables. EPS can be extruded using conventional equipment to form continuous sheet. This sheet may later be formed (e.g., using vacuum forming, pressure forming) to produce articles such as fruit trays), etc. (Fig. 5.3).
- It is widely used material to produce food service containers like drink cups, food trays and clamshell containers.

5.5 Modified Atmosphere Packaging (MAP)

Modified atmosphere packaging is commonly known as MAP technique. This is also considered as one of the kind of smart packaging. MAP is a packaging technique based on the replacement of the ambient air (78.09% nitrogen (N_2), 20.95% oxygen (O_2), 0.03% carbon dioxide (CO_2) and 0.93% Argon (Ar) plus others) inside the package with a single gas or a specific mixture of gases that can lead to produce shelf life extension. Gas mixtures containing low O_2 (15%), high CO_2 (10–20%) and N_2 as make-up gas are normally used [22–28].

Modified atmosphere packaging (MAP) is also known as gas flushing, protective atmosphere packaging or reduced oxygen packaging. Modified atmosphere enables fresh and minimally processed packaged food products to maintain visual, textural and nutritional appeal. The controlled MAP environment enables food packaging to provide an extended shelf life without requiring the addition of chemical preservatives or stabilizers.

Fig. 5.3 Fresh fruits packed in polystyrene trays

5.5 Modified Atmosphere Packaging (MAP)

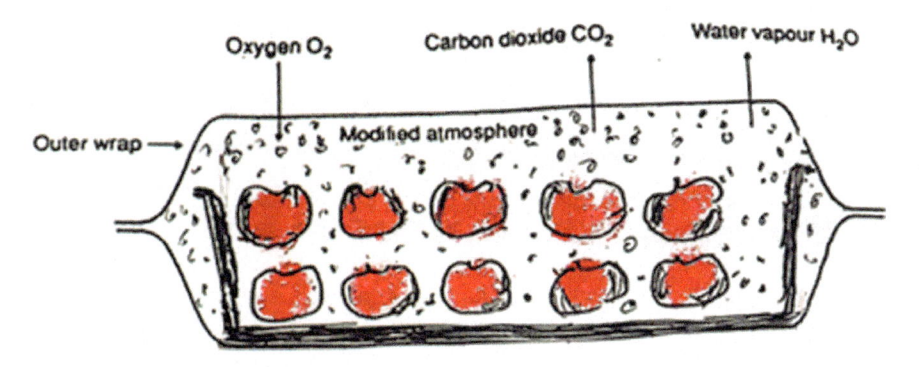

Fig. 5.4 Schematic diagram of MAP

5.5.1 Role of Gases in MAP

Oxygen: Low O_2 is desirable as it slows down respiration rate, ethylene production, enzymatic browning and growth of aerobic microorganisms. However, the amount of O_2 needs to be carefully controlled since concentrations of this gas below 1% lead to produce fermentation and growth of harmful microorganisms, such as clostridium botulinum.

Carbon di oxide: CO_2 has direct antimicrobial capacity, but excessive levels of CO_2 can cause a reduction in pH, flavor tainting and drip loss.

Nitrogen: Nitrogen is used as a make-up gas to avoid the collapse of the package, and by displacing O_2, it has an indirect antimicrobial capacity.

Other gases (e.g., Ar, ozone, other noble gases, carbon mono oxide, etc.) and gas mixtures (e.g., 95% O_2 + 5% N_2) have been proven to enhance produce quality and safety as well.

The most adequate gas or mixture of gases is selected based on the type of produce to be packaged, due to the differences in respiration rate, ethylene production and/or sensitivity, etc., between different fruits and vegetables. Schematic diagram of MAP is shown in Fig. 5.4.

5.5.2 Classification of MAP [21, 22]

Depending on how the ambient air inside the package is replaced with a desired gas composition, MAP is classified into two types: active MAP (AMAP) and passive MAP (PMAP).

5.5.2.1 Active Modified Atmosphere Packaging (AMAP)

Under this technique, the ambient air in the package is replaced with a desired mixture of gases or a single gas by direct flushing prior to package sealing.

5.5.2.2 Passive Modified Atmosphere Packaging (PMAP)

In this technique, air free of contaminants, such as medical air, replaces the ambient air inside a package by direct flushing, and then after package sealing, the composition of the air is modified.

Both continuous and microperforated films can be used in PMAP to control gas exchange. Microperforated films perform better than continuous films since they allow a faster entrance of O_2 into the package that mitigates the high/low concentrations of CO_2/O_2 that develop in continuous film packages containing produce. A desired gas composition can easily be obtained with microperforated films by varying the number, area and length of the microperforations and thereby optimizing the gas exchange.

AMAP is used to extend the shelf life of a variety of fresh-cut leafy greens, vegetable salads, sliced apples and sliced peaches, among others, and PMAP with microperforations to commercialize produce including baby spinach, ready-to-eat blueberries and sliced apples. AMAP is widely used to extend the shelf life of bulk produce during warehouse storage and transportation, including strawberries, blueberries and cherries. While MAP in combination with refrigeration can definitively delay produce deterioration and health risk.

A schematic diagram of AMAP and PMAP is given in Fig. 5.5.

5.5.3 Principle of MAP Technique

Modified atmosphere packaging is essentially a dynamic system where respiration of the packaged commodity and gas permeation through the packaging film take place simultaneously. During respiration, packaged commodity takes oxygen from the package atmosphere, and the carbon dioxide, produced by the commodity, is given away to the package atmosphere. As a result, there will be depletion of oxygen level inside the package and accumulation of carbon dioxide within the package.

Consequently, the composition of package air changes. Initially, the composition of package air soon after the sealing of package will remain nearly same as that of ambient air. Hence, respiration rate of the packaged produce will be high as that in ambient condition. However, as the time goes, the concentration of oxygen inside the package air reduced and that of the concentration of carbon dioxide increases, and thus, the rate of respiration declines. On the other hand, there will be development of a gradient of oxygen and carbon dioxide gases between the package atmosphere and ambient atmosphere due to the decrease of concentration of oxygen gas and

5.5 Modified Atmosphere Packaging (MAP)

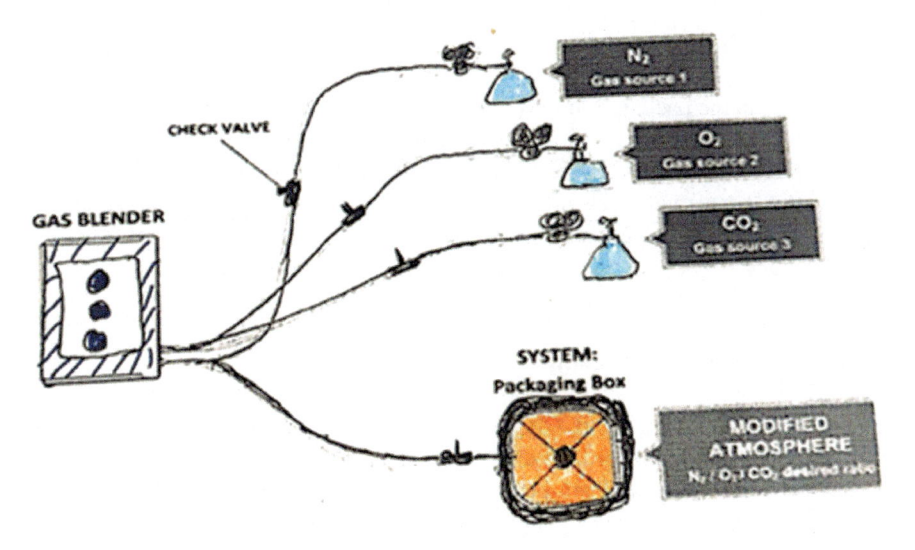

Fig. 5.5 Schematic diagram of AMAP and PMAP

increase of concentration of carbon dioxide gases inside the package resulting into the penetration of oxygen gases from the atmosphere and the liberation of carbon dioxide gas from the package. However, this phenomenon will also depend upon the permeability of oxygen and carbon dioxide gases of the packaging film, surface area of the package available for gaseous exchange and the temperature.

As the package produce continues to respire though at a decreasing rate, the concentration gradients continue to increase which would lead to more permeation of gases from atmosphere into inside of the package. This process would continue till it reaches to an equilibrium stages. At equilibrium stage, the rate of oxygen permeating (ingress) becomes equal to the rate of oxygen consumption (respiration), and the rate of carbon dioxide permeating (egress) will also become equal to the rate of carbon dioxide evolution (respiration). This equilibrium state is known as steady state, and the time taken by the package to reach equilibrium state is termed as equilibrium time or transit state or unsteady state.

5.5.4 Techniques of MAP

There are two basic techniques by which air can be replaced in MAP package; gas flushing and compensated vacuum.

5.5.4.1 Gas Flushing

Air presence in the package can be replaced by passing a stream of gas. Due to this, gas gets diluted, and the pack is sealed. In this process, dilution of air is a continuous process, and thus, the system is very fast. Gas flushing is normally done with the use of vertical or horizontal form–fill–seal machines. In such case, the concentration of oxygen gas inside the package may be up to 2.5%. However, in our country, nitrogen gas flushing is very common for ready-to-eat snacks and fried items.

5.5.4.2 Compensated Vacuum

In this technique, vacuum is first applied to remove the air, and then, the desired gas or gas mixture is incorporated. Since it involves an extra step, this process is slower. In this technique, there will not be any residual oxygen, and hence, it has its own advantage to package oxygen-sensitive food products.

5.5.5 Advantages of MAP Technique

- In MAP system, the produce is sealed in a retail size packages unlike bulk storage system. Hence, in the event of any infection, the chances of its spread over the bulk of produce are extremely low.
- MA packages obviate the use of bulky sophisticated equipment to maintain the required optimum air composition. Due to this, this technology is very cost effective and also simple which can be adopted easily.
- MAP packages are easy to handle as this technology is employed on the retails packages.
- In MAP system, the packaging films are almost impermeable to aromatic compounds. Hence, MAP package can retain the aroma of packaged commodity and also protect the same from contamination of any kind of foreign odor. This helps to share the different commodity in packaged condition together at retail store and also during transportation.
- MAP system is available in retail packages, and hence, it facilitates to display the information about the packaged commodity.
- Normally, once the equilibrium is established in MAP package, the optimum air composition remains constant inside the package till the package is cut-open by the consumer.

5.5.6 Limitation MAP Technique

- In this system, the establishment of optimal air composition inside the MAP package depends on many parameters as there is no external means like CAP to control and monitor the package air composition as and when required. Due to this, there is probability of variation in package air composition, in the event of substantial variation in ambient atmosphere.
- Most of the packaging materials used for MAP techniques are plastic film which is manufactured in combination of different polymeric materials, thus, these materials may not be recyclable, and hence, the packaging materials would be non-compliance to environmental regulation.

5.5.7 Application of MAP

MAP is used to extend the shelf life and retain the quality of a wide variety of fruits and vegetables. A study indicates about shelf life of fruit and vegetables in MAP package and exposed to low temperature, given in Table 5.1.

Table 5.1 Recommended mixture of gases for fruits and vegetables

Product	Recommended O_2 percent	Recommended CO_2 percent	Temperature in °C	Status
Avocado	2–5	3–10	5–13	Good
Apple	1–3	1–5	0–5	Excellent
Banana	2–5	2–5	12–15	Excellent
Grape	2–5	1–3	0–5	Excellent
Lemon	5–10	0–10	10–15	Good
Mango	3–5	5–10	10–15	Fair
Orange	5–10	0–5	5–10	Fair
Peach	1–2	3–5	0–5	Good
Pear	1–3	0–3	0–5	Excellent
Plum	1–2	0–5	0–5	Good
Asparagus	Air	5–10	0–5	Excellent
Beans	2–3	4–7	5–10	Fair
Cabbage	2–3	3–6	0–5	Excellent
Cucumber	3–5	0	8–12	Fair
Onion	1–2	2–5	0–5	Good
Pepper	3–5	0	8–12	Fair
Spinach	Air	0–20	0–5	Good
Tomato	3–5	0–3	12–20	Good

5.5.8 Packaging Material Used for MAP [27]

When selecting packaging films for MAP of fruits and vegetables, the main characteristics to consider are gas permeability, water vapor transmission rate, mechanical properties, transparency, type of package and sealing reliability. Traditionally used packaging films like low-density polyethylene (LDPE), polyvinyl chloride (PVC), ethylene–vinyl acetate (EVA) and oriented polypropylene (OPP) are not permeable enough for highly respiring products like fresh-cut produce, mushrooms and broccoli. As fruits and vegetables are respiring products, there is a need to transmit gases through the film. Films designed with these properties are called permeable films. Other films, called barrier films, are designed to prevent the exchange of gases and are mainly used with non-respiring products like meat and fish.

MAP films developed to control the humidity level as well as the gas composition in the sealed package are beneficial for the prolonged storage of fresh fruits, vegetables and herbs that are sensitive to moisture.

5.5.9 Application of MAP Technique

- Fresh fruits
- Leafy vegetables
- Processed foods like bakery items and snack foods
- Fresh fishes and meat
- Cooked food (Fig. 5.6).

Fig. 5.6 Fruits and vegetables packed in MAP

5.6 Controlled Atmosphere Packaging (CAP) [28–30]

In controlled atmosphere packaging storage chamber, the optimum levels of carbon dioxide and oxygen and temperature are maintained for extending shelf life of stored produce. Controlled atmosphere (CA) storage involves altering and maintaining an atmospheric composition that is different from air composition (about 78% N_2, 21% O_2 and 0.03% CO_2); generally, O_2 below 8% and CO_2 above 1% are used. Atmospheric modification should be considered as a supplement to maintenance of optimum ranges of temperature and RH for each commodity in preserving quality and safety of fresh fruits, ornamentals, vegetables and their products throughout post-harvest handling.

The efficiency of this technique can be achieved by continuous monitoring and controlling of storage air composition throughout the period of storage with the help of external means like nitrogen flushing and carbon dioxide scrubbing. In addition, a refrigeration unit is employed for maintaining the storage temperature. In general, liquid nitrogen generator is commonly installed to flush the chambers with liquid nitrogen as and when required for maintaining the optimum level of oxygen inside the storage chamber. Similarly, chamber air is circulated through carbon dioxide scrubber frequently to maintain the optimum level of carbon dioxide in the chamber. The shelf life of the fresh produces can be increased by 2–4 times as compared to normal life.

In case of CA, stored products would deteriorate rapidly when exposed to normal ambient condition, especially for marketing of these commodity. Moreover, in India, fresh fruits and vegetables are commonly transported, retail store and transported in ambient conditions or sometimes, in refrigerated condition. Due to this fact, this kind of bulky and sophisticated equipments with CA technology during the transportation of fresh fruits and vegetables and also continuous monitoring as well as controlling the air composition make the technology very cost-intensive. Due to this, CA storage technology is only restricted to the bulk storage of fruits and vegetables with high commercial value.

5.6.1 Advantages of CA Technique

- Retardation of senescence (including ripening) and associated biochemical and physiological changes, i.e., slowing down rates of respiration, ethylene production, softening and compositional changes, reduction of sensitivity to ethylene action at O_2 levels <8% and/or CO_2 levels >1%
- CA technique can have a direct or indirect effect on post-harvest pathogens (bacteria and fungi) and consequently decay incidence and severity. For example, CO_2 at 10–15% significantly inhibits development of botrytis rot on strawberries, cherries and other perishables. Low O_2 (<1%) and/or elevated CO_2 (40–60%)

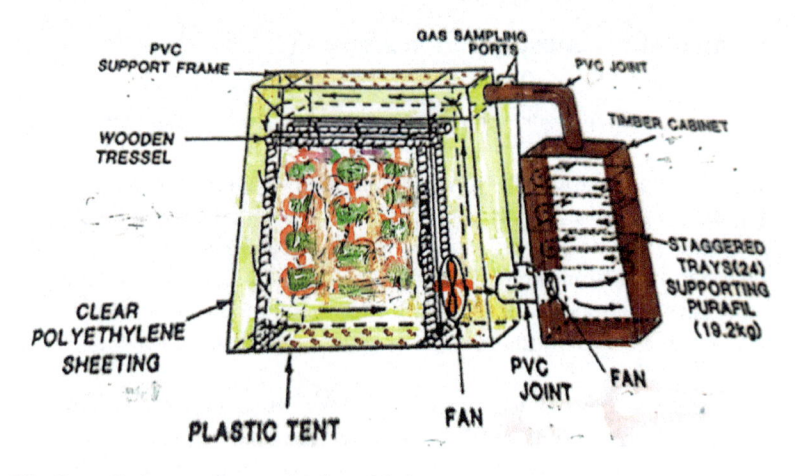

Fig. 5.7 Controlled atmosphere packaging of fruits

can be a useful tool for insect control in some fresh and dried fruits, flowers, vegetables and dried nuts and grains.

5.6.2 Disadvantages of CA Technique

- Initiation and/or aggravation of certain physiological disorders such as internal browning in apples and pears, brown stain of lettuce and chilling injury of some commodities
- Irregular ripening of fruits, such as banana, mango, pear and tomato can result from exposure to O_2 levels below 2% and/or CO_2 levels above 5% for >1 mo
- Development of off-flavors and off-odor.

Low cost plastic tent made from clear polyethylene sheeting can be used for controlled atmosphere storage of fruit. A small fan serves to circulate the C.A. storage air (2% O_2 and 5% CO_2) through a chamber of potassium permanganate on aluminum oxide. Ripening is delayed as ethylene is scrubbed from the storage air. The shelf life of bananas under these conditions is four to six weeks at ambient temperatures (Fig. 5.7; Table 5.2).

5.6.3 Application

It is used for perishable foods like different fruits and vegetables, e.g., apple, pears, okra, spinach, etc. CAP can also be called as "*Gas Flushing*" because a mixture of different gases is flushed or inserted into the food package and replaces the air which extends the shelf life of fresh produce.

Table 5.2 Controlled atmospheric requirements of fruit and vegetables

Fresh produce	Oxygen %	Carbon dioxide %	Temperature °C
Apricot	2–3	2–3	0–5
Banana	2–5	2–5	12–16
Grapes	2–5	1–3	0–5
Guava	2–5	0–1	5–15
Lemon	5–10	0–10	10–15
Lichi	3–5	3–5	5–12
Mango	3–7	5–8	10–15
Orange	5–10	5–10	0–5
Papaya	2–5	5–8	10–15
Pineapple	2–5	5–10	8–13
Plum	1–2	0–5	0–5
Cabbage	2–3	3–6	0–5
Cauliflower	2–3	3–4	0–5
Cucumber	1–4	0	8–12
Okra	Air	4–10	7–12
Onions	1–2	0–10	0–5
Radish	1–2	2–3	0–5
Spinach	7–10	5–10	0–5
Tomatoes	3–5	2–3	12–20

5.7 Active Packaging [31, 32]

Active packaging is also called as functional packaging or "Interactive Packaging". It is also a type of "Smart packaging". Active packaging refers to the incorporation of certain additives into packaging film or within packaging containers with the aim of maintaining and extending product shelf life. Active packaging includes additives or freshness enhancers that are capable of scavenging oxygen; adsorbing carbon dioxide, moisture, ethylene and/or flavor/odor taints; releasing ethanol, absorbants, antioxidants and/or other preservatives and/or maintaining temperature control. All the fruits and vegetables have a unique deterioration mechanism that must be understood before applying this technology. The shelf life of packaged food is dependent on numerous factors such as the intrinsic nature of the food, e.g., acidity (pH), water activity (aw), nutrient content, occurrence of antimicrobial compounds, redox potential, respiration rate and biological structure, and extrinsic factors, e.g., temperature, relative humidity (RH) and the surrounding gaseous composition.

Mechanism of active packaging

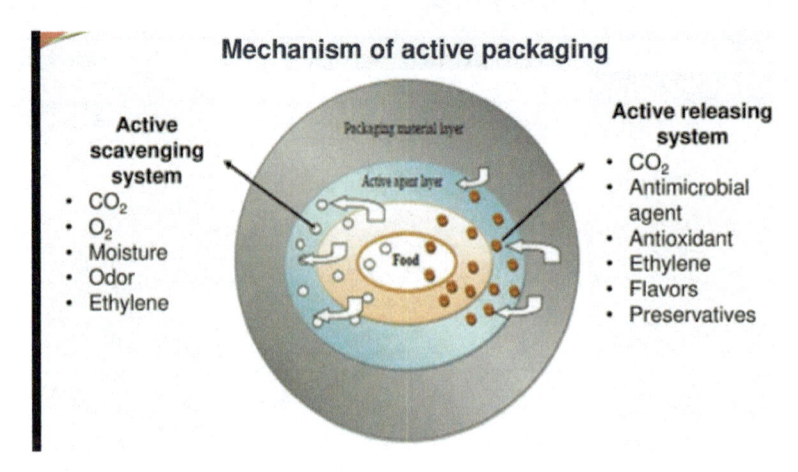

Active scavenging system
- CO_2
- O_2
- Moisture
- Odor
- Ethylene

Active releasing system
- CO_2
- Antimicrobial agent
- Antioxidant
- Ethylene
- Flavors
- Preservatives

Fig. 5.8 Mechanism of active packaging

These factors will directly influence the chemical, biochemical, physical and microbiological spoilage mechanisms of individual food products and their achievable shelf lives. By carefully considering all of these factors, it is possible to evaluate existing and developing active packaging technologies and apply them for maintaining the quality and extending the shelf life of different food products.

5.7.1 Principles of Active Packaging

The concept of active packaging has been developed to correct the deficiencies in passive packaging. For example, if a polymeric film is having good barrier to moisture content but not to oxygen gas barrier like polyethylene, the film can still be used safely with the use of oxygen scavenger to exclude oxygen from the pack. This is active, interactive or functional packaging where the package protects the contents inside the package besides its own classical functions of any packaging. Oxygen scavengers or sachets are popularly used today as an active agents. A schematic diagram is given in Fig. 5.8.

5.7.2 Active Packaging Principles and Its Application to Perishables

A wide variety of materials are normally used for the purpose of active packaging, and the manner in which these materials "act" is given in Table 5.3.

5.7 Active Packaging ...

Table 5.3 Principles of active packaging versus its application

Sl. No.	Principles	Applications
1	Inorganic/organic oxidation	Oxygen scavengers, carbon dioxide generators and ethylene scavengers
2	Ethylene catalysis	Oxygen scavengers
3	Adsorption	Taint removal, oxygen scavengers and ethylene scavengers
4	Hydrolysis	Sulfur dioxide release
5	Acid/base reaction	Carbon dioxide absorption, odor absorption
6	Organic reactions	Ethylene removal, oxygen barrier
7	Polymer permeability control	Temperature compensation, gas compensation balance

5.7.3 Different Types of Active Agents

The different active agents can be used to modify the headspace inside the package and thereby extend the shelf-life of fresh produce.

- Oxygen scavengers
- Carbon dioxide absorbers/emitters
- Ethanol emitters
- Ethylene absorbents
- Odor absorbents (Fig. 5.9).

Active compounds are normally incorporated in the package in the following manner:

- Sachets
- Label

Fig. 5.9 Chlorine dioxide pouches placed inside fruit-packing boxes kill pathogens but do not damage fruit

Table 5.4 Active agents in food packaging

Oxygen scavengers	Carbon di oxide emitters	Ethylene scavengers	Antimicrobial	Antioxidant
Iron	Sodium bicarbonate	KMnO$_4$	Metals	Essential oils
Palladium	Citric acid	Metal oxides	Nisin	Phenolic compounds
Ascorbic acid	Ferrous carbonate	Activated carbon	Essential oil	Plant extract
Pyrogallol		Titanium dioxide	Lactoferrin	Lignin
Gallic acid		MOFs	Lysozyme	Alpha tocopherol
Glucose oxidase			Chitosan	

- Film
- Coating.

However, the current trend of active agents used in food packaging is given in Table 5.4.

5.7.3.1 Ethylene Scavengers

In fresh produce, ethylene scavengers are used as active agent because they respire even after harvesting also. Ethylene (C_2H_4) is a plant growth regulator which accelerates the respiration rate and subsequent senescence of horticultural products such as fruit, vegetables and flowers. Many of the effects of ethylene are necessary, e.g., induction of flowering in pineapples, color development in citrus fruits, bananas and tomatoes, stimulation of root production in baby carrots and development of bitter flavor in bulk delivered cucumbers, but in most horticultural situations, it is desirable to remove ethylene or to suppress its negative effects. Activated carbon-based scavengers with various metal catalysts can also effectively remove ethylene. They have been used to scavenge ethylene from produce warehouses or incorporated into sachets for inclusion into produce packs or embedded into paper bags or corrugated board boxes for produce storage. KMnO$_4$ oxidizes ethylene to acetate and ethanol and in the process changes color from purple to brown and hence indicates its remaining ethylene scavenging capacity. KMnO$_4$-based ethylene scavengers are available in sachets to be placed inside produce packages or inside blankets or tubes that can be placed in produce storage warehouses.

5.8 Intelligent Packaging [32]

Intelligent packaging is an extension of the communication function of traditional food packaging and communicates information to the consumer based on its ability to sense, detect, or record changes in the product or its environment.

Intelligent packaging systems can be classified into three categories as follows:

1. Indicate product quality, for example, quality indicators, temperature and TTIs and gas concentration indicators.
2. Provide more convenience, for example, during preparation and cooking of foods.
3. Provide protection against theft, counterfeiting and tampering (Fig. 5.10).

5.8.1 Quality or Freshness Indicators

In this application of intelligent packaging, quality or freshness indicators are used to indicate if the quality of the product has become unacceptable during storage, transport and retailing and in consumers' homes. Intelligent indicators typically undergo a color change that remains permanent and is easy to read and interpret by consumers.

5.8.2 Time–Temperature Indicators

TTIs are devices which integrate the exposure to temperature over time by accumulating the effect of such exposures and exhibiting a change of color (or other physical characteristic). Many devices which can be attached to food packages to integrate the time and temperature to which the package is exposed have been developed. TTIs can be divided into two categories: partial history indicators, which do not respond unless some predetermined threshold temperature is exceeded, and full history indicators,

Fig. 5.10 Image of intelligent packaging

Fig. 5.11 Time–temperature indicators

which respond continuously to all temperatures (within the limits of the functioning temperature range of the TTI) (Fig. 5.11).

5.8.3 Self-adhesive Labels

These are used on packages of perishable products to assure consumers for freshness. It is a full history indicator whose working mechanism is based on the color change of a polymer.

5.8.4 RFID

Radio frequency identification (RFID) is advanced form of data carriers that can trace and identify the product. RFID is the use of radio frequencies to read information at a distance with few problems from obstruction or disorientation on a small device known as a tag or transponder that can be attached to an object (commonly a pallet or corrugated box) so that the object can be identified and tracked. Almost all conventional RFID devices contain a transistor circuit employing a microchip, an antenna and a substrate or encapsulation material (Fig. 5.12).

5.9 Antimicrobial Packing

Antimicrobial packing is a form of active packaging in which the packaging acts to reduce, inhibit or retard the growth of microorganisms that may be present in

Fig. 5.12 Image of RFID tag on banana

the packaged food or packaging material itself. To control undesirable microorganisms on foods, antimicrobial substances can be incorporated in or coated onto food packaging materials.

5.9.1 Microbial Agents Are of Different Types

- Natural antimicrobial agents—It includes extracts from spices like cinnamon, allspice, clove, thyme, rosemary, oregano and other plant extracts like onion, garlic, radish mustard and horseradish.
- Derived from microbes—These are derived from substances produced from fungal and bacterial action like polypeptide nisin, natamycin, pediocin, and various bacteriocins.
- Chemical antimicrobial agents: benzoic acid, sorbic acid
- Derived from animals: Chitosan.

5.9.2 There Are Different Methods of Incorporating Antimicrobial Agents into the Film

- Addition of sachets-pad
- Incorporation directly into polymer
- Coating or adsorption on polymer surface
- Immobilization of antimicrobials by covalent linkage
- Antimicrobial polymers.

5.9.3 Properties of Antimicrobial Packing

- No need for preservatives
- Reduce the cost of refrigeration and chilling.

- Protect against antibiotic resistance and heat resistance pathogens.
- Environment-friendly, non-toxic and do not leach and release chemicals.

5.9.4 Antimicrobial Package Material Can Be Classified into Two Types

- Those containing antimicrobial agents that migrate to the surface of the packaging material and are effective against surface microbes without migration of the active agent to the food
- The incorporation of antimicrobial agent to product formulations may result in partial inactivation of active compounds by food constituents.

The antimicrobial food packaging material has to extend the lag phase and reduce the growth phase of microorganisms in order to extend the shelf life and shelf life safety.

Edible films and coatings are carriers for AM agents. The matrices for edible films are polysaccharides-chitosan, alginate, κ-carrageenan, cellulose ethers, high-amylose product and starch derivatives and protein-wheat gluten, soy, zein, gelatin, whey and casein. AM agents such as benzoic acid, sorbic acid, propionic acid, lactic acid, nisin and lysozyme have been added to edible composite films and coatings. Silver nanoparticles (AgNP) are used to develop antimicrobial packaging. AgNPs possess antifungi, anti-yeast and antiviral properties. These nanoparticles can be mixed with edible and non-degradable polymers.

Application

The major potential food applications for antimicrobial agents include meat, poultry, bread, cheese, fruits and vegetables.

References

1. Didone M, Saxena P, Brilhuis-Meijer E, Tosello G, Bissacco G, McAloone TC, Howard TJ et al (2017) Moulded pulp manufacturing: Overview and prospects for the process technology. Packag Technol Sci 30(6):231–249
2. Soltani M, Alimardani R, Mobli H, Mohtasebi SS (2015) Modified atmosphere packaging: a progressive technology for shelf-life extension of fruits and vegetables. https://scholarworks.rit.edu/cgi/viewcontent.cgi?article=1032&context=japr. Cited on 9/9/2020
3. Kitinoja L, Kader AA (2020) Small-scale postharvest handling practices: a manual for horticultural crops, 4th edn. Post-harvest Horticulture Series No. 8E July 2002, Slightly Revised in November 2003. University of California Control Atmosphere Packaging cited on 9/9/20. http://www.fao.org/3/ae075e/ae075e19.htm
4. Robertson GL (2016) Food packaging: principles and practice. CRC Press
5. https://www.trioworld.com/en/products-solutions/hygiene-film/products/breathable-microporous-film/#:~:text=In%20the%20hygiene%20industry%2C%20polyethylene,film%20to%20create%20micro%20holes

References

6. https://www.extrememeterials-arkema.com/en/markets-and-applications/chemical-industry-and-general-industry/breathable-films/
7. https://www.marketresearch.com/MarketsandMarkets-v3719/Breathable-Films-Type-Polyethylene-Polypropylene-12805449/
8. https://www.permachink.com/resources/what-is-a-breathable-film?item=
9. https://patents.google.com/patent/WO2017151463A1/en
10. https://patents.google.com/patent/WO2017011341A1/en
11. https://patents.google.com/patent/US6953510B1/en
12. https://www.jeffjournal.org/papers/Volume2/Wu.pdf. J Eng Fibers Fabr 2(1) (2007). http://www.jeffjournal.org
13. https://patents.google.com/patent/CN102205654A/en#:~:text=The%20production%20process%20of%20the,stretching%20the%20base%20film%3B%20heating%2C
14. J Eng Fibers Fabr 2(1) (2007). http://www.jeffjournal.org
15. https://patents.google.com/patent/WO2016100699A1/en
16. https://www.marketsandmarkets.com/Market-Reports/breathable-film-market-208493214.html#:~:text=What%20are%20the%20different%20types,polypropylene%2C%20polyurethane%2C%20and%20others
17. Paine FA, Paine HY (2012) A handbook of food packaging. Springer Science & Business Media
18. Coles R, McDowell D, Kirwan MJ (eds) (2003) Food packaging technology, vol 5. CRC Press
19. https://omnexus.specialchem.com/selection-guide/expanded-polystyrene-eps-foam-insulation
20. https://www.corrosionpedia.com/definition/2473/expanded-polystyrene-eps
21. https://www.engineeredfoamproducts.com/news/101-applications-for-expanded-polystyrene-polypropylene/
22. Han JW, Ruiz-Garcia L, Qian JP, Yang XT (2018) Food packaging: a comprehensive review and future trends. Compre Rev Food Sci Food Saf 17(4):860–877
23. Dordi MC, Bandekar S (2005) Plastics for food packaging. Packaging of Fresh fruits and Vegetables
24. Kumar P, Mishra HN (2002) Beverage and food world. Application of membrane technology in water processing
25. Satish HS (1998) Modern food packaging book. Packaging of fresh fruits and vegetables—international practices
26. Prasad M (1998) Modern food packaging book. Modified atmosphere packaging of fresh fruits and vegetables
27. Beverage and Food world, Sept 2002, pp 45, 46
28. Shorter AJ et al (1987) Controlled atmosphere storage of bananas in bunches at ambient temperatures. CSIRO Food Res Q 47:61–63
29. OBrian D (2017) Chlorine dioxide pouches can make produce safer and reduce spoilage. AgResearch Mag. Retrieved 2 Oct 2020
30. Hu B, Li L, Hu Y, Zhao D, Li Y, Yang M, Jia A, Chen S, Li B, Zhang X (2020) Development of a novel Maillard reaction-based time–temperature indicator for monitoring the fluorescent AGE content in reheated foods. RSC Adv. https://doi.org/10.1039/D0RA01440K **10**(18):10402–10410
31. https://onlinelibrary.wiley.com/cms/asset/284bf297-bd47-44b3-a208-e6ec05cb027d/pts2148-fig-0010-m.jpg
32. https://evigence.com/wp-content/uploads/2017/06/food_in_package.png

Chapter 6
Packaging Techniques for Processed Food Products

N. C. Saha and Meenakshi Garg

6.1 Introduction

The word "food" can be defined in many ways, that is a substance which is consumed for the growth, repair and maintenance of life of organism. It is also explained that any substances or materials containing nutrients such as carbohydrates, proteins and fats can be ingested by a living organisms and metabolized into energy in body tissue.

In other words, food is a material which has nutrients like protein, carbohydrates and fat. It is used by an organism for growth, repair of tissues, for functioning of vital processes and for energy to do various work. Food products can be divided into various categories like.

- Basic food: Cereals, pulses and legumes
- Fresh fruits
- Fresh vegetables (leafy, root)
- Underground vegetables (potato, onion)
- Spices and condiments
- Stimulating foods: Tea, coffee
- Oil seeds (mustard, ground nut, sesame)
- Meat, fish and poultry
- Table eggs
- Processed foods.

"Eat healthy think better" it is rightly said, a healthy food can only pave way to wealthy and bright mind to think innovative things in our daily life. Food scientists and packaging technologists are doing their best to simplify the things that humans consume daily into more simple product which can fulfill the daily requirements. Apart from bringing changes in the food one eat, they are also coming out with mind blowing ideas of packaging that are having qualities like: easily handling, biodegradability, easily disposability, edible packaging, packaging sustainability, free from breakage, high mechanical strength, etc. The field of packaging has changed its

Table 6.1 Categories of processed food products

Aerated drinks	Extruded foods (noodles)	Indian sweets
Bakery products	Edible oil	Malt-based products
Beverages	Fruit-based products (jam, jelly, marmalade)	Papad
Breakfast foods	Fast foods	Spices (ground and whole)
Concentrates	Ghee	Snack foods
Confectionery	Ice cream	Stimulating products (tea, coffee)
Dairy products	Infant foods	Vegetable oil
Dry fruits	Instant mixes	Winning foods

faces from milk in glass bottles to milk in aseptic carton of having six layers of structures for protection and longer shelf life.

A way back of eight years, coconut water in aseptic carton was just a dream but now, it is available in our market in paper-based multilayered composite cartons and PET bottles. This has become possible due to constant research undertaken by packaging technologists to meet the requirements of newly developed innovative processed food products by the food scientists.

In today's scenario, different types and variety of processed foods are made available in our market, which have made humans life too easy and comfortable. In fact, one may not be even knowing that there are numerous varieties of processed foods which are being consumed regularly by the modern consumers. The different types of processed foods are shown alphabetically in Table 6.1.

Interestingly, the critical factors of different types processed foods are varying from each other. Due to this, the requirement of packaging for variety of processed would also vary. For example, biscuits are having crispiness due to low moisture content, and hence, biscuits need packaging materials with high barrier against moisture content to avoid the absorption of moisture from the environment at ambient condition to retain its crispness. But in case of potato chips, the chips are normally deep fried in oil, and thus, product would get oxidized in presence of atmospheric oxygen which leads to rancidity resulting into create off odor and bitter taste. To avoid the oxidation process, there is a need to have packaging materials with high oxygen gas barrier properties, but at the same time, potato chips are also having low moisture content, and in order to retain its crispiness, the materials should also have high barrier to moisture content. Due to this, one has to choose a packaging materials which would provide both barrier against the ingress of moisture content and also the permeation of oxygen gas. Considering this critical nature of this type of processed food, plastic-based laminate structure, made of a combination of polyester and polyethylene film, would be recommended packaging material, where polyethylene would provide moisture barrier and polyester would contribute barrier against oxygen gas. In short, food packaging is need-based technology, and the structures of packaging materials are developed by considering the critical factors of different types of processed foods.

6.1 Introduction

Considering the above, the choice of materials used in packaging is based on the requirement of processed foods to avoid the spoilage and contamination of food products with limited shelf life. Food spoilage, in general, means the adverse changes in quality of food caused by the action of any specific conditions or agents that induce physical and chemical changes due to the effect of microorganisms, insects, bird and rodent, pests etc. In addition, mechanical damage during handling, storage and transportation would also lead to deterioration of quality of processed foods due to chemical and biological hazards. The primary causes of food spoilage include the following:

- Physical: Physical damages of fresh fruits and vegetables include breakage, bruises, crushing and cut or otherwise dismembered surfaces.
- Chemical: These include enzymatic or nonenzymatic reactions.
- Biological: These include microorganisms like bacteria, yeasts and mold and other agents like insects, rodents and birds.

In order to avoid the deterioration of food products due to contamination either by physical, chemical or biological agents resulting into the spoilage of processed and fresh foods and to enhance the shelf life of food products, many innovative packaging techniques and methods have been developed and employed to store the processed food products for longer duration and to deliver the products in fresh and acceptable condition by maintaining the nutritional quality to the consumers.

Over the years, a number of packaging techniques are developed and discussed below.

6.2 Canning

The canning of processed food products in tinplate containers was discovered in the early nineteenth century as a revolutionary technique of preserving foods, and since then, this particular packaging technique has given unblemished service by developing and adapting to the needs of the time so that it remains as a modern innovative method as it was nearly two centuries ago. But the novelty brings its own appeal even though new is always not better.

Canning is a technique by which the food products are packed in a tin container and then processed under high-temperature and pressure to completely kill the pathogens, and the container is closed to make it hermetically sealed, and thus, the food products will have longer shelf life. In short, this food preservation method uses hermetic sealing of food product, and after that, the sterilization or pasteurization of the food product is done by using thermal energy. No preservatives are required to prevent the growth of microorganisms to stop food spoilage [1, 2].

6.2.1 Methods of Canning

The different methods of canning techniques are normally used.

6.2.1.1 Boiling Water Bath Method

This method is used to process foods which contain high acid and can be processed at 100^0 C, which can be obtained in water bath canner at sea level. During the time of loading of food cans/jars, the water in the canner must not be boiling, and it should be hot if the jars have been raw-packed. The can should be covered with water by 1–2 inch. above the height of can for the full processing time so that food will be thoroughly heated from top, bottom and sides. The water should boil steadily for the full recommended time so that food can be sterilized properly.

The boiling water bath canner method is suggested for canning of fruits like tomatoes, foods with added vinegar and fermented foods like pickles. Jams, jelly, marmalades, conserves and preserves can also be processed by using the water bath canner method [1].

6.2.1.2 Steam Pressure Canner Method

It is used to process low-acid foods under pressure at a temperature of 115–121 °C. A steam pressure canner is used to process food at pressures greater than one atmosphere. The steam pressure canner is used for canning foods in the low-acid group like vegetables (except tomatoes), meat, poultry, fish, mushrooms, soups, etc.

Food canning operation is shown in Fig. 6.1 [2].

6.2.2 Material Used for Canning

Commercial canning uses glasses, plastic and metal cans for packaging. Glass container is fragile and heavy and presented challenges for transportation. Hence, glass jars were replaced in with tin can or tin free canisters. Cans are cheaper and quicker to make and much less fragile than glass jars. Glass jars are still in use for some high-value products. Nowadays, tin-coated steel or chromium-coated steel is most commonly used. Aluminum cans are also used for beverages. Bi-metal cans are made of aluminum bodies and steel lids are also used [1].

6.2 Canning

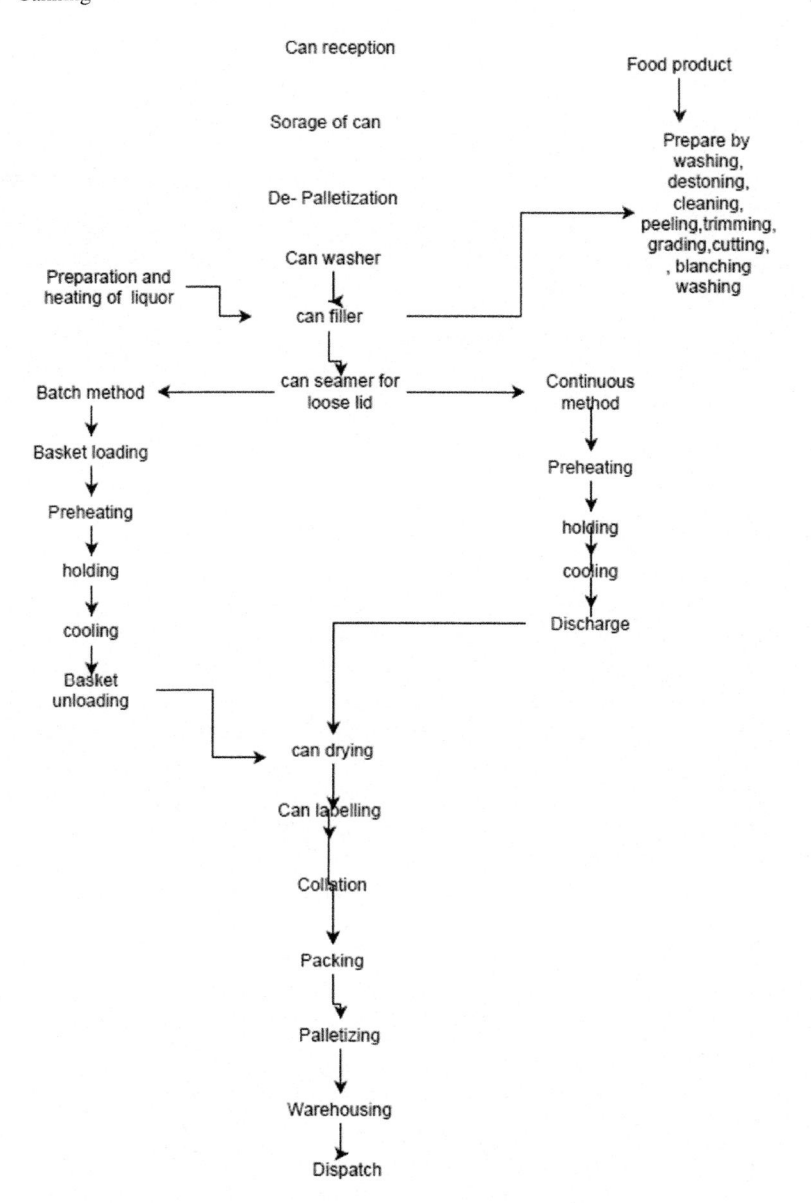

```
                          Can reception                    Food product
                                 │                               │
                          Sorage of can                           ▼
                                 │                        Prepare by
                                 ▼                         washing,
                          De- Palletization               destoning,
                                 │                         cleaning,
                                 ▼                    peeling,trimming,
                           Can washer                  grading,cutting,
  Preparation and                │                        , blanching
  heating of liquor              ▼                          washing
         └──────────────────► can filler ◄──────────────┘
                                 │
                                 ▼
  Batch method ◄──────── can seamer for ────────► Continuous
         │                 loose lid                method
         ▼                                            │
  Basket loading                                      ▼
         │                                         Preheating
         ▼                                            │
  Preheating                                          ▼
         │                                         holding
         ▼                                            │
  holding                                             ▼
         │                                         cooling
         ▼                                            │
  cooling                                             ▼
         │                                         Discharge
         ▼                                            │
  Basket                                              │
  unloading                                           │
         └──────────────► can drying ◄────────────────┘
                                │
                                ▼
                          Can labelling
                                │
                                ▼
                            Collation
                                │
                                ▼
                             Packing
                                │
                                ▼
                           Palletizing
                                │
                                ▼
                          Warehousing
                                │
                                ▼
                            Dispatch
```

Fig. 6.1 Food canning operation

6.2.3 Application

Tin cans are used for storing vegetables, fruit, meat and dairy and processed foods. Aluminum cans are used for soft drinks and other beverages. These are lighter in weight and do not rust [1] (Fig. 6.2).

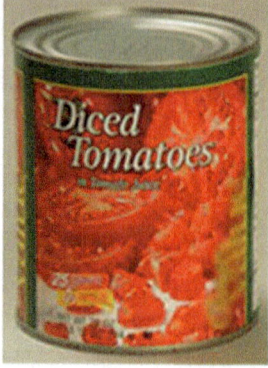

Fig. 6.2 Canned food

6.3 Aseptic Packaging

Aseptic packaging is the method of filling commercially sterile product into sterile containers under aseptic/sterile conditions and then hermetically sealing the containers so that chances of reinfection is prevented. Aseptic means the absence or removal of unwanted microorganisms from the food product, packaging environment and container. Hermetic sealing (air tight) is used to remove the entrance of microorganisms in the container and water vapor and air from the package or from the environment. The term commercially sterile means 99.9% sterile but there is probability of presence of 0.1% microorganisms but not capable of reproducing in the food under room temperature conditions of storage and distribution; thus, it can be said that the absolute absence of microorganisms is almost impossible [3] (Fig. 6.3).

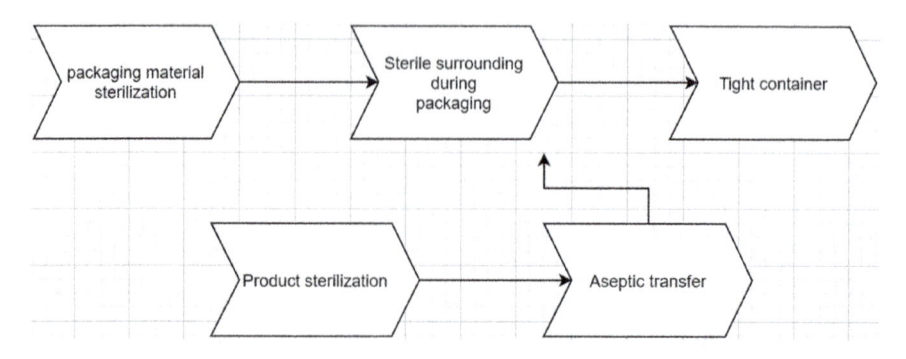

Fig. 6.3 Aseptic packaging

6.3 Aseptic Packaging

Main application for aseptic packaging is:

(1) Packaging of different varieties of presterilized and sterile product like milk, dairy products, puddings, soups, sauces, desserts, beverages and products with particulates.
(2) Packaging of different varieties of non-sterile products like *dahi, lassi, yoghurt, charanamrit, kanji,* etc. to avoid infection by microorganisms [4].

Main reasons for using aseptic packaging are:

i. To extend the shelf life of food product at room temperature
ii. To use those packages which cannot tolerate pasteurization or sterilization temperature or medium [5] (Fig. 6.4).

There are many sterilization processes in aseptic packaging for packaging material which can be used either individually or in combination.

i. **Irradiation**—Interior of empty container or package can be sterilized using gamma rays from cobalt-60 or cesium-137. Generally, material which cannot tolerate the temperatures required for thermal sterilization because of their material, shape, etc. can be conveniently sterilized using irradiation.
ii. **Pulsed light**—Aseptic packaging material can be sterilized using broad spectrum white light (200–1000 nm) by storing electrical energy in a capacitor and

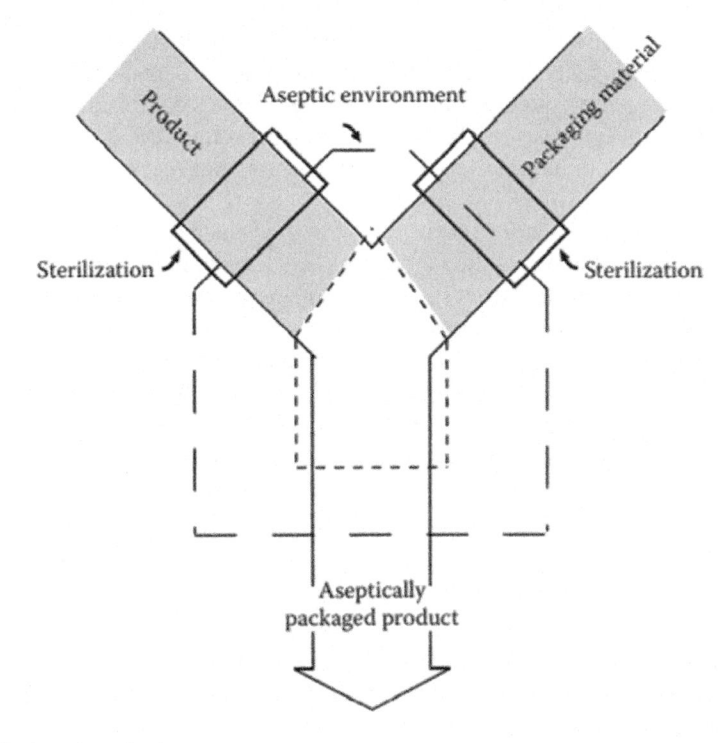

Fig. 6.4 Aseptic packaging

releasing it in short pulses. High peak power levels can be generated using this process. The flashes are applied at a rate of 1–20 flashes/second, and the duration of pulse is from 1 μs to 0.1 s.

iii. **UV-C radiation**—UV radiation is effective in killing microorganisms in range of wavelength between 248 and 280 nm, but it can be used from a wavelength of 200–315 nm. Its optimum effectiveness is at 253.7 nm. It is used mostly in combination with (H_2O_2) hydrogen peroxide.

iv. **Heat**—In heat sterilization processes, either moist heat (steam) or dry heat can be used. Moist heat is pure water in the form of vapor with no air. Dry heat is hot air which does not contain any moisture or water molecules. The sterilization effect of dry and moist heat depends on time and temperature used for sterilization. In steam sterilization process, material is sterilized at 121 °C for 20 min, and in dry heat sterilization process, material is sterilized at 170 °C for 60 min. This shows that steam is more effective in sterilization than dry heat.

v. **Saturated steam**—Steam is the most reliable sterilant. It is used under pressure to sterilize mainly plastic containers. For example, immediately after deep drawing, molded polystyrene cups and foil lids are subjected to steam at 165 °C temperature 600 kPa pressure for 1.4 s (cups) and 1.8 s (lids). After this in order to reduce the heating effect on the internal surface of the cups, the exterior surface of the cups is cooled immediately. This process has been shown to achieve a five to six decimal reduction in Bacillus subtilis spores.

vi. **Superheated steam**—Superheated steam method was used for the sterilization of tinplate and aluminum cans and lids in aseptic canning process. The cans were passed continuously through 220–226 °C saturated steam at an atmosphere pressure for 36–45 s, depending on the construction material given that aluminum cans have a shorter heating time because of their higher thermal conductivity.

vii. **Hot air**—Paper board laminated cartons are usually sterilized by hot dry air at a temperature of 315 °C so that a surface temperature of 145 °C for 180 s can be maintained. However, such a system is apparently only suitable for acidic products having a pH < 4.5.

viii. **Hot air and steam**—Inner surface of cups and lids made from polypropylene are sterilized by use of combination of hot air and steam. These polypropylene cups are stable up to 160 °C. In this process, hot air is blown into the cups through a nozzle in such a way that the base and walls of the cup are uniformly heated [3, 4].

6.3.1 Chemical Treatments

i. **Hydrogen peroxide**—Hydrogen peroxide is used in combination with dry heat to sterilize the surface of packaging material specially paperboard laminate. Resistant spores of microorganisms can be destroyed at a temperature 70 °C, at 30% H_2O_2 concentration and at a minimum time of 6 s. The concentration of H_2O_2 should not be greater than 500 ppb at the time of filling so that it can reduce to approximately 1 ppb within one day.

ii. **UV irradiation and hydrogen peroxide**—When UV irradiation and H_2O_2 are used together on packaging material they act synergistically. By this process, less than 5% concentration of H_2O_2 can be used instead of 30–35% with heat together). Use of less amount of peroxide can reduce the problems of atmospheric contamination. It also reduces the level of residual peroxide in the filled product also.

iii. **Peracetic acid (PAA)**—It is also called peroxyacetic acid. It is a liquid sterilant and is effective against spores of microorganism. It is produced by the oxidation of acetic acid with hydrogen peroxide and this solution is effective against microbial spores. Major drawback of PAA is that it causes corrosion of metals especially iron, copper and aluminum. Concentration of PAA in closed containers greater than 60% can cause explosive decomposition. Hence, it is generally stored at 40% concentration [3, 4].

6.3.2 Packaging Materials Used

The aseptic system can be employed either in paper-based laminated composite carton or in plastic container. In India, Tetra Pack Company has introduced this aseptic package by using different combination of packaging materials. The packaging is available in three formats.

- Tetra Brick Carton
- Tetra Hedron
- Gabletop Carton.

A schematic diagram of the structure of packages is given in Figs. 6.5, 6.6 and 6.7.

The brick-shaped carton is made with the combination of six layered structure of packaging materials. However, three important layer of packaging materials with three bonding layers make the carton.

Paper Board: Paper board is used to provide the stiffness to the package.

Aluminum foil: It provides the barrier properties like moisture, oxygen gas, light, odor and heat resistance.

Polyethylene: It helps for sealing of the carton.

Bonding layers: Additional two layers of polyethylene act as bonding layers in between the layers of paper board and aluminum foil and the aluminum foil and the sealant layer. The top layer of polyethylene provides abrasion resistance to the print surface [5].

Fig. 6.5 Brick Carton

- polyethylene
- polyethylene
- aluminum
- polyethylene
- paper
- polyethylene

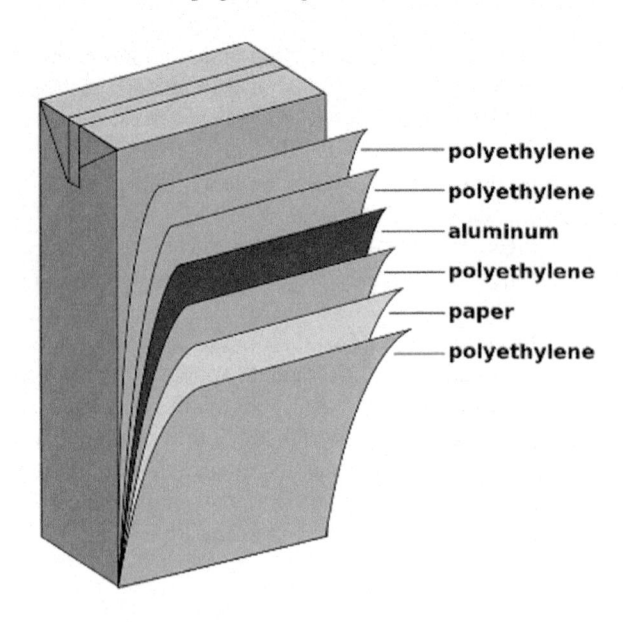

Fig. 6.6 Tetra Hedran

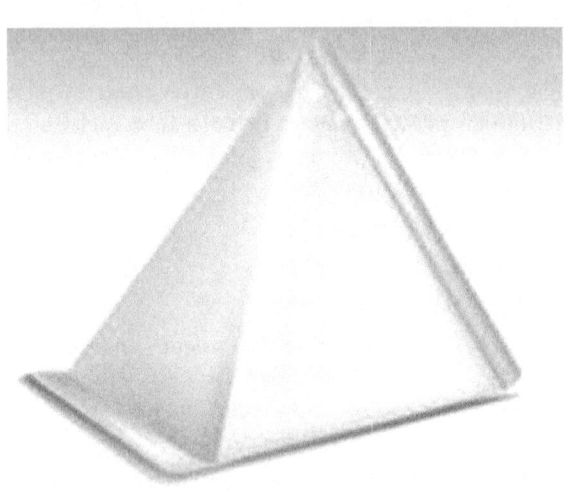

6.3.3 Main Features of Aseptic System

An aseptic packaging system is comprising of the following criteria:

- Before filling strelization of packaging materials/package/containers and closures
- Before filling strelization of the food product
- Sterilization of UHT unit, lines for products, air, filter and machine zones and maintaining sterility in the total system during operation

Fig. 6.7 Gabletop Carton

- Able to carry out all the filling, sealing and transfer operations in a sterile environment and should be cleaned properly after use [4].

6.3.4 Advantages of Aseptic Packages of Food Products

i. Production of hermetic package
ii. Can be stored at ambient condition
iii. Can be able to provide shelf life up to six months without refrigeration
iv. Easy to handle and dispose off
v. The used packages can be reprocessed to manufacture value added goods like composite board-based doors, furniture, desks, pallets, etc.

6.3.5 Application

Liquid processed foods like ready to serve beverages, fruit juices, soups, milk, alcoholic beverages, etc. [3].

6.4 Retort Pouch

'Ready to eat food' is possible because of the development in packaging materials and thermal processing techniques. The retort pouches are made from laminates of aluminum foil, plastic and paper. These pouches are consumer friendly and can be called "heat and eat" products as food material in the pouch can be consumed only by keeping it in boiling water prior to serving. These pouches are usually rectangular, flexible and must be sealed hermetically from all the four sides and thermally processed at high temperature. These can be used to pack highly viscous "curries," dry vegetables in oil, sweets, etc. which are consumed in the Indian households extensively. Retort pouches are used globally. These are light in weight, safe, have high quality, convenient, shelf stable, durable and used commercially for a wide variety of food stuffs. These can be kept at room temperature without refrigeration for year also. These are also called as "flexible can."

6.4.1 Characteristics of Materials Used for Retort Pouch

i. Retort pouches are made from aluminum foil/plastic laminates sometimes foil-free plastic laminate films are also used where food visibility is the requirement.
ii. Retort pouches should be chemically inert, heat sealable, dimensionally stable and heat resistant at least up to 121 °C temperature so that product can be sterilized properly according to nature of product.
iii. These pouches should be physically strong, must have long life and have low oxygen and water vapor barrier properties.
iv. The outer layer of pouch is made from polyester film which provides strength and toughness, while the inner layer of pouch is made from polypropylene polymer which provides good heat sealing properties, strength, flexibility and compatibility to all food products stored in these pouches.
v. Sometimes oriented polyamide (nylon) film layer is also laminated which increases the durability of the pouch, used where more strength is required.
vi. AlOx (metallized) coatings are used to improve barrier properties against moisture and air. Another type of coating is of SiOx which also provides high barrier properties against oxygen and moisture, and it is also suitable for microoven heating.
vii. Thermostable, food compatible, adhesives are used to laminate different layers of packaging material together in retort pouch. Generally, different layers have different thickness, e.g., PET 12 μm, PA 15–25 μm, Al foil 7–9 μm and CPP 70–100 μm.
viii. Print never comes in contact with food. Usually, pouches are reverse printed in a wide range of graphics on the PET film before lamination to provide safety [2].

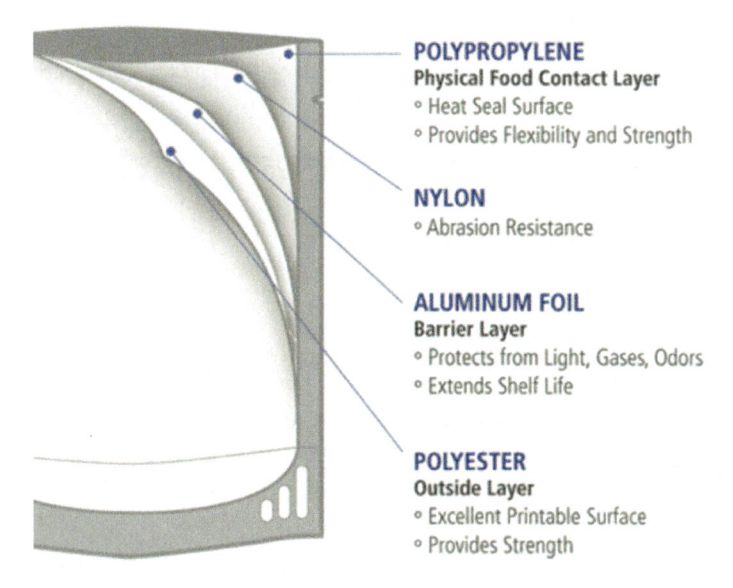

Fig. 6.8 Structure of retort pouch

POLYPROPYLENE
Physical Food Contact Layer
° Heat Seal Surface
° Provides Flexibility and Strength

NYLON
° Abrasion Resistance

ALUMINUM FOIL
Barrier Layer
° Protects from Light, Gases, Odors
° Extends Shelf Life

POLYESTER
Outside Layer
° Excellent Printable Surface
° Provides Strength

A schematic diagram of structure of retort pouch is given in Fig. 6.8.

6.4.2 Advantages of Retort Packaging Technique

The important advantages are given below:

i. Pouch making is easy and less energy consuming process compared with cans
ii. Transport of empty pouches is cheaper and 85% less space is required than cans.
iii. Retort pouches filling lines can be easily changed to a different size according to requirement.
iv. Nutrition and flavor can be retained better in pouches due to faster processing and rapid heat penetration.
v. In retort pouches, refrigeration is not required and the food content are shelf stable at room temperature.
vi. Pouches used for processed foods are compact and require 10% less shelf space as compared to can.
vii. Pouches are cheaper to transport. They require less amount of brine or syrup and are lower in mass.
viii. In pouches, no need to clean after use, fast reheating of contents only by dipping in hot water.
ix. These can be opened easily by tearing or cutting.
x. These are ideal for serving size control.

xi. Material of retort pouch is noncorrosive and does not produce any toxins.
xii. These are convenient for outdoor leisure activities and for military rations use.

6.4.3 Disadvantages of Retort Packaging Technique

The important disadvantages are given below:

i. A major investment is required for equipment for filling and processing and unit.
ii. Production speed for single filler or sealer is less than half as compared to can seamers used in industries.
iii. New handling techniques are difficult to adopt and introduce.
iv. Heat processing in these pouches is more complex and critical.
v. Pouches have limitations to retain rapid heat penetration throughout the pouch in all directions and corners.
vi. Individual outer wrapping increases the cost.
vii. Some fruits which are nonrigid lose their shape.
viii. Knowledge to the consumer regarding its correct storage and proper use is required during marketing [6].

6.4.4 Application

i. Retort pouches can be used for solid meat packs like sliced meat in gravy, high-quality entrees, fish, sauces, soups, vegetables, fruits, drinks, baked items, sweets, etc. Retail packs up to capacity of 450 g can be used for home and other outdoor activities. High visibility products like vegetables can be packed in foil-free pouches for short shelf life up to six months.
ii. These pouches can be used for fruit juices, other type of beverages, soups, etc. Pouches for the institutional trade having capacity of 3.5 kg which is equivalent to the A 10 size can, suited for prepared vegetable products such as carrots, peeled potatoes and potato chips can be made available.
iii. These pouches can be disposed easily after use. This quality offers an advantage to the catering and institutional markets.
iv. High heat sensitive products which are not processed by canning can be packed in retort pouches and nutrient and flavor can be retained better because of low heat exposure.
v. A wide variety of products like vegetables, fruits, soups, curries, stews, hashes, prepared meats, fish in sauce and other popular dishes are packed in it. Few photographs are shown in Fig. 6.9 [6].

Fig. 6.9 Retort pouches

6.5 Ultra-High-Temperature Processing (UHT)

Ultra-heat treatment is a food processing technology that sterilizes liquid food like milk, juice, etc. by heating it above 135 °C (275 °F) temperature for 2–5 s. This time temperature combination is required to kill bacterial endospores. UHT is most commonly used in milk production, but this technique is also used for processing of fruit/vegetable juices, cream, soy milk, yogurt, wine, soups, honey, etc.

Ultra-high-temperature processing is carried out in a production plants, which have many stages of processing and conduct several unit operations automatically in a sequence:

- Flash heating and cooling
- Homogenization
- Aseptic packaging.

Flash heating stage—The liquid which is to be processed is preheated first to a noncritical temperature (70–80 °C for milk) and then heated quickly to the temperature required by the ultra-high process.

There are two types of heating technologies: *direct*, where the product is kept under direct contact with the steam and the steam is mixed with liquid and *indirect*, where the product and the heating medium/steam are in indirect contact, means there will be a barrier and will not be mixed together and remain separated by the equipment's contact surfaces. This process helps to maintain the product quality as well as maintain the high product temperature for the shortest period possible. This technology also ensures that the temperature is evenly distributed throughout the process [7].

- **Direct heating systems**

In direct systems, the product is mixed with heating medium and held at a high temperature for a shorter period of time. It reduces the thermal damage for the sensitive products such as milk but if steam is used as a heating medium then it may dilute the product. There are two groups of direct systems:

- Injection-based system: In this system, the high-pressure steam is injected into the liquid food product. It is suitable only for few products. There may be chances of overheating of product.
- Infusion-based system: In this system, the liquid food product is passed through a nozzle into a chamber containing high-pressure steam at a low concentration so that all the particles of food product come in contact with steam, providing a large surface contact area. This method provides instantaneous heating and cooling and even distribution of temperature. Overheating can be avoided by this method. It is suitable for low and high viscous liquid foods.

- **Indirect heating systems**

In indirect systems, the product is heated in a heat exchanger without contacting the food surface. A higher temperature and pressure is applied to prevent boiling. There are three types of exchangers in use:

- Plate heat exchangers
- Tubular heat exchangers
- Scraped-surface heat exchangers.

These exchangers are highly efficient and prevent fowling, steam is used as the medium for heating and the scrappers are attached to remove any adherent material which may reduce the efficiency of exchanger. It also contains the regeneration unit which allows reuse of the heat and helps in energy saving.

- **Flash cooling**

Flash cooling is a process in which hot liquid food is passed through the holding tube and then to a vacuum chamber, where it loses its heat and the temperature reduces. This process reduces the risk of overheating/thermal damage and also helps in removing water obtained by the contact with steam. It also removes some of the volatile compounds which is a draw back and affects the quality of product. Cooling rate is determined by the level of vacuum which must be carefully handled and should be calibrated properly.

- **Homogenization**

Homogenization is a mechanical treatment to prevent formation of cream on top of layer of milk. In this process, fat globules are broken down into smaller particles and the number of globules, and their total surface area increases. This process enhances the stability of milk and make the milk more palatable to the consumers.

6.5.1 Advantages of UHT Technique

i. The food product processed from UHT method is of high quality because reduction in processing time due to increase in temperature and the minimal come-up and cool-down time leads to a higher quality product.
ii. Shelf life of UHT processed product can be expected greater than six months without refrigeration.
iii. UHT process allows filling of large containers which are required by food manufacturers, food service providers, food sellers, etc. because processing conditions are independent of container sizes.
iv. Packaging material cost, storage cost and transportation cost are less as compared to other techniques [7].

6.5.2 Disadvantages of UHT Technique

i. Sterility should be maintained throughout the aseptic packaging system. Equipment and plant are needed to sterile between processing and packaging. It includes packaging materials, pipework, tanks, pumps, etc. Sterility must be maintained through aseptic packaging.
ii. Large particle size in food product may cause danger of overcooking of surfaces.
iii. Sometimes, overprocessing may occur because of lack of equipment for particulate sterilization.
iv. Flavor deterioration may take place due to presence of heat stable lipases or proteases. Gelation of milk may be possible when stored for long time, and cooked flavor is also more pronounced in UHT-treated liquids [4].

6.5.3 Application

UHT treatment is used very efficiently now a days and requires both a sterilizer and an aseptic unit for packaging the food products. It can be used for low-acid foods having pH above 4.6 like UHT milk, UHT flavored milk, UHT creams, soya milk, vegetable stock and other dairy products. This process is also used to sterilize other processed foods like soups, sauces, desserts, juices, baby food, etc. [7].

6.6 High-pressure Processing

It is a nonthermal technology used to develop microbiologically safe foods without bringing any undesirable changes in physicochemical, microbiological and nutritional properties of food. This technology is highly accepted by consumers. HPP is a technology in which product is processed under high pressure to inactivate microorganism and enzymes present in food [8].

It can inactivate parasites, plant cells, vegetative microorganisms, fungal spores, food borne viruses. Macromolecules can change their confirmation but flavor is generally unaffected. Generally, 200–800 MPa pressure is required to inactivate enzymes like alkaline phosphatase, proteases, etc. For commercially processed foods, optimum pressure is 600 MPa. However, up to 1000 MPa can be used. Food products are usually compressed uniformly from all directions and later food products, after releasing pressure can return to their original shape. It interrupt the cellular function and kill the microorganisms without bringing any changes in nutrient content, texture, taste, flavor and color of food. Product should contain sufficient amount of moisture to prevent damage or any physical change at macroscopic level [9].

6.6.1 HPP Systems Are of Two Types

- Batch process
- Semi-continuous process.

Step wise procedure of working of high-pressure processing technique is shown in Fig. 6.10 and functioning of this technique is also shown in schematic diagram in Fig. 6.11.

6.6.2 Foods Suitable for HPP

- Low/medium moisture containing semisolid or solid foods-cured or cooked meat, cheese, fish, other sea food, ready to eat meals, sauces
- High moisture containing solid food: Marmalades, jelly, preserve, jams, vegetables
- High moisture containing liquid foods: Dairy products like lassi, butter milk, juices and other beverages
- Solid foods: Sweets, bread, cakes, canned food, spices, dry fruits, powder. Different foods suitable for HPP along with temperature and pressure are mentioned in Table 6.2.

Packaging material used in HPP: Various type of packaging material can be used in HPP like plastic laminated material, nylon/coextruded EVOH, nylon/PP,

Fig. 6.10 Stepwise procedure for HPP

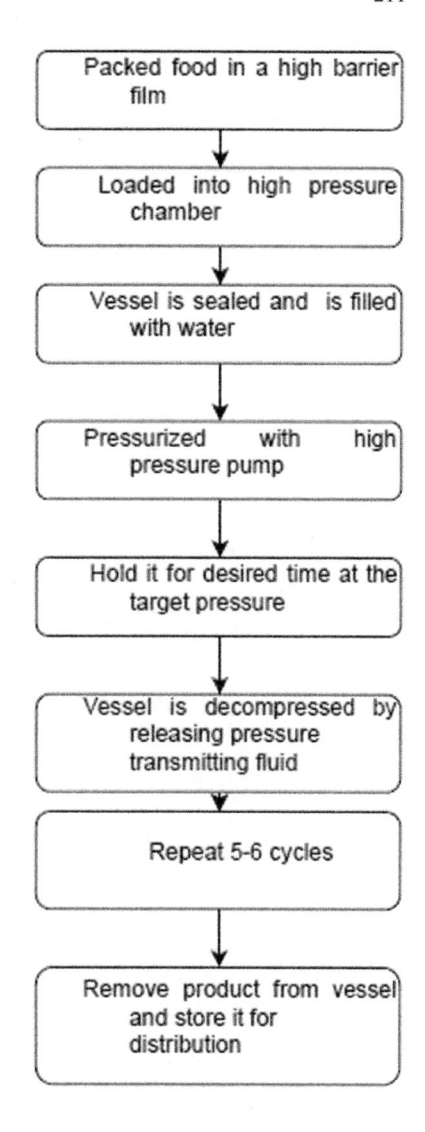

```
┌─────────────────────────────┐
│ Packed food in a high barrier│
│ film                         │
└──────────────┬──────────────┘
               ▼
┌─────────────────────────────┐
│ Loaded into high pressure    │
│ chamber                      │
└──────────────┬──────────────┘
               ▼
┌─────────────────────────────┐
│ Vessel is sealed and is filled│
│ with water                   │
└──────────────┬──────────────┘
               ▼
┌─────────────────────────────┐
│ Pressurized with high        │
│ pressure pump                │
└──────────────┬──────────────┘
               ▼
┌─────────────────────────────┐
│ Hold it for desired time at the│
│ target pressure              │
└──────────────┬──────────────┘
               ▼
┌─────────────────────────────┐
│ Vessel is decompressed by    │
│ releasing pressure           │
│ transmitting fluid           │
└──────────────┬──────────────┘
               ▼
┌─────────────────────────────┐
│ Repeat 5-6 cycles            │
└──────────────┬──────────────┘
               ▼
┌─────────────────────────────┐
│ Remove product from vessel   │
│ and store it for             │
│ distribution                 │
└─────────────────────────────┘
```

nylon/aluminum oxide/CPP, PET/PE, aluminum foil laminated pouches, PET/Al/PP, nylon/Al/PP, nanocomposite materials, etc. These flexible packaging should have more than 15% volume contraction, must have extra tight seals, edges should be round and reinforced, head space should be minimum, must have good tear strength and puncture resistance. Surface should be smooth [8, 9].

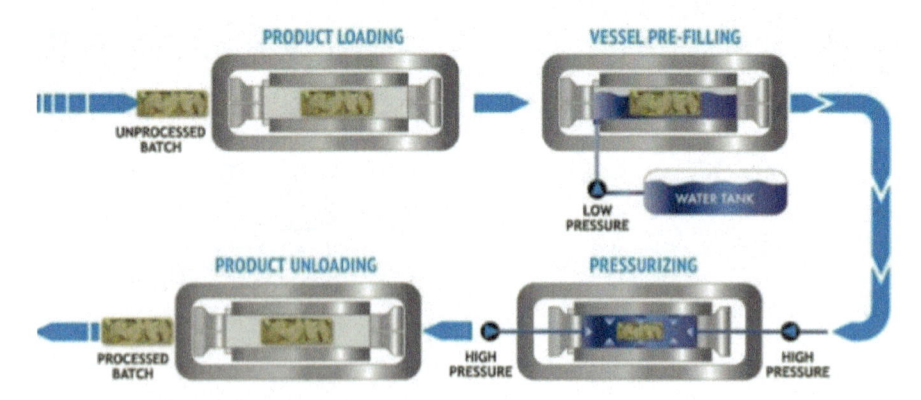

Fig. 6.11 Schematic diagram of high-pressure processing of food product

Table 6.2 Pressure, time and temperature requirement of different food products

Product	Pressure (Mpa)	Holding time (min)	Temp. °C
Carrot/potato cubes	400	15	5–50
Orange juice	350	1	30
Vegetable juice	300	10	35
Cauliflower, spinach	370	10	35
Skim milk	310	0.05	25
Raw milk	100–400	10–60	20
Apricot nectar	600–900	1–20	20

6.7 Individual Quick Freezing (IQF)

It is a freezing method used in food processing industry to freeze individually smaller pieces of food products like berries, diced/sliced fruits and vegetables, seafood—shrimps and small fish, meat, poultry, pasta, cheese, grains, etc.

6.7.1 Benefits

The main benefit of this technology is that this freezing process takes only a few minutes to freeze the product. The time and temperature depend on the type of IQF freezer and the quality, size and shape of the product. Quick/fast freezing helps in formation of small ice crystals in the product's cells, which does not destroy the

Fig. 6.12 IQF mixed vegetable

membrane structures at the molecular level. This helps in maintaining shape, color, smell and taste of food product even after defrosting also.

IQF technology separates each unit of the product like peas, carrot slices, bean pieces etc. during freezing, which produces higher quality product. This is important for food sustainability. The consumer can defrost the frozen food and use the required quantity only [10] (Fig. 6.12).

6.7.2 Methods

There are two main IQF technologies:

- Mechanical IQF freezers
- Cryogenic IQF freezers.

Mechanical IQF freezers: This type of freezers works on the principle of cold air circulation. Air flows from lower side of the bed plate or transportation belt; fans help the circulation of air in chamber. The cold air passes through the pieces of product in circular motions and freeze the product which can be moved out from freezer toward the exit end. This technology is being suited for an increasing range of products [10].

Cryogenic IQF freezers: These type of freezers immerse the food product in liquid nitrogen at very low temperatures, continuously move the product to avoid lump or block formation and freeze it rapidly. Major drawback of this method is higher processing costs per kg of product because liquid nitrogen is very costly.

6.8 Freeze-Drying

Freeze-drying method is used to preserve perishable materials like fresh fruit, vegetable, etc. to extend their shelf life or to make the material more convenient for transport. Working of this method is based on the principle that first freezes the material, then reduces the pressure and adds heat to allow the frozen water in the material to sublimate. It also known as lyophilization or cryodesiccation.

This freeze-drying process is in contrast to dehydration that evaporate water using heat. Freeze-drying provides high-quality product because low temperature is used in food processing. The original shape, color and texture of the product is maintained and quality of the rehydrated product is much better than other processing methods. This method maintains the structural integrity of the product and preserves the flavor. Freeze-drying is an expensive process, it is mainly used with high-value and high-quality products like seasonal fruits and vegetables because of their limited availability, coffee because of highly hygroscopic and for preservation of aroma, and foods used for military rations, astronauts/cosmonauts, etc. [11].

There are four stages in freeze-drying process.

Pretreatment
Freezing and annealing
Primary drying
Secondary drying.

1. **Pretreatment**

This stage means treating the product prior to freezing. It includes following treatments like concentrating the product, formulation revision which means addition of components to increase stability, or to preserve appearance or for better processing or decreasing a high-vapor pressure solvent, or increasing the surface area. Food pieces usually before freeze-drying are IQF treated so that they can flow freely.

2. **Freezing**

During this stage, the food product is cooled below its triple point. (Triple point—the lowest temperature at which the solid, liquid and gas, all three phases of the food can coexist). This ensures that process of sublimation will occur later on. Formation of larger ice crystals is preferred to facilitate faster freeze-drying. The big ice crystals make a network within the food product which promotes faster removal of water vapor during sublimation process. To produce larger crystals, the food product should be frozen slowly and slowly. Speed of reconstitution, duration of freeze-drying cycle, product stability and appropriate crystallization all these will depend on the freezing method. Hence, freezing phase is the most critical phase. Usually, the freezing temperatures for food products for freeze-drying are in between -50 and $-80\ °C$. Freeze-drying process is shown in Fig. 6.13.

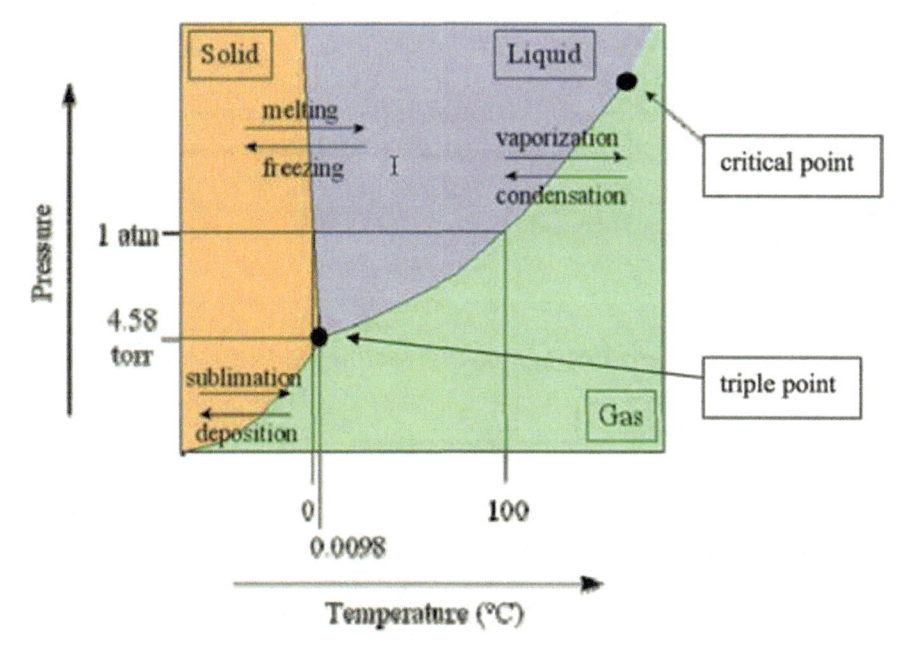

Fig. 6.13 Freeze-drying process

3. **Primary drying**

During the primary drying phase, the pressure is reduced, and heat energy is supplied to the food material for the ice to sublime. In the initial drying phase, about 95% of the water in the material is sublimated. This phase may be slow because, if large amount of heat is added, the material's structure could be altered.

In this phase, pressure is controlled by applying partial vacuum. The vacuum speeds up the sublimation. Condenser provides a surface for the water vapor to reliquify and solidify on. Heat is brought mainly by conduction or radiation, due to low air density the convection effect is negligible.

4. **Secondary drying**

During secondary drying phase, unfrozen water molecules are removed, and ice was removed earlier in the primary drying phase. This part of the process is because of the material's adsorption isotherms. The temperature is raised more than in the primary drying phase and can even be above 0 °C. This is done to break physicochemical interactions between the water molecules and the frozen material. The pressure is reduced in this stage to encourage desorption. The vacuum is usually broken with an inert gas, such as nitrogen after completion of freeze-drying process before the material is sealed. The final residual water content in the product is around 1–4%.

6.8.1 Packaging Requirement

Freeze-dried food products must be sealed in airtight containers to prevent absorption of moisture from the air. Several types of containers may be used:

- Plastic laminated foil pouches
- Metal and plastic cans, or metal
- Fiber drums for bulk packaging.

6.8.2 Advantages of Freeze-Drying Technique

Freeze-dried food has many advantages.

- About 98% of the water content has been removed. Hence, the food is extremely lightweight, which reduces the cost of shipping.
- It requires no refrigeration; hence, shipping and storage costs are reduced further.
- In freeze-dried food, yeast and harmful bacteria cannot survive. Hence, food is free from contamination.
- The physical structure of the food product is not altered while freeze-drying process, the food retains its natural color, shape, texture and flavor.
- Freeze-dried food is attractive to consumers as compared to food preserved by some other methods.
- Freeze-drying causes less damage to the food components as compared to other dehydration methods while using higher temperatures. Heat sensitive nutrients can be preserved by this method.
- Freeze-drying usually does not cause shrinkage or toughening of the food material being dried. Flavors, and nutritional content usually remain unaltered during this process.
- Freeze-dried products can be rehydrated easily because of microscopic pores present in material.

6.8.3 Disadvantages of Freeze-Drying Technique

Major disadvantages of freeze-dried food are.

- The equipment required for this process requires expensive; the process is time consuming and labor intensive.
- Water is not the only chemical capable of sublimation, and the loss of other volatile compounds such as acetic acid and alcohols can yield sometimes undesirable results [11].

6.8.4 Application

- Liquids milk, thin portions of meat, and small fruits and vegetables can be freeze-dried easily. Almost all fruits and vegetables can be freeze-dried, including beans, corn, peas, tomatoes, berries, lemons, oranges and pineapples. Even items like olives and water chestnuts can be processed this way.
- Coffee is the most common freeze-dried liquid. Flavor and aroma are developed due to the Maillard reaction during roasting can be retained by freeze-drying.
- Chunks or slices of shrimp, crab, lobster, beef and chicken can be freeze-dried. They are often mixed with vegetables as part of soups or main course entrees.
- Freeze-dried products are popular and convenient for hikers, as military rations, or astronaut meals. A greater amount of dried food can be carried compared to the same weight of wet food. In replacement of wet food, freeze-dried food can easily be rehydrated with water if desired and shelf life of the dried product is longer than fresh/wet product making it ideal for long trips taken by hikers, military personnel or astronauts. The development of freeze-drying increased meal and snack variety for consumers.
- Some fruits like berries can degrade in quality as their structure is very delicate and contains high levels of moisture. They have the highest quality when freeze-dried. It retained color, flavor and ability to be rehydrated. A photograph is given in Fig. 6.14.

6.9 Microwave Heating

Microwave processing heating destroys microorganisms via thermal effects. Frequencies of 950 and 2450 Hz are used to excite polar molecules, which produces thermal energy and increases temperature. Rapid heating may improve the quality of foods that are sensitive to thermal degradation. Domestic microwave has a greater impact on the food industry. A wide range of foods are available in microwave reheat able packaging. Nowadays, desirable browning and crisping is also possible in microwaves [12] (Fig. 6.15).

Fig. 6.14 Freeze-dried strawberries

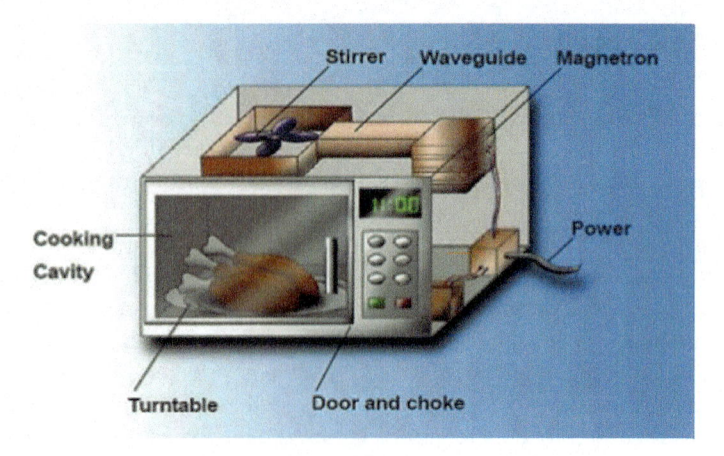

Fig. 6.15 Design of microwave

6.9.1 Packaging Materials Used for Microwave Heating

Compared to most other types of food packaging, the uniqueness of microwave packaging is that the package should be part of the microwave heating cycle, like a wok in conventional cooking. Consumers actually may consume the foods in the package after it is heated. It can serve as a part of heating configuration and a dinning device as well [13]. Meeting above requirements, many materials can be used for microwave packaging, such as

- Glass and ceramics
- Paper, paper board and structured film
- Polymer trays
- Metalized paper
- Metal trays
- Edible materials

6.9.2 Advantages of Microwave Oven

The following are some advantages of the microwave oven.

- Microwave heating is very fast and takes less time to heat food. The main reason behind the cooking is the vibration of water molecules by the reflected microwaves. These microwaves vibrate the water molecules for about a million times in a second. So this results in fast cooking in a microwave oven [5].
- No changes in taste, flavor and nutrient content of food occurs because of microwave heating as it is a fast cooking method. It does not burn the food.

- Microwaves are easy to clean because the oven makes only the food to heat and does not make the food to stick around utensils. So, it is easy to touch and clean as there is a smooth glass finish inside the oven. It requires only a damp towel to wipe off the spills.
- In microwave oven food can be defrosted easily. It melts and cooks the frozen food within a short span of time without any changes in taste.
- It consumes less energy when compared to normal ovens. The wattage of the microwave oven is around 600–1200 W.
- It does not produce smoke and heat in the kitchen and have child lock facility also. These are safe for children.

6.9.3 Disadvantages of Microwave Oven

- The leakage may occur when there is damage of any type like holes in the surface of the cooking chamber, improper functioning of the door. The microwaves are short radio waves, the continuous exposure to the microwave radiation leakage can cause some health problems like burns, damage to the eyes (while looking into the oven with damaged protective mesh), decreases immunity, change in heart rate, cancer cells formation etc.
- Standing waves pattern in the microwave oven is responsible for occurrence of hot and cold spots in food. This causes food poisoning. Formation of these spots depends on the shape and size of the food because the microwaves cannot penetrate into the food up to 1.5 inch. So, it also important to maintain the shape and size of the food, in order to avoid uneven cooking.
- Cooking of food by microwave is dependent on water molecules, and if water content is less, it will results in the formation of dry food. So, there must be enough water content [12].

6.9.4 Applications

Commercial applications of microwave processing include the following.

- Continuous microwave cooking systems are used to cook meats. Microwaves penetrate deep and internal cooking is possible, reduces cooking time by 50%.
- Use of microwave in drying is very popular. It is used in for snacks, as well as spices and other ingredients.
- Microwaves can generate heat rapidly, and defrosting is possible in minutes rather than hours/days, even for large blocks of food product. Tempering can also be performed directly inside the package. This process results in a significant reduction in drip loss and reduces product deterioration due to microbial growth. It is commonly used to temper deep-frozen beef, pork, poultry and fish [12].
- Microwave sterilizes products shorter process time and provides improved quality.

6.10 Irradiation

Food products can be treated with radiations by the use of gamma rays (with Co-60 or Cesium-137 radioisotope), electron beams (high energy of up to 10 meV), or X-rays (high energy of up to 5 meV). Radiations interaction result in the formation of energetic electrons at random throughout the matter, which cause the formation of energetic molecular ions. These ions may be subject to electron capture and dissociation, as well as rapid rearrangement through ion–molecule reactions, or they may dissociate with time depending on the complexity of the molecular ion [14]. Factors which effect the radiation on matter are.

- Type of the radiation
- Energy level
- Composition
- Physical state
- Temperature
- Atmospheric environment of the absorbing material.

Proper application of irradiation can eliminate or reduce microbial load and insect infestations along with the foodborne diseases and helps in improving the safety of foods as well as extends their shelf life.

6.10.1 Packaging Material

Food is usually packaged prior to irradiation to prevent recontamination. Hence, radiations also affect the food-packaging materials so while evaluating the safety of irradiated foods, packaging material must be considered. Irradiation can affect integrity of material which act as a barrier to microbial contamination. Irradiation can cause radiolysis that could affect odor, taste, and the safety of the food.

Many food-packaging materials are made of polymers. Radiation effects on polymers are the result of competing crosslinking or chain scission, i.e., degradation, reactions. Hence while selecting a polymer for a particular application, the effect of radiation on the overall stability of the material must be examined. Radiation sterilization has been successfully applied to products and their packaging, which is made of thermoplastics and includes polyesters, polystyrenes, polyethylene, nylon, and cellulose and their copolymers; the typical dose used on food is 10 k Gy [14] (Fig. 6.16).

6.10.2 Advantages of Irradiation Technique

- It extends the shelf life of product and reduces the need for preservatives.

6.10 Irradiation

Fig. 6.16 Irradiated food

- Food irradiation reduces sprouting in onions, potatoes, etc.
- Food irradiation destroys/kill insects their larvae, eggs, etc., otherwise many health problems may occur.

6.10.3 Disadvantages of Food Irradiation

- There is no definitive way of killing all of the organisms that can come into contact with our foods. Some organisms are resistant to the mild radiation
- Irradiated foods may cause health problems and safety issues in animals like allergies, chromosome abnormalities, premature deaths, etc.
- Both vitamins and minerals may be destroyed by irradiation. Sometimes, the textures and flavors can also be altered.

6.10.4 Application

- Irradiation extends the shelf life of the food and therefore reduces the spoilage of food.
- This process reduces the risk of foodborne diseases in humans.
- Doses of irradiation-food irradiation may be achieved using low-dose, medium-dose or high-dose levels of radiation.
    i. Low-dose irradiation (< 2 k Gy) is used to delay sprouting of vegetables and aging of fruit
    ii. Medium dose (between 1 and 10 k Gy) is used to reduce the levels of pathogenic organisms
    iii. High dose (> 10 k Gy) is used to achieve sterility of the product [14].

6.11 Pulse Electric Field

Pulsed electric field (PEF) processing technology is a nonthermal food preservation method. It uses high intensity pulsed electric fields (10–80 kV cm^{-1} in a short pulse of 1–100 μs) to control the spoilage. It enhances the shelf life of the food by eradicating the spoilage microorganisms by creating electromechanical instability of the cell membrane based on electroporation phenomena. It alters the functional group of quality-degrading enzymes and make them inactive without the application of heat. It preserves the quality attributes of food products [15]. Mechanism of PEF is shown in Fig. 6.17.

6.11.1 Advantages

PEF kills vegetative cells of microorganisms and make them unavailable for spoilage of food. Colors, flavors and nutrients of products are not affected by this treatment. This technology is safe, have relatively short treatment time. It is suitable for liquid foods like fruit juices, soups, liquid egg and milk. PEF does not produces any negative effect on environment.

6.11.2 Disadvantages

Major drawbacks of PEF technology are.

- It has high initial cost of set up.
- It has no effect on enzymes and spores of microorganism.
- PEF alone is suitable only for liquids particles in liquids.

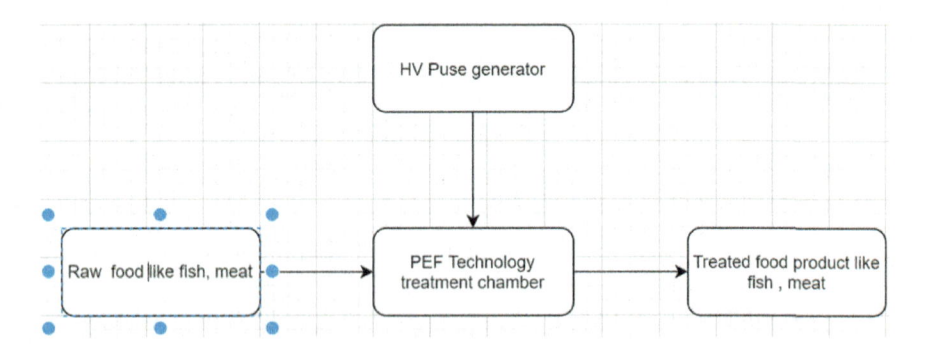

Fig. 6.17 Preservation of food by pulse electric field

6.11 Pulse Electric Field

6.11.3 Applications

PEF is a nonthermal alternative to conventional pasteurization methods. The PEF method has no known detrimental effect on heat-labile components like vitamins in foods. PEF has been mainly used to preserve the quality of foods like bread, milk, orange juice, liquid eggs, apple juice, vegetable juices, etc. [15].

6.12 UV Sterilization

UV light food sterilization does not add radiation to food. Hence, food does not become radioactive. It is a safe process of sterilizing food using UV lamps which emit wavelengths in the UV-C range. UV light is divided mainly into three parts.

- UVA—It uses 320–400 nm (nm) wavelength which is the longest one.
- UVB—It uses 280–320 nm wavelength.
- UV-C—It measures between 100 and 280 nm which is the shortest wavelength.

The use of UV radiations is a part of food manufacturing process since long time. Irradiations cannot be used in organic food. Many food products like juices, apple cider, grains, cheese, baked items, frozen foods, fresh fruits and vegetables, liquid egg products, beverages, etc. are processed using UV-C. UV-C eliminates or reduces *E. coli*, Salmonella, Listeria and other foodborne pathogens. Effectiveness of the treatment depends on the exposure time, level of irradiation and the type of technology used. The UV radiation is very effective on juices and 4000 L of juice can be treated in less than 30 s [16]. UV-C irradiated strawberries and green leaves are shown in Fig. 6.18.

6.12.1 Advantages

- Germs on fresh produce can be killed using UV rays to reduce food borne microbial load.
- UV treatment is better as compared to other processing technology in terms of cost, labor and the need for technically trained personnel for operation.
- Pasteurization can change the color, taste and flavor of juices but UV rays do not show any negative effect on color, taste and flavor of product. Boiling.

6.12.2 Disadvantages

- Suspended particles are a problem in liquid because microorganisms buried within particles are shielded from the UV light and pass through the unit unaffected.

Fig. 6.18 UV irradiated food

- Folds or fissures present in the food may protect the microorganisms from UV rays and these cannot be killed.
- UV rays are applicable mostly on liquid foods and only on surface of solid food.
- Efficacy of UV light depends on the microbial exposures. Sometimes, spores of bacteria may persist [16].

References

1. Featherstone S (ed) (2015) A complete course in canning and related processes: volume 3 processing procedures for canned food products. Woodhead Publishing
2. Coles R, McDowell D, Kirwan MJ (eds) (2003) Food packaging technology, vol 5. CRC Press
3. von Bockelmann B (1998) Aseptic packaging, modern food packaging. Indian Institute of Packaging
4. David JR, Graves RH (1996) Aseptic processing and packaging of food and beverages: desktop reference for food industry practioners. CRC Press
5. Robertson GL (2016) Food packaging: principles and practice. CRC Press

6. Al-Baali AAG, Farid MM (2007) Sterilization of food in retort pouches. Springer Science & Business Media
7. Burton H (2012) Ultra-high-temperature processing of milk and milk products. Springer Science & Business Media
8. Paine FA, Paine HY (2012) A handbook of food packaging. Springer Science & Business Media
9. Huang HW, Wu SJ, Lu JK, Shyu YT, Wang CY (2017) Current status and future trends of high-pressure processing in food industry. Food Control 72:1–8
10. Steinka I, Barone C, Parisi S, Micali M (2017) Technology and chemical features of frozen vegetables. In: The chemistry of frozen vegetables. Springer, Cham, pp 23–29
11. Ciurzyńska A, Lenart A (2011) Freeze-drying-application in food processing and biotechnology-a review. Polish J Food Nutr Sci 61(3):165–171
12. Datta AK (2001) Handbook of microwave technology for food application. CRC Press
13. Microwave Food Packaging. Available from https://www.researchgate.net/publication/312192924_Microwave_Food_Packaging. Accessed 18 Sep 2020
14. Arvanitoyannis IS (2010) Irradiation of food commodities: techniques, applications, detection, legislation, safety and consumer opinion. Academic Press
15. Kumar Y, Patel KK, Kumar V (2015) Pulsed electric field processing in food technology. Int J Eng Stud Techn Approach 1(2):6–17
16. Koutchma T (2008) UV light for processing foods. Ozone Sci Eng 30(1):93–98

Chapter 7
Food Packaging Design—Its Concept and Application

7.1 Introduction: Significance of Design for Packaged Food [1, 2]

Packaging, once known as the "silent salesman" can be heard the loudest now [1].

Take a walk down a modern supermarket aisle or look at the crowded store shelves in the neighborhood bazaar? For that matter, just glance at the overflowing garbage bins? *Modern packaging is not silent.* It is loud and wears brightly colored trade dress. It shouts out from the shelves: *"New"; "Fresh"; "Save"; "Now With"; "Extra"* and so on and on. It glitters in the colors of gold or silver. It sparkles and catches the eye in bright colors even while decaying in a garbage dump!

With the gradual disappearance of conventional advertising communication into online and electronic media, pesky "noisy" packaging has emerged to intrusively make a constant pitch for our attention. Packaging is being designed to aggressively grab eyeballs. Each product is on a mission impossible—to be the first to get our attention within the 8–10s available to successfully activate our buying impulse.

As we all know, the consumer product markets today are fiercely competitive. There is little to differentiate among competing products in terms of real quality attributes. Most retail platforms have moved to self-service or online buying where there are no sales people to push sales. In such an environment, the role of branding and persuasive on-pack messaging is critical to forcefully activate the customer's buying intent.

The rules of persuasive messaging have changed drastically in our increasingly digital, bits and bytes environment of media and noise. The messaging of the past, which worked then, seems woefully ineffective today. Several schools of thought have evolved over time to learn from insights into behavioral and neurological research about what will really grab attention and kick-start the buying intent. Such insights and market research help to emphasize how important it has become for packaging

---

With contributions by "Deepak Manchanda."

to help products break through the "white noise" of the market and engage the head, as well as the heart of the customer.

Getting to the heart of the customer turns out to be an even more stiff challenge when, not just the market environment, but the nature of consumer society has altered. The willful millennial is here. Social media wildfires rage around the world. Ideas are expressed in characters. Feelings have become emoticons. Issues have become memes. Packaging can no longer stand silently outside this raging maelstrom of voices. Due to the intimacy with which "our packaging" shares our lives, it must also speak up in a way that really reaches our heart—preferably with topicality and humor!

7.1.1 First Moment of Truth [2]

In this context, it is also important to recall the terms FMOT, ZMOT and SMOT coined by P&G in 2005, which had become very popular. FMOT refers to the first 3 to 5 seconds when a shopper notices an item—invariably due to its pack design—amidst a crowded shop shelf and makes a decision whether or not to purchase the item. It is believed that there are possibly three stimuli that will cause the customer to notice the packaging. First, it may be a TV commercial; a radio advertisement; a magazine or newspaper advertisement; digital or social media or an outdoor banner. The second stimulus may be due to the customer searching the web to locate the product or service or due to a visit to the retail outlet. The third, critical stimulus is the moment the customer actually finds the product in the store, or online and experiences the so-called first moment of truth. The buying decision at that moment is due to the cumulative media, packaging and point-of-purchase experience (or online experience) of the customer.

Subsequently, after the customer takes the product home and starts using it, the SMOT—or second moment of truth—occurs and it determines the customer's brand perception and future buying decisions. Much later, in 2011, the term ZMOT—or zero moment of truth—was coined by Jim Lecinski, VP Sales, Google, to describe the perceived changes in consumer behavior due to the advent of digital media. Hence, ZMOT refers to the consumer's journey of discovery of the product or service via online search, word-of-mouth referrals or traditional publicity, promotions or media campaigns.

From function and protection to communication and convenience

Clearly the nature of food and beverage packaging is continuously evolving according to needs and concerns of people, their life styles, as well as global events (e.g., wars and pandemics) and technological discoveries or patents. Over the past couple of decades, there has been global concern about the environmental impact of packaging, and consequently, the packaging industry is on the defensive in order to develop processes and materials to reduce environment impact. A mix of such issues will

7.1 Introduction: Significance of Design for Packaged ...

continue to impact the development of packaging designs as we move toward the future.

Take Away Points

1. Once thought to be the silent salesman, modern packaging is no longer silent and in fact reaches out, by design, to our head and heart to make an emotive connect.
2. The term FMOT was coined in 2005 to describe the event when a consumer encounters a media promoted packaged product in the retail environment for the first time.
3. Functionally, packaging design focuses on carrying, protecting and displaying the product. However, by careful design, style is added to the package to create desire for the product and deliver on the brand promise. By this means, package design and POP enrich the user experience.
4. By exploiting insights into basic human nature as well as learning from the world of nature persuasive packaging design helps to make products successful.

7.2 Role of Package Design as a Marketing Tool [3]

A creative convergence of technology and mass communication methods

From the above, it would be apparent that the role of packaging goes much beyond containment, transportation or safety of the product being packaged. In the modern marketplace—be it retail or online—there is a constant rush of new product launches. This makes it critical for the packaging to create differentiation for the purpose of brand recognition. Such differentiation inevitably needs to be created by innovation in terms of the materials, the ease of use or the form and graphic communications presented by the package. This requires a creative convergence of the use of technological processes with behavioral science and marketing communication methods.

In order to achieve such results effectively, it is therefore important to define clearly the objectives of a pack design process in terms of: targets, needs, requirements, budgets, timelines and materials. In marketing terms, such an outline of the packaging requirements is known as "the brief." It is observed that whenever a brief for packaging innovation is clearly and tightly defined and goes deeper than just stating pack size; quantity and timeline—the outcome is likely to be more innovative and commercially successful.

For these reasons, it is important to consider deeply the following roles that all packaging needs to perform effectively, in a more or less way, depending on the nature of the product. For example, packaging for a daily use food staple item would entail all the listed considerations but to a much lesser extent as compared to say, a luxury confectionery product.

The various roles that all packaging to be designed [4] must therefore entail.

7.2.1 Disruption and Differentiation [4]

The primary role of all packaging in the market needs it to be disruptive (or "hatke"—as said colloquially) such that it creates "cut-through" on crowded shop shelves. Though, it must be admitted that most often a disruptive role is adopted by the market leader while the me-too, less ambitious brands often take a follower approach and simply imitate the design cues of the leading pack.

The differentiation needs of the packaging need to be met by:

- Delivering a unique user experience by offering a desirable product in a functional and efficient style.
- Conveying the targeted brand image of the product as, for example, from a global brand to being local—viz by being "global."
- Making best possible use of materials, shape, structure, size, texture and print design to create a successful image.
- Satisfying the special needs of niche customer groups such as making products "senior friendly" or "child resistant."

7.2.2 Convenience and Desirability [4]

The desirability of a package increases with how easy it is to open and use or store and handle. A well-designed package must offer consumer convenience along with safety of the contents from spoilage or contamination. For consumers with special needs, such as the aged, the package must allow easy access. For the modern, fast-moving world always on the go, the packaging needs to be designed for immediate gratification or relief.

7.2.3 Communication and Branding [4]

As mentioned earlier, the package and brand identity work best when merged into each other by design. The "brand personality" is integrated across various media as well as the point of sale. For example, design for Gen X brands would aim to address their aspirations and lifestyle.

7.2.4 Brand Integrity and Consumer Safety [4]

Consumer safety and protection needs to be ensured by design in such a way that the package cannot be tampered with or counterfeited. For this purpose, various pilfer-proofing methods are used to protect the contents and ensure the authenticity of the product. Barcodes, QR codes or other forms of code marking the package also help to achieve this purpose.

7.2.5 Process Feasibility [4]

Effective package design ensures that the method of processing and production of the package is feasible, technically efficient and scalable to meet fluctuating sales demand. The design must also be capable of being produced within the process capability of the plant where it is to be manufactured.

7.2.6 Supply Chain and Logistics [4]

For products that are to be distributed across the globe or across various terrains and conditions, it is important for the package to be suitable for the supply chain and logistics involved.

7.2.7 Commercial Viability [4]

Packaging costs and format play an important role in making an affordable product and this has to be factored in while designing pack SKUs that will meet consumer desired price-points or household budgets.

7.2.8 Sustainability and CSR [4]

Across the world, there has been great awareness of the harmful effects of discarded packaging that ends up in landfills or littered packaging that defaces the environment. Many "Go-Green" movements have forced governments to enact laws to ensure the producer of the packaging is responsible for its responsible disposal or recycling. This has come to be known as EPR, or extended producer responsibility. Brands and corporates are obliged to meet EPR norms under their CSR, or corporate social responsibility.

Table 7.1 Key package design considerations

Design consideration	Order of importance for Daily Need Products	Order of importance for Discretionary Use Products
Disruption and differentiation	Low	High
Convenience and desirability	High	High
Communication and branding	Medium	Very High
Brand integrity and consumer safety	High	Very high
Process feasibility	High	High
Supply chain and logistics	High	High
Commercial viability	High	Low
Sustainability and CSR	High	Low
Special needs	Low	High

7.2.9 Special Needs [4]

In addition to all above, many global brands have led campaigns to meet their obligations toward consumers with special needs, such as the blind or physically impaired. With increasing life expectancy, there is also a growing population of aging consumers, globally, toward whom specific products and packs are targeted. In addition, developments in IT and data management have enabled packs to be capable of a human to digital interface (Table 7.1).

7.3 Branding and Integrated Brand Identity [4]

Brand logos and packaging relate together like one large family.

Creating an intimate connection between a manufactured, packaged product and its consumers is the perennial goal of all product marketers. For this reason, packaging and branding are always seen together. But is it possible to achieve successful brand identity just by placing a brand logo on the packaging?

Familiar branded packaging in our homes, in our offices, while traveling, while working or while relaxing is as reassuring as a dear friend. Finding a recognizable branded package in unfamiliar surroundings while traveling abroad can be as happy an experience as meeting a loved member of the family. Reminiscing about "long gone brands" of the good old days can be as nostalgic as thinking about old friends.

For such reasons, a lot of attention needs to be given to the design of the pack. Since all product promotion activity finally culminates in the pack, the packaging is considered to be half promotion and half product. In fact, for products that do not, or cannot, advertise the pack itself is the only source of promotion. The pack design of the brand becomes the main manifestation of the product and "is" the brand.

Influencing customer choice therefore is the result of the creation of an orchestrated brand identity across media, retail environment and packaging platforms. It is often termed as "establishing brand consistency and continuity in a larger perspective—across advertising and visual merchandising." It may also be described as "conveying the brand." In the latter case, packaging elements are often used to convey advertising cues about the product to enhance customer experience. This is why some customers may feel that "the same ice cream tastes better from a tub than from a carton"—for completely subjective reasons. Such an experience is termed *sensation transference* and is frequently observed in food category products.

It is believed that a company's brand identity matters more than the brand being sold and that is created in a world outside the packaging. While different categories of products may require varying levels of brand management, it is unfortunate that every manufacturer in India does not even think very consciously about branding. Packaging design in India has historically been dominated by advertising agencies which tend to give pack design a second place and often do the design work as an afterthought to the advertising campaign. Sometimes even large brands get their print suppliers to do the design work free of cost. In such cases, the brand identity invariably suffers and the brand personality is compromised.

Very often, it is believed that creating a unique form of packaging alone will get it the attention it needs on the retail shelves. In the same way, it is believed that easy open dispensing or convenience dispensing features built into the pack will unlock the market. There are different opinions about this. While it can be argued that in some cases—Coca Cola, for example—form can be "highly differentiating," it is felt that in most cases package form or structure, convenience or product attributes alone are "rarely differentiating." On the other hand, it is believed that a successfully branded package must be a "calculated gamble" of creating a cost effective, easy to handle and stack and distribute package with exclusive shelf appeal.

Take Away Points

1. Packaging and branding are almost always seen together.
2. Successful branding must achieve an emotive consumer connect with the customer.
3. Packaging is the culmination of all advertising and visual merchandising and therefore considered "half packaging and half advertising."
4. The choice of pack format strongly influences customer perception of the product and the resulting customer experience is termed "sensation transference."
5. Packaging design in India is often developed by advertising agencies, almost as an afterthought and does not get the attention it deserves.
6. Unique looks or features alone will not make a pack successful unless it is a "calculated gamble" of cost effectiveness, ease of use, easy to distribute and exclusive shelf appeal.

7.4 Design and Sensation Transference [5]

Winning organizations use high-grade packaging finish to produce champion products

Louis Cheskin, scientific researcher, clinical psychologist and pioneering marketing innovator spent most of his life, in the early 1930s, investigating how design elements impacted people's perceptions of value, appeal and relevance. He discovered that most people could not resist transferring their feelings toward the packaging to the product itself. He observed that people's perception of products and services were directly related to aesthetic design. This relationship was termed *"sensation transference."*

By the 1950s, Cheskin had already implemented the concept that brands, messages, and offerings could be coordinated and delivered through multiple contexts and media. His groundbreaking book *"Color for Profit,"* published in 1951, initiated a scientific approach to color and design. His marketing philosophy propounded around *three core concepts* appears just as relevant today, more than half a century after it was first posited.

1. Good taste has little to do with how well a design sells.
2. Asking customers what they think of a package design is not a useful way to measure effectiveness. Surveys and polls do not measure unconscious reactions; and what consumers—do, not what they say, is what matters. Research shows that most people who claim advertising does not affect them tend to buy widely advertised products.
3. Colors have symbolic meanings: "We associate red with festivity, blue with distinction, purple with dignity, green with nature, yellow with sunshine. Pink is generally associated with health… White is a symbol of purity. Black expresses evil." Preferences for pure colors are often associated with the poor. The rich tend to prefer tints. And while women generally prefer tints and men deep shades, both are attracted to flesh tones.

From: *The Louis Cheskin Animal Coloring Book* by Jack Szwergold.

One famous study by Cheskin involved the testing of identical deodorants in different packages. The deodorants were mailed to a sample set of users and they were told that the formulations were different. However, the only difference between them was their packaging—which was in three different color schemes. Interestingly, the trials showed that customers preferred one or the other versus the others. Some even perceived one sample as so threatening that they reported rashes and trips to dermatologists and yet had no trouble with the same formula in a different package.

Today Cheskin's pioneering work is acknowledged as having contributed to much of modern marketing folklore.

In addition, Macolm T Gladwell, the famous author of the bestseller *"Blink: The Power of Thinking Without Thinking"* (2005) has used Cheskin's work to illustrate how "companies have learnt to manipulate our first impressions." The point simply put is that when we put something in our mouth (from a package) and in that blink of an eye decide whether it tastes good or not, we are reacting not only to the evidence from our taste buds and salivary glands but also to the evidence of our eyes and memories and imaginations, and it is foolish of a company to service one dimension and ignore the other. (See: http://www.getrichslowly.org/blog/2007/08/13/malcolm-gladwell-on-the-power-of-marketing/).

7.4.1 Packaging is the Future of Finishing [5]

In contemporary packaging, high interaction products such as the wine and champagne categories are among the most extensive users of *finish* to support the brand proposition. In such products, appearance and feel are extremely important and a variety of finishing effects are used to create the impressions. These may include foil blocking, varnishes, embossing, de-bossing, die-cutting or laser etching. In fact the print derived packaging finish options are so versatile and so popular that many in the printing industry have begun to believe that packaging is the future of "print finishing." Don Piontek, writing his blog, Pi-World, points out that while the print industry may be in negative growth, almost every consumer product we buy needs to be packaged. It is this demand for varying finishes on a wide variety of packaging substrates that assures and sustains print finishing. (See: http://www.piworld.com/blog/is-packaging-finishings-future-don-piontek).

7.4.2 Quality and Perception—Neck Finish [6]

For rigid packaging—be it glass or plastic—the neck finish is considered to be the single largest culprit contributing to poor packaging finish. This is due to a variety of reasons with which packaging technologists are familiar. A variety of processes from design to mold-making to converting are involved which demand strict adherence to a set of neck dimension standards laid down by the Glass Packaging Institute (GPI) and the Society of Plastics Industry (SPI). A cross-referencing of the standards between GPI and SPI is involved as in many cases there may be plastics closures applied to glass bottles. The details of achieving the best possible finish for a wide range of bottle or jar necks are too vast to be covered here but it may suffice to note that a micron level mismatch between bottle neck and cap thread dimensions can ruin the reputation of the best brands.

7.5 Pack Structure Design

Art and sculpture have an intimate relationship with packaging to make it distinctive

From time, immemorial mankind has been known to have a fascination for art and sculpture. Stone Age men painted representations of animals and life around them in their caves. Sculpture found a place in our lives in many different forms. The desire for art and fine objects continued to evolve with civilization. In the late nineteenth century, with Art Nouveau, art was considered a way of life and artists aimed to combine the fine arts with applied art even for utilitarian, everyday objects of daily life. During this period, the world of art and business became closely related. This was reflected in the industrial design and advertising of the day. It was inevitable that packaging would also be strongly influenced by the art and sculpture trends of the day. By applying the aesthetic principles of art, designers quickly discovered that it was possible to design packaging that could surpass "mere need and go beyond—to arouse desire." Designers know this well and it is for such reasons that many products are intentionally made with "sexy" curves or shapes? Indeed, packaging that focuses on being "easy on the eyes" and stands out on the shelf for its sculpturesque form instead of harsh lines and aggressive communication can create vital space of its own. However, the leverage that sculptured packs can provide to brand marketing is not often shared by packaging purists. The race to create more artistic packaging often leads to its own difficulties. While sophisticated 3D visualization software has made it easier to visualize and design shapes of increasing complexity the same complexity leads to difficulties in mold design and efficient production. Production slows down. Rejections increase. Filling line difficulties arise. End-of-line handling automation is affected. Secondary pack design is inefficient. Freight volume increases. In fact in many cases, even the customer feels cheated when the curves of the pack prevent full use of the product inside. The Holy Grail in sculpture inspired packaging is therefore, to successfully blend sculpture with functionality. The simplicity with which that can be achieved adds to the beauty and appeal of the pack in such a way that it takes on a personality of its own. Such packs stand out from competition. Such packs capture popular imagination. Over time such packs become iconic.

What does it mean to be iconic?

The Oxford dictionary defines it as "to be an icon a person or thing needs to be regarded as a symbol of a belief, nation, community, or cultural movement." The Bloom-London website (www.bloom-london.com) lucidly lists iconic packaging as having the following five features:

1. **It is a part of culture**: associated with popular memories and nostalgia.
2. **It is timeless**: to have a unique identity and have its own visual vocabulary.
3. **It dares to change rules**: in such a way that it improves our relationship with the object.
4. **It is fun to use**: invites to be picked up and played with.
5. **It has an idea at its heart**: that takes it beneath surface beauty to offer a deeper emotional connect.

Over time, the world has seen many examples of iconic packaging. Among these perhaps the one that stands out universally is the bottle of Coca Cola. While it has seen continuous subtle change in shape and size over the decades its basic iconic sculpturesque form has remained untouched. It is difficult to think of a similar example of iconic packaging from India though there have been several that have attempted to do so. The liquor and perfumery industry are usually the most prolific in creating iconic packs. Many homes in India proudly display bottles of whisky and perfume in showcases. One quirky example however that comes to mind of a "stand out" (maybe) iconic pack is the sipper pouch of Paper Boat—the "drinks and memories" from Hector

Beverages, Gurgaon. Modern design tools and technology is making sculpturesque packs accessible to all categories of products. The art of packaging with care and craftsmanship has moved up to the art of marketing to appeal to more focused cross sections of customers. Everyday packs have taken on their own unique appeal. It may not be long before a walk down the super market aisle will resemble a walk down an art gallery.

7.6 Package Structure Design Process [6]

Typically, the package structure design process, when approached professionally, should consist of five key stages, consisting of eleven different parts. These can be listed as:

Stage 1: Understanding and specifying the requirement

- Part 1: Outlining the creative brief
- Part 2: Researching benchmark products
- Part 3: Brainstorm and preparation of mood-boards

Stage 2: Creative conceptualization

- Part 4: Preparing initial, hand drawn 3D sketches
- Part 5: Developing the design approach in 3D software like Solid Works, Rhino, 3D Max or others.
- Part 6: Refining the design approach

Stage 3: Making mock-ups

- Part 7: Preparation of design mock-ups by hand crafting or 3D printing
- Part 8: Preparation of pilot production if needed

Stage 4: Taking feedback

- Part 9: Presenting the design to the team
- Part 10: Making revisions as suggested

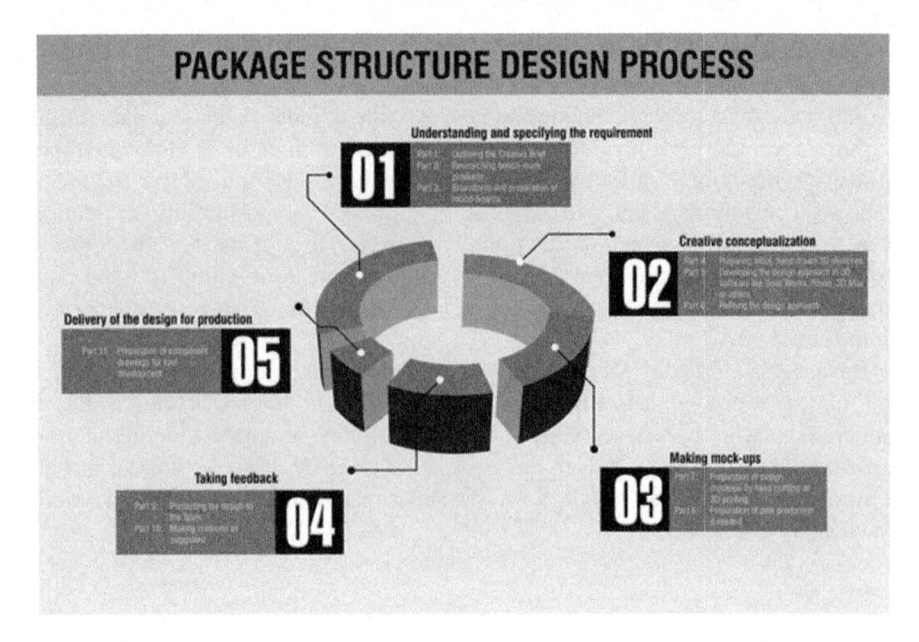

Fig. 7.1 Typical package structure design process

Stage 5: Delivery of the design for production
- Part 11: Preparation of component drawings for tool development.

Such a process can be applied to almost every kind of design project (Fig. 7.1).

7.7 Product and Pack Compatibility for Design Decisions [6]

All product formulations need to be lab-tested rigorously for product to pack compatibility prior to a finalizing the packaging production. Detailed methods of testing food contact safety of all materials that will come into contact with the product have been stipulated by the relevant standards organizations or many brand leaders follow their own standards.

Key tests that need to be performed are:
- Food contact safety
- Shelf life study.

7.8 Graphic Design Significance and Process [7]

Creating strong brands with art, photography and text

Once a package format and shape has been designed in 3D, the surfaces offered by it become prime real estate for marketing communication. This is critical because the product must have a strong shelf appeal in order to be able to stand out among the thousands of different items found in a typical shop or super store. It is generally believed that more than 75% of food purchase decisions are made in store. The human brain is primarily visual. It is said to be able to process images 60,000 times faster than text. In a big super store with thousands of packages on display, a package design may get only about 2–3 s to communicate what product it is and why it should be bought. According to market studies, around 64% consumers may try a new product because the package catches their eye. Subsequently, almost 41% such consumers will continue to buy the product if they prefer the packaging.

For this reason, the graphic design and print appearing on a package has become a professional discipline, and there are several design agencies of repute that specialize in creating strong, iconic pack designs.

7.8.1 Graphic Design Process [7]

Typically, the package design process, when approached professionally, should consist of four key stages, consisting of eight different parts. These can be listed as:

Stage 1: Understanding and specifying the requirement

- Part 1: Outlining the creative brief
- Part 2: Researching graphic design approaches
- Part 3: Brainstorm and preparation of mood-boards

Stage 2: Creative conceptualization

- Part 4: Preparing initial sketches
- Part 5: Developing the design approach
- Part 6: Refining the design approach

Stage 3: Taking feedback

- Part 7: Presenting the design to the team
- Part 8: Making revisions as suggested.

Stage 4: Delivery of the design for production
Such a process can be applied to almost every kind of design project.

As may be apparent, in the above process, the Stage 2 of creative conceptualization is central to the quality of the design development. While a lot may be said about "artistic freedom" and "creative inspiration," in professional graphic design projects,

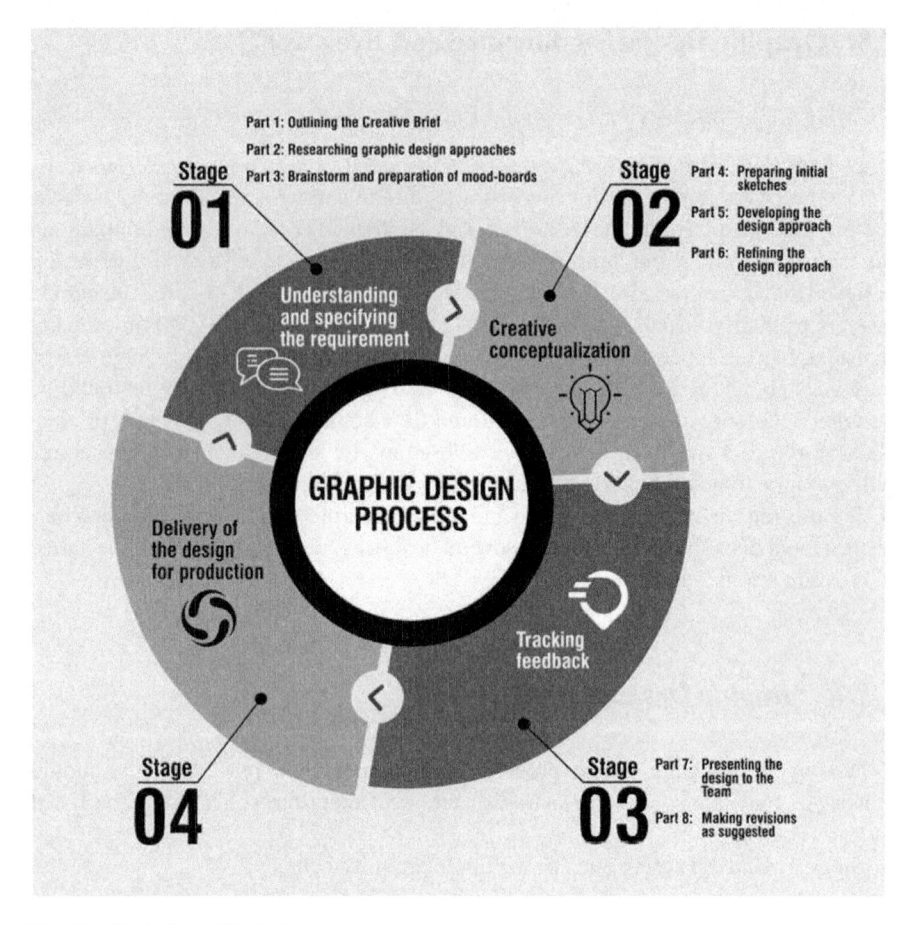

Fig. 7.2 Typical graphic design process

there are generally accepted design norms, which, if followed, help to deliver quality outcomes (Fig. 7.2).

7.8.2 Visual Hierarchy [8]

The principle of visual hierarchy is a set of design norms which designers use to influence the order in which we notice the text and graphics on a pack. These universal rules help to create designs that are pleasing to the eye as well as attract attention to themselves. For food and beverage brands, this aspect may be ranked as one of the most critical pieces in the overall marketing puzzle. Emerging or small and medium brands can leverage good design to compete and be noticed. Unlike advertising, which requires a constant flow of spending, strong packaging design can help a

7.8 Graphic Design Significance ...

brand to grow in sales over longer periods without needing additional spending. The main aspect that helps to make a strong design is known as the "visual hierarchy" of the design (Figs. 7.3 and 7.4).

The visual hierarchy of a design refers to the order in which the flow of communication on the package is prioritized. It is made up of a combination of the brand identity; appetizing food photographs or illustrations; product descriptions and statutory information; claims; batch coding and other information that the shopper needs to know.

The hierarchy of information becomes important in order to be able to convey the brand's story and unique selling proposition in a flash and in the right order of priority. A creative designer varies the weight and importance of the visual elements on the

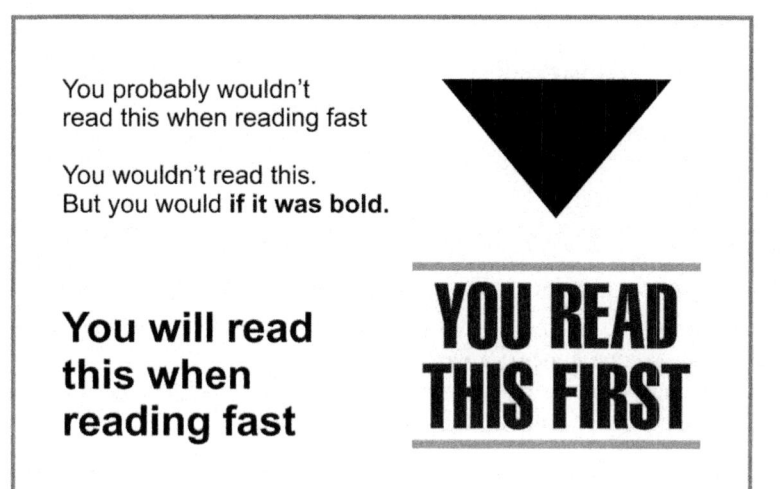

Fig. 7.3 Example of visual hierarchy in text layout

Fig. 7.4 Example of text without visual hierarchy

pack so that a center of interest is established. Around that center the information is arranged in an organized aesthetic manner such that there is a desired viewing pattern, which communicates the most important information first. An important aspect of this arrangement of visual elements is to reduce visual clutter and manage empty space on the design such that it allows prominence to the key design elements, which must get the required attention. Unlike print advertising, packaging design is viewed in 3D and may be viewed from the side or back instead of just the front. For this reason, the viewing pattern should ideally form an infinite loop that tells the brand story effectively from whichever angle it is viewed. Inevitably, designers are known to make use of big design elements, bright colors and bold lines to draw the viewer's attention to specific portions of the label. It can be observed that the stronger the contrast, the higher in visual hierarchy the design element will rise.

Take Away Points

1. A product must have strong shelf appeal to be able to stand out among the thousands of items found in a modern retail store.
2. More than 75% food purchase decisions are made in the blink of an eye by the customer and more than 64% customers choose to buy a product if the packaging catches their eye.
3. Any professionally run package design process must consist of four key stages, viz: Understanding the requirement; creative conceptualization; taking feedback and finally design delivery for production.
4. The visual hierarchy of a design refers to the order in which the flow of communication on the package is prioritized.
5. A creative designer varies the weight and importance of the visual elements on the pack so that a center of interest is established.

7.9 Evolution of Print Processes [9]

Designer's vision made possible only by better print capability

In the comparatively brief period of just about six decades, the printing industry has undergone such revolutionary change that it appears like science fantasy in contrast to its centuries old origins. Till about the mid-eighties, artworks for commercial print needed to be manually prepared painstakingly. The text was prepared with the help of photosetting and prints were pasted skillfully according to the design layout. Illustrations were processed separately and the artwork was "combined" photographically. Hence prepress processes were laborious and time consuming. The outcome depended a lot on the skill and craft of the prepress "artists" who had an eye for color and retouching of photographs. Color photographs had to be "separated" into the standard 4-color, Cyan, Magenta, Yellow and Black set to be prepared for printing by separate "plates." Color "separations" were prepared with the help of filter lenses attached to large size process cameras. The color separated negatives of the color illustration were converted to "film positives" that often needed to be retouched with great expertise before being used for making the printing plates or blocks. A study of the equipment, technology and skills needed to produce quality print on packaging even as late as the 1970s will reveal how difficult and expensive it was.

All this began to change with the development of "drum scanning" technology by Dr Ing Rudolf Hell in Germany. The technology arrived in India by the early 80s. Photographic or art illustrations could be scanned electronically but the data could not be stored digitally because these scanners were not linked to computers and available memory devices could not offer enough memory capacity. Hence it took some time before scanning could move from analog to digital data storage.

This may not be so evident nowadays but it should be remembered that the availability of digital scanning and data storage has revolutionized preprint process technology. The entire workflow nowadays is completely integrated and ensures a high level of fidelity and accuracy that leads to delightful print outcomes.

Color Management

A critical aspect of package printing is to ensure the repetitive reproduction of brand colors identically every time the job gets printed and across the various media and processes that are used for the print production. For this purpose, meticulous color management is needed to ensure that color appearance on the pack is matched as closely as possible between the input design to the output from the printer on the substrate. Inadequate color management often leads to disputes between the brand owner and the printer and needs to be avoided by careful planning.

The need for color management arises because the different devices used for capturing digital images these days use slightly different CMYK or RGB formulas to create a color. Digital cameras and scanners or displays use additive color and the range (gamut) can vary between devices. On the other hand printers rely on a subtractive color system like CMYK or RGB and the different inks and papers used also affect the outcome. For such reasons, professional prepress color management equipment needs to be continuously calibrated and checked to ensure that the color display is reliable and accurately works across the entire workflow. This requires meticulous profiling of all the equipment in the workflow and constant comparison and conversion from the digital data in the original file. An important aspect of this process is now called "soft proofing" and requires a good quality display monitor which can display the image at the correct color temperature and gamma (brightness) settings.

In view of the numerous variables and complexity of managing the entire color workflow process, printers and customers often choose to take shortcuts and make manual adjustments during each print run. To do this color measurement devices such as colorimeters or spectrophotometers are relied upon. An important tool for communicating color shade accurately between the designer and printer is the Pantone color shade system, or its equivalents. At the end of it, in spite of best efforts, since the process inevitably produces unavoidable print shade variations, an LSD chart of light, standard and dark versions of the print are agreed upon between printers and brand owners in order to accept or reject supplies.

While it is not possible to go into too much detail about this vast subject here, it is important to have an understanding of the various print processes that are available to produce print designs on packaging. Broad outlines of the print processes commonly used to produce packaging are listed below.

Offset Lithography

This has been the most popular process to print on flat surfaces of absorbent materials like paper or boxboard in sheet, or even reel form. Later, with the development of heat or UV curable inks, the process could be extended to print on plastic and tinplate sheets too. The process of offset lithography involves transferring the image to be printed on to an aluminum plate in such a way that the portions to be printed can hold the ink (which is oily in nature) while the nonprintable area is prevented from receiving ink by a process of wetting with water. The inked plate then transfers the image to an offset "blanket" across which the substrate to be printed is passed. The blanket transfers the image in this way to the printing substrate.

In a typical 4-color offset print process, a plate for each process color—CYMK— is made and the sheet passes sequentially across each color plate to pick up the color and finally become the original combined image. Nowadays, with the growth of technological sophistication, it has become possible to increase the number of color plates, to print up to 6 colors in one pass, such that the printed pack can feature customized "spot" colors recognizable as the brand's identity. For example, the typical red color of Colgate is printed as a spot color and is identifiable as its "brand color." It is also possible to apply transparent or textured spot lacquers, or metallic inks by this process to premium-ise the design.

Flexography

For flexo printing, the image needs to be transferred to a rubber, or silicon-nylon compounded material plate, and etched in relief such that only the raised portions of the plate can pick up ink when passed across an ink roller. The substrate to be printed is then passed across the inked plate and the image gets directly transferred in a manner like "rubber stamping." The inks are designed to be quick drying for nonabsorbent surfaces like polymer films, or aluminum foils and air-drying for kraft paper, etc.

This process is versatile and popular for narrow web printing such as for label-stock although it has limitations to the extent of color tonal values that can be reproduced by it. A specific problem of the flexo process is the "dot gain" that occurs due to the resilience of the soft rubber plate squeezing the ink dots under pressure on the print substrate. For this reason, the design work needs to be customized to avoid having very fine tonal shades or tightly print registered areas.

Gravure Printing

Unlike offset or flexo processes the printing image for rotogravure needs to be engraved into a metal plated printing cylinder, around its circumference. The engraving is such that the image area of the printing cylinder picks up and holds liquid ink till the print substrate passes across it and the ink gets transferred from the engraved cells. This process originally evolved for high volume work like newspapers and magazines but later adapted to printing multilayer flexible packaging films in reel form.

In India, gravure printing became hugely popular in flexible packaging convertor units due to the capability to print up to 8 colors in one pass by this process at high output speed without any print distortion. This runs contrary to packaging production processes in the USA or Europe where flexography is more popular.

Other Processes

In addition to the above, there are other print processes that are available to printers but seldom used for packaging unless it is for creating special effects like foil stamping, bronzing, flocking, embossing or de-bossing, etc. Some of these other processes can be listed as:

- **Letterpress**—The image is in relief and gets inked over by a roller to be able to stamp the impressions on the print surface.
- **Screen Print**—The pasty ink is spread over a finely meshed screen to create an image in the area of the screen through which the ink can pass.
- **Thermal ink transfer**—The image is transferred from a printed ribbon by application of heat and pressure.
- **Ink-jet printing**—The print image is created by squirting tiny droplets of ink through a nozzle.
- **Xerography**—The image is formed on a metal cylinder by applying a charge to the image areas. The charged area attracts the ink particles, which get transferred to the substrate where they are "fused" to "fix" the image.

Another print process that is fast gaining popularity because of its versatility is the growing technology of digital printing. This process is increasingly replacing short-run jobs that were done earlier by offset, flexo or screen-printing.

Digital Printing

Development of workflow integrated digital print proofing equipment has opened up the possibility of creating short-run, customized packaging labels, cartons and even flexible packs or collapsible tubes for pilot scale production. A range of equipment for this purpose is available from the likes of Xerox, HP, Canon and Konica Minolta.

7.10 Types of Food and Typical Pack Formats

Packaging designers need to bear in mind a few specific design considerations that affect different types of food. Designs made with an awareness of these considerations help to streamline the process of development and prevent rework or market failures due to poor package quality or spoilage.

Frozen Food or Ice Cream

Such products are packaged in plastic tubs, boxboard cartons or multilayer flexible pouches. Print surfaces of such packs have to withstand continuous dampness or getting frosted over while they are on display in freezer shelves. For this reason, the design is usually made with bright colors that can standout on the shelf. Attractive images of the food product packed inside help to make this packaging look appetizing.

Fried Snack Food

Such products are consumed "on the go" and preferred by younger customers with hectic lifestyles. Popular formats for such packs are a range of multilayer laminate pouches or cardboard composite tubes. In some cases even small tins with easy open ends are offered. The designs need to be trendy and bright to suit the preferences of Gen X customers.

Confectionery

Hard-boiled candy, eclairs, toffee and gum in India is sold across tiny kiosks from cities to remote rural areas. Many such kiosks are around the roadside or in the open air and exposed to a range of temperature and humidity extremes. Depending on the target customers, packaging for such products needs to be lowest cost yet robust to deliver dependable shelf life. The design appeal must be largely visual and iconic to cut across multilingual barriers.

On the other hand, chocolates and wellness bars aimed at a more sophisticated customer segment must be designed for more protected environments and the designs could be more subtle or oriented toward children.

Beverage

This is a very diverse and dynamic segment stratified by class of offering, ranging from aerated soft drinks in PET or glass bottles to aluminum cans to Tetrapak and spout pouches for juices and ready-to-drink segment. Designs for this segment need to be focused toward the user group, and the market continues to be sliced into narrower segments based on identifiable niche groups. The pack design in such cases often needs to be aligned closely with media advertising in the manner Coke and Pepsi bring out special edition packs to commemorate sporting tournaments.

Processed Food

This is seen as a high potential segment with vast market growth opportunities in the ready-to-eat or ready-to-cook categories. Popular design formats for such products range from the traditional tin packs and glass jars to plastic tubs, boxboard cartons and bag in box with retort pouches. Product offerings in this category can range from items of daily need or refreshment to gourmet food products with luxury gifting packs with fine quality print decoration and embellishment.

Dairy and Meat

From the humble milk pouch or bottle fulfilling, a daily staple need to specialty milk and dairy products or processed meat products, packaging must be designed to ensure high barrier properties and protection from contamination. Design of such packaging usually tries to communicate freshness, purity and health and nourishment. Very often the design graphics include agrarian scenes of pastures from the product supposedly originates. Pack formats for such products range from glass jars and bottles with quarter-turn caps or tins with easy open ends or plastic tubs, lined cartons or laminate pouches.

7.11 Design Guidelines [9]

For all types of packaging formats that are printed or labeled the starting point for design is the preparation of key-line drawings for all the components to be decorated. This applies equally to rigid, flexible or semi-rigid packaging in metal, glass, paperboard or plastic.

Key-Line Drawing (KLD)

The KLD is prepared in consultation with the packaging producer based on machine limitations of the printable area within which the design graphics can extend. It is based on limitations of the print process as well as the package manufacturing processes. For example, package surface areas that are likely to be overlapped at the sealing or seaming areas are identified and marked out from the KLD. For collapsible tubes the printable area is defined as 3 mm away from the shoulder and the rear seal. In some cases however, designers take liberties within these limitations to stretch the possibility of design on a case-to-case basis. The KLD also sets down limits on the minimum typeface font sizes that would print well with a chosen print process.

In addition to the various packaging formats possible for a given product designers need to have a broad understanding of the manufacturing processes of different types of packaging. Although it is not possible to go into much detail here to discuss the various pack formats and their production processes, a few pointers are listed here for later reference (Figs. 7.5, 7.6, 7.7, 7.8, 7.9 and 7.10).

Glass Containers

Molten glass from a furnace is poured into molds to form glass containers. This process was traditionally done manually by glass blowing (and is still done in Firozabad, UP) but has been replaced by semi-automatic and automatic processes. Different grades of glass quality are used depending on the end use. Glass intended for pharmaceutical or perfumery use is generally USP Type 1. For bottles, there is a standardized set of neck or mouth "finishes" in order to regulate the choices of caps and closures. Glass containers are specified by "weight group" and brimful capacity. With the emergence of modern glass manufacturing methods it has become possible to reduce the weight of glass required for a container volume to get the benefit of "light weighting." This is achieved without losing the physical strength of the glass

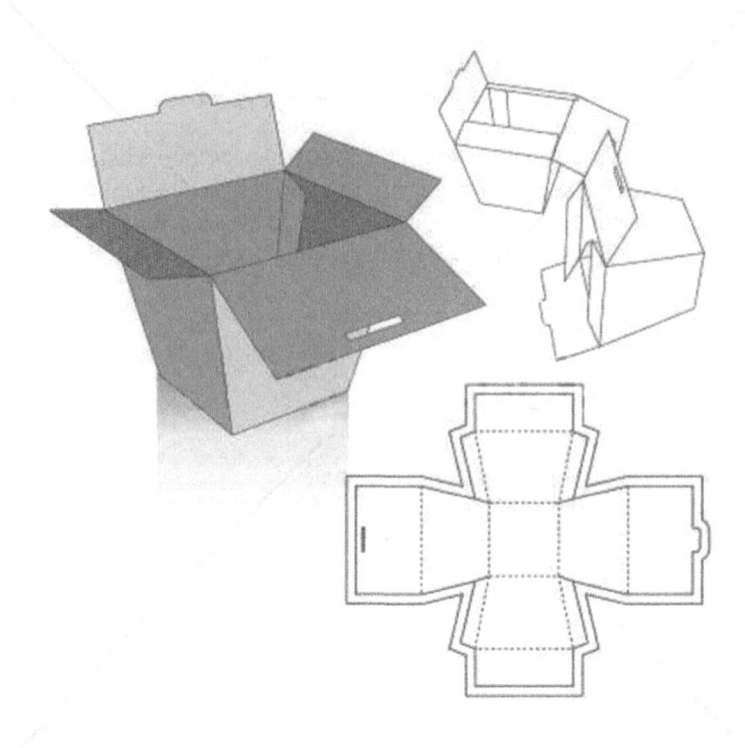

Fig. 7.5 Example of die-line drawing for a tub-shapeed boxboard carton for confectionery

to withstand vertical load or transit vibrations. For good labeling or print decoration, the glass container needs to have good verticality and cylindercity.

Metal Containers

Tinplate, three-piece cans are made from steel sheet coated with tinplate. For nonfood application, such cans are known as general line cans whereas for food applications they are known as Open Top Sanitary (OTS) cans. Both types of cans are produced from tinplate in sheet form which is printed by offset and cut to size and curled and side seam welded to form a cylinder. Internal lacquer may be applied to the can to prevent reaction between the can and the food. Top and bottom ends are seamed on by folding the edges together. There is a standardized set of OTS can dimensions which is offered by can manufacturers in order to regulate the variables.

On the other hand, two-piece aluminum cans are produced by impact extrusion from aluminum disks and then thinning the walls by a process of "ironing." For this reason the cans are known as Drawn and Wall Ironed (DWI) cans. Such cans are printed after being formed into a cylinder by a dry offset print process.

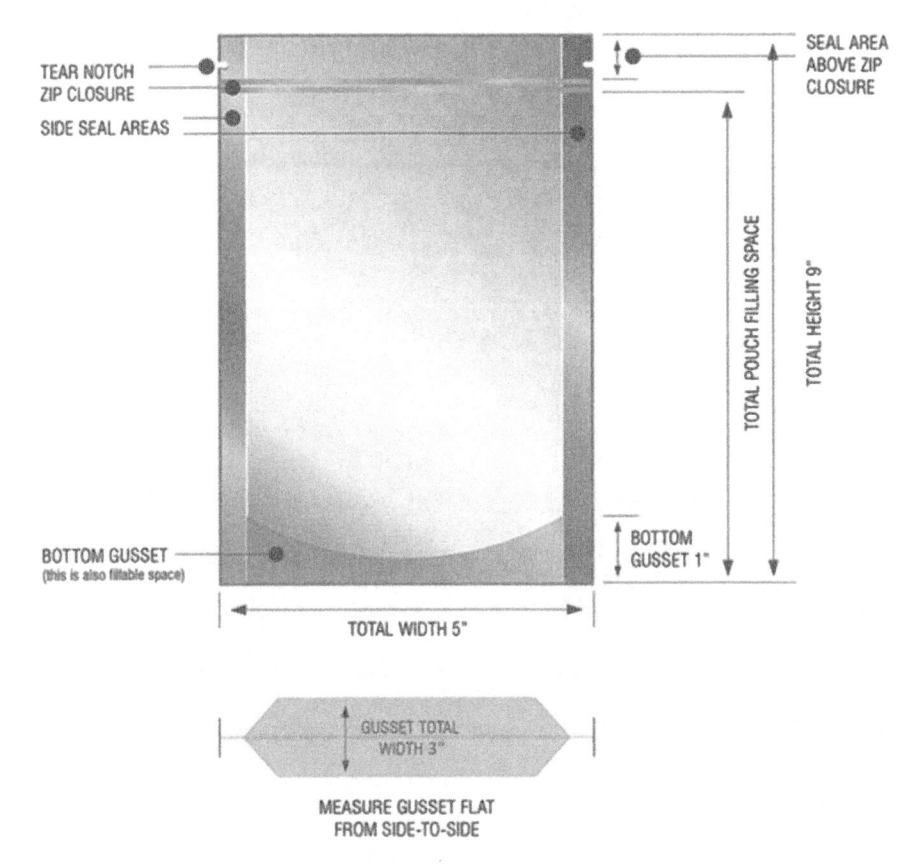

Fig. 7.6 Example of stand-up pouch dimensional drawing and seal areas

7.12 Statutory Requirements [10]

Regulating freedom to produce and sell packaged food

Till not long ago food entrepreneurs were free—or loosely controlled—by a number of ambiguous or outdated standards, to produce, label and sell food in whatever manner they chose.

However the regulatory environment has evolved and keeping pace with it while designing food packaging is critical. Consequently, a number of statutory regulations have had a bearing on the labeling and marking of food. In a bid to simplify the multiplicity of regulations the Food Safety and Standards Authority of India (FSSAI) was established in 2006. The objective of FSSAI was to lay down science based standards for articles of food and to regulate their manufacture, storage, distribution, sale and import so that only safe and wholesome food is offered for human consumption. With FSSAI a *"single line of command"* and a single reference point for all matters

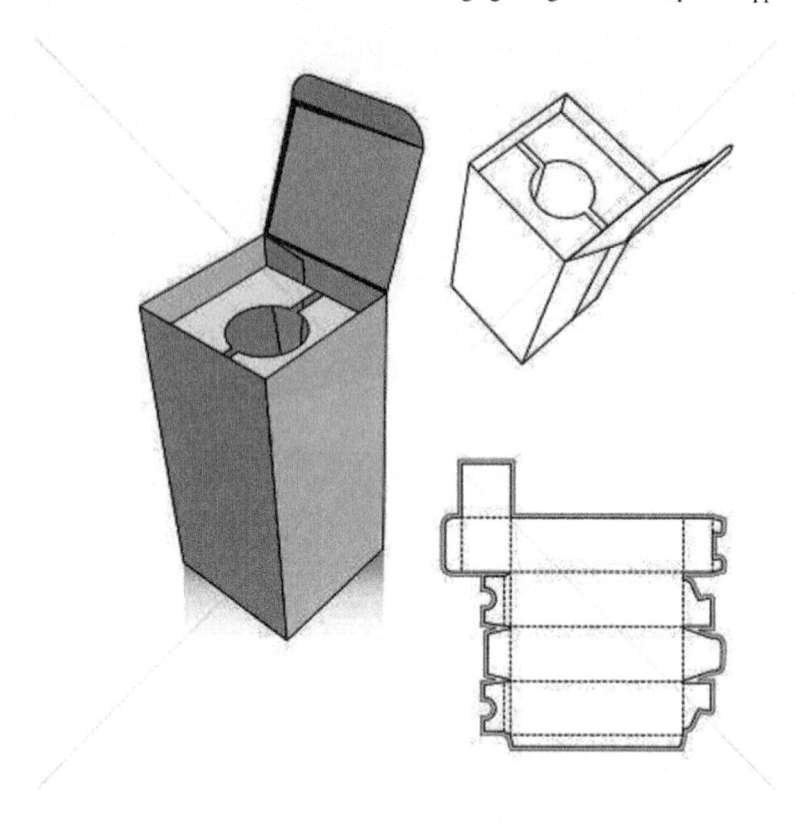

Fig. 7.7 Example of die-line for boxboard carton for a bottle with provision for internal neck support

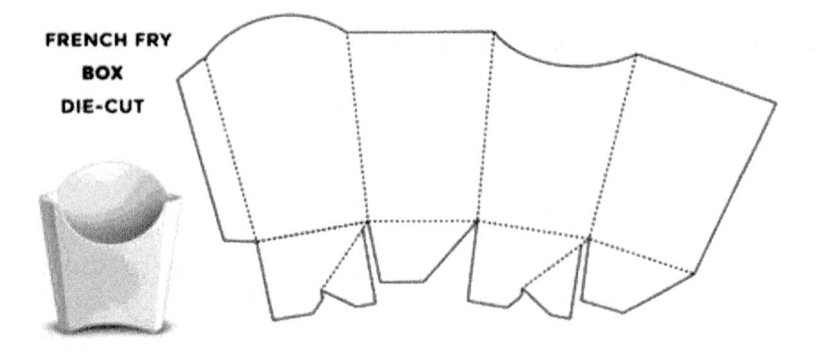

Fig. 7.8 Example of die-line for a French fry tub boxboard carton

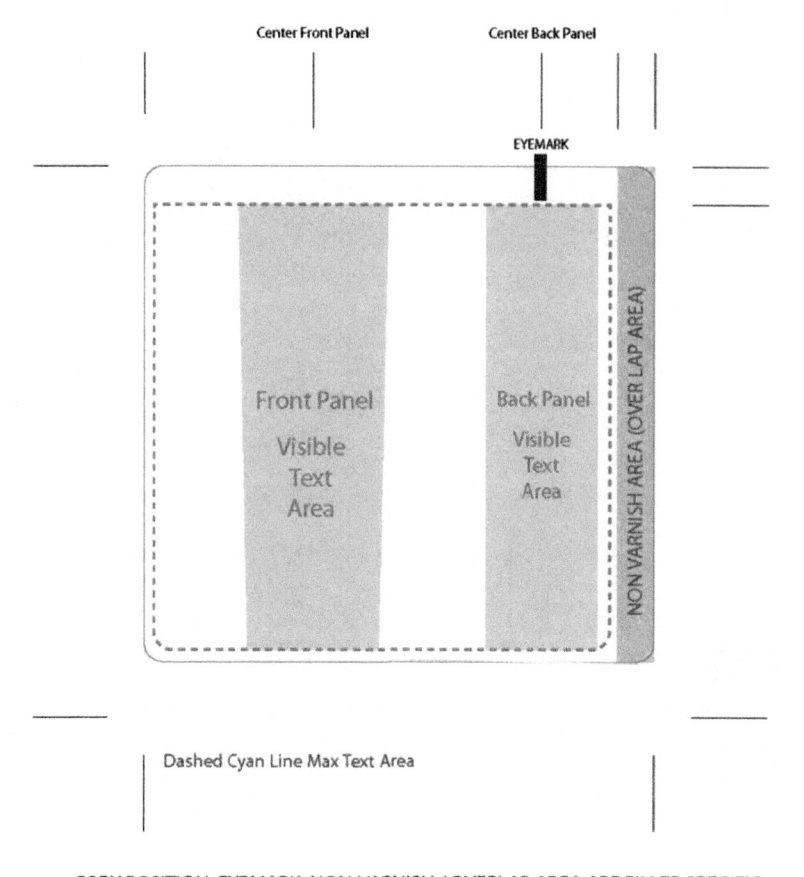

7.12 Statutory Requirements ... 251

Fig. 7.9 Example of key-line drawing for a FFS polylaminate pouch

relating to food safety and standards has been established instead of a hitherto multiplicity of levels and departments overseeing fragmented aspects of food safety. As a consequence many outdated Acts and Orders dating as far back as 1947 or 1954, relating to Food Adulteration, Fruit or Edible Oil or Dairy or Meat production were repealed and replaced by the FS&S Act 2006.

Hence all packaging and label designs for the Indian market need to be governed by the packaging and labeling regulations notified periodically by FSSAI in their website. Important marking requirements in this regard may be listed as:

- Nutrition table and "traffic signal" listing of ingredients
- Vegetarian/nonvegetarian marking
- Batch coding, MRP and best before marking
- Manufacturing and marketing address and license number marking
- Customer care marking.

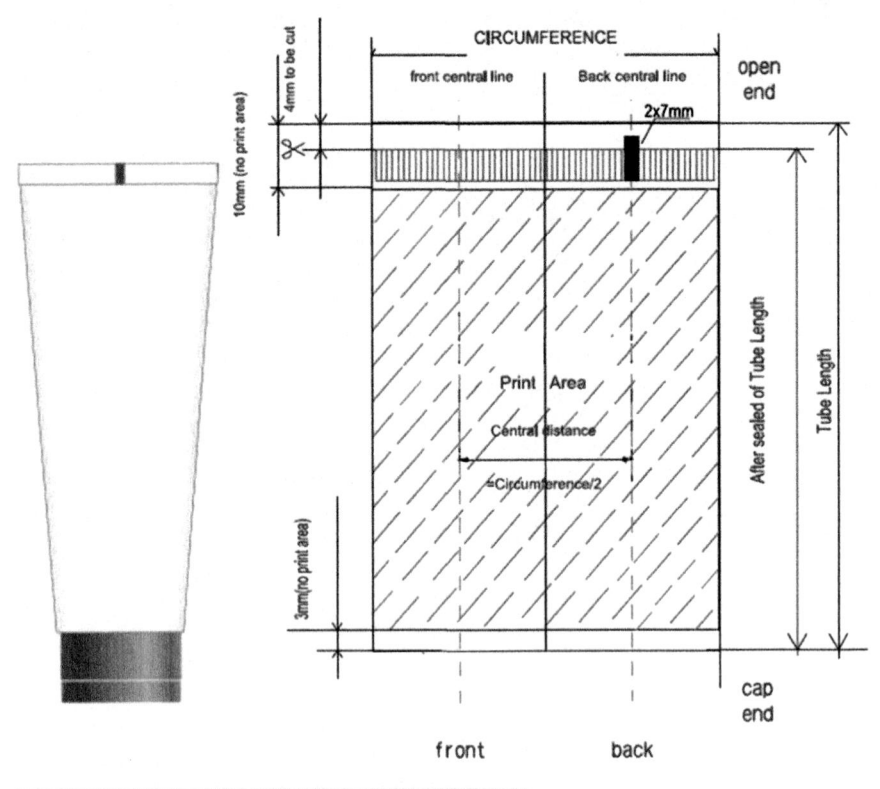

Diameter (mm)	Circumference (mm)	Center distance (mm)
16mm	49.24mm	24.62mm
19mm	58.66mm	29.33mm
22mm	67.08mm	33.54mm
25mm	76.50mm	38.25mm
30mm	92.50mm	46.25mm
35mm	107.9mm	53.95mm
38mm	117.32mm	58.66mm
40mm	123.60mm	61.80mm
45mm	139.37mm	69.685mm
50mm	151.86mm	75.93mm

Fig. 7.10 Example of key-line drawing for seamless plastic tube

7.12 Statutory Requirements ...

- Bar coding and product authentication options
- Regulations regarding pack size and weights and measures.

Example of typical text layout for nutrition facts

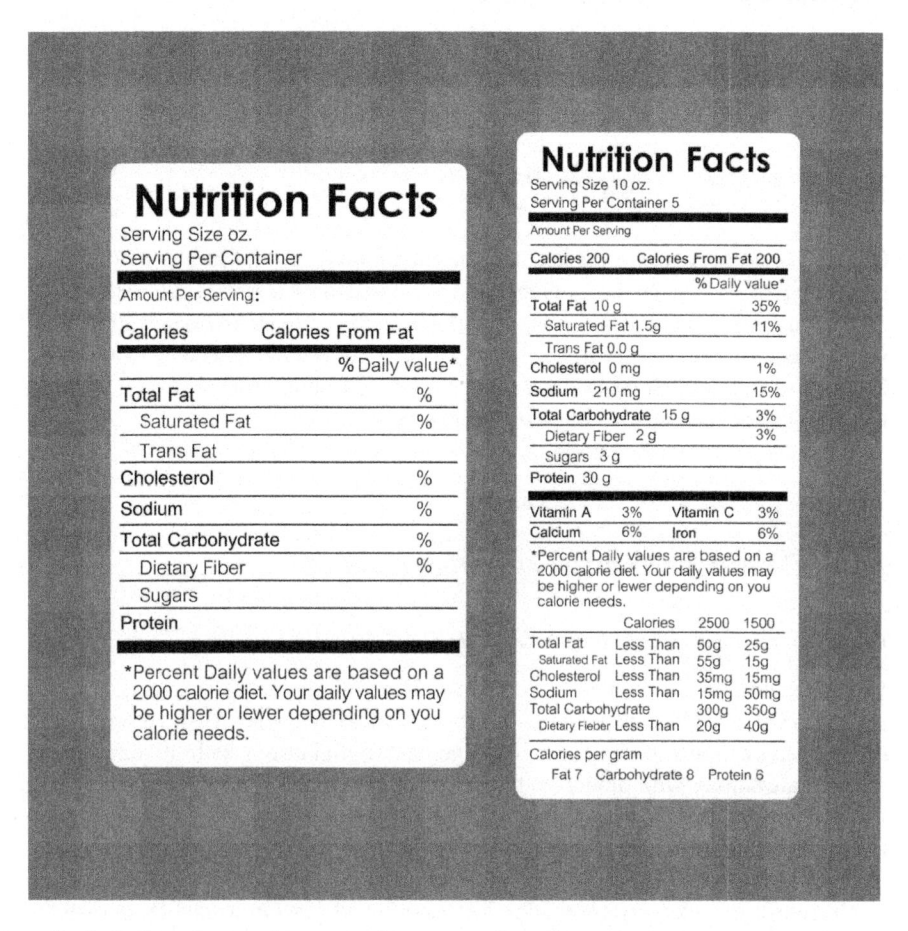

Nutrition Facts

Serving Size oz.
Serving Per Container

Amount Per Serving:

Calories	Calories From Fat	
		% Daily value*
Total Fat		%
Saturated Fat		%
Trans Fat		
Cholesterol		%
Sodium		%
Total Carbohydrate		%
Dietary Fiber		%
Sugars		
Protein		

*Percent Daily values are based on a 2000 calorie diet. Your daily values may be higher or lewer depending on you calorie needs.

Nutrition Facts

Serving Size 10 oz.
Serving Per Container 5

Amount Per Serving

Calories 200	Calories From Fat 200	
		% Daily value*
Total Fat 10 g		35%
Saturated Fat 1.5g		11%
Trans Fat 0.0 g		
Cholesterol 0 mg		1%
Sodium 210 mg		15%
Total Carbohydrate 15 g		3%
Dietary Fiber 2 g		3%
Sugars 3 g		
Protein 30 g		

Vitamin A	3%	Vitamin C	3%
Calcium	6%	Iron	6%

*Percent Daily values are based on a 2000 calorie diet. Your daily values may be higher or lewer depending on you calorie needs.

	Calories	2500	1500
Total Fat	Less Than	50g	25g
Saturated Fat	Less Than	55g	15g
Cholesterol	Less Than	35mg	15mg
Sodium	Less Than	15mg	50mg
Total Carbohydrate		300g	350g
Dietary Fieber	Less Than	20g	40g

Calories per gram
   Fat 7 Carbohydrate 8 Protein 6

Pack designs featured here are from www.freepik.com.

7.13 Brand Launch, Pilot Production and Pre-testing [11]

Software capability nowadays enables design finish to be pilot produced and minutely pretested

Increasing costs of new product launch, widely dispersed markets and high risks of failure have charted the path of developing and pretesting modern pack design. Visualizing the subtleties of print finish, such as foil blocking or embossing, without actually doing print trials was a major handicap for pack designers in earlier times. Nowadays, this task has been greatly simplified by the development and availability of professional design software such as Esko's visualizer. This software has the capability to simulate the printing and finishing operations one by one, in the correct order, and on the right substrate even for complex luxury packs. It allows the designer to see the packaging as it would look when produced and even place it virtually on the shop shelf amidst its competing brands to measure shelf appeal.

7.13.1 Finish or Finesse [12]

Clearly, it is imperative for brands to create champion level packaging. In order to do this, high performing business organizations must not only learn to get into the minds of their customers but also build champion level teams within their organizations. The technology options exist and are constantly evolving. The consumer and her tastes are evolving too. A number of virtual tools to study this vital interface between finishing technology and the consumer's mind are now also readily available.

We must be prepared to cross that vital finish line as champions—*with finesse!*

7.14 Consumer Connect and Digitalization

Creating customer delight by customization

Creating a unique user experience of the product with the help of packaging design and the point of sale is now a major objective of brand managers. Rapid development of 3D visualization software; color management digital print capability on a variety of substrates along with plotter-cutter mock-up making and 3D printing have made it possible to produce low cost, short-run packaging of just about any kind. This has made it feasible to be able to offer customized packaging design for a range of niche customer groups to be able to achieve even more customer connect and delight.

Customized packaging is "made-to-measure" and can be printed on-demand to include personalized graphic designs and specially designed gift boxes intended to enhance brand recognition.

All of this has been made possible over the last two decades largely because of the development of various prepress proofing systems known as the Approval Digital Imaging System. These systems allow customers to view three-dimensional mock-ups of packages, for example, in cardboard, metal (including aluminum), glass, plastic or shrink-wrapping. Approval Digital Imaging System workflows are supported by Kodak Proofing Software (KPS), Kodak (HQ-1), Prinergy, FlexRIP and EskoAtworks. Other specialized prepress software workflows for this purpose interface via Open Front End (OFE) or Approval Interface Toolkit (AIT).

Specialized companies capable of offering services of brand engagement or short and medium run packaging production have grown to occupy this space.

7.15 Challenge of Online Retail

From shelf display to customer experience

The increasing popularity of online retailing appears to be leading to a perceptible paradigm shift in the way products are packaged and branded. Traditional retail store packaging relies heavily on the quality of its "shelf appearance." On the other hand, products intended for online sale are often displayed without outer packaging (with labels only) and need peripheral brand engagement and media support to get customer attention.

Competition in the online space is severe and customer engagement depends not only on the primary packaging but extends to the full range of user experience—from the website to order placement and delivery to the receipt of the package in its transport pack. Attention to details such as transport box, filler materials and adhesive tape in this context can hugely enhance the overall user experience and help to create customer loyalty.

7.16 Supply Chain and Secondary Packaging

Include logistics considerations in the packaging design process

A modern supply chain demands speedy and efficient distribution of the packaged product. In a vast country like India the package in transit will be subject to widely varying handling, transit and temperature conditions. Statutory marking and labeling conditions to be met within the targeted jurisdictions may be varied too. It is for these reasons that the pack needs to be designed with a holistic perspective. It must take into account the available market knowledge about the supply chain conditions.

In the prevailing supply chain conditions within India, secondary packaging design must ensure:

- A case sizes that is easy to lift and carry for manual labor handling.
- Dimensional accuracy of the case to avoid product movement (and breakage) within it.
- Adequate compliance with statutory marking needs to declare product packed, batch number, MRP and quantity packed.
- Markings like Fragile, This Side Up, etc. for warehouse handling.
- Flap gumming and taping to prevent product pilferage or spillage in transit.

7.17 Visualizing the Future

Thinking beyond the box

Leading from all the above and the rapid pace of design, technology and market development in packaging it is tempting to imagine what everyday packaging will look like a decade ahead from now. Evidence of many likely developments to come is already available. Some of the notable developments that are likely to influence packaging design in the near future could be listed as below:

Printed Electronics

Printed electronic circuits and components will make it possible to create packaging could glow when touched. Smart tags would allow better shelf life management of products. Brand integrity would be ensured with the help of covert and overt printed counterfeit protection.

Full EPR Compliance

In deference to consumer interest groups as well as statutory stipulations, brands will need to move to full Extended Producer Requirement (EPR) compliance. This would ensure that packaging is designed to be reusable or recyclable or simply "edible" as is being attempted for some product categories.

Augmented Reality

In a unique fusion of the real world with the virtual world it will be increasingly possible to delight customers with AR capability printed into the label or box. With AR capability the customer will be immersed in a mix of reality and virtual experience that will be designed for maximum impact on the users senses.

7.18 Examples of Modern Packaging Design

Pack designs featured here are from www.freepik.com.

Pack designs featured here are from www.freepik.com.

References/Suggested Reading

1. https://www.packagingsouthasia.com
2. https://www.monash.edu/business/marketing/marketing-dictionary/s/second-moment-of-truth-smot
3. https://99designs.com/blog/packaging-label/6-rules-of-great-packaging-design/, https://en.wikipedia.org/wiki/Codd-neck_bottle, https://www.smithsonianmag.com/smart-news/father-canning-knew-his-process-worked-not-why-it-worked-180961960/
4. www.bloom-london.com
5. http://www.piworld.com/blog/is-packaging-finishings-future-don-piontek, http://www.getrichslowly.org/blog/2007/08/13/malcolm-gladwell-on-the-power-of-marketing/
6. https://depersico.com/package-design-hierarchy/, http://jamandco.com.au/7-step-packaging-design-process/
7. https://filestage.io/blog/graphic-design-process/
8. https://99designs.com/blog/tips/6-principles-of-visual-hierarchy/, https://www.advancedlabelsnw.com/blog/simple-visual-hierarchy-for-labels-explained
9. http://www.piworld.com/blog/is-packaging-finishings-future-don-piontek, https://www.prepressure.com/printing/processes, https://www.printingforless.com/Packaging-Box-Dimension-Guidelines.html, http://www.thedieline.com/blog/2015/1/27/packaging-101-part-i
10. http://www.logopeople.in/blog/fssai-guidelines-for-product-packaging-label-design-approval/
11. https://bizongo.com/blog/5-techniques-customized-packaging/, http://www.esko.com/en/Products/Overview/studio/modules/visualizer
12. https://en.wikipedia.org/wiki/Approval_proofer, https://www.piworld.com/article/print-connect-customers-digital-age/

Chapter 8
Testing and Quality Evaluation of Packaging Materials and Packages

N. C. Saha

8.1 Introduction

In general, the term, "Quality" is normally understood as an index or measurement. It is the degree of excellence, a degree of conformation to standard. Quality is the distinctive inherent features, property and virtue. In other words, it is the totality of features and characteristics of a product or service that bear on its ability to satisfy stated or implied needs. In short, quality is the fitness for the purpose. According to dictionary meaning, the term "Quality" could be explained as follows [1]:

- It is a peculiar or essential Character.
- It is a distinctive, inherent features, property and virtue.
- It is an inherent or intrinsic excellence of character or type: superiority in kind.

But "quality control" or "quality evaluation" is the operational techniques and activities that are used to fulfill requirements for quality. And the quality assurance indicates about all planned and systematic actions necessary to provide adequate confidence that a product or service will satisfy given requirements for quality.

In order to achieve the quality evaluation, "Testing" is considered as a tool for the measurement of different qualitative parameters to assess the performance and ensure the quality of packaging materials and packages. The important packaging materials and packages like paper and paper board, corrugated fiber board boxes, plastic films and laminates, plastic containers, metal Containers, glass Containers, Composite containers, plastic multisqueezable tube, plastic pouches are normally used for food products. The testing and quality evaluation of these materials are discussed below:

8.2 Paper and Paper Board

Conditioning: Paper and all types of paper board including corrugated board are being hygroscopic in nature, these samples are required to be conditioned under a specific condition in order to have equilibrium moisture content in the sample. As per Indian Standard IS:1060 (part-1), the samples are conditioned by exposing the samples at 27 ± 1 °C and $65 \pm 2\%$ RH for 24 h. The quality of paper and paper boards are assessed by considering the following types of tests:

8.2.1 Physical tests
8.2.2 Mechanical tests
8.2.3 Optical tests

8.2.1 Physical Tests

(i) **Subsatnce/Grammage/Basis weight** (gm/sq.mt) Standard: IS: 1060 (Part-1)

The test specimen is cut to specified size and then weighed. The weight divided by area gives grammage of each sample. The average of these values will grammage of paper [2, 3] (Fig. 8.1).
Calculation:

$$\text{Substance in grammage per meter square} = \frac{10,000 \times M}{A \times B}$$

Fig. 8.1 Weighing balance and template

8.2 Paper and Paper Board

where, A = length in cm of test specimen.

B = Width in cm of test specimen.

M = Weight in gm of test Specimen.

(ii) **Thickness** (mm): (Reference test method is TAPPI T 410) Caliper is the measure of the thickness of a sheet of paper or paper board. Number of sample of specified size is cut and their thickness is measured by using calibrated micro-meter. The average reading is taken as the caliper of paper of board [3] (Fig. 8.2).

(iii) **Density** (gm/cubic centimeter): The density of paper is its specific gravity of paper which is most important property of paper. Density is related to the porosity, rigidity, hardness and strength of paper. In fact, density influences the physical and optical property of paper except weight. Density or specific gravity tends to increase with basis weight. The density of paper is controlled by the type of fiber, amount of calendaring rosins, fillers etc. which affect the density of paper.

$$\text{Density(cc)} = \frac{\text{Substance of paper gms/mt}^2}{\text{Thickness of paper(mm)}}$$

(iv) **Bulk:** Ratio of the volume of paper measured at standard atmospheric conditions to the volume of an equal weight of water at 4 °C and is calculated from the [2]

Fig. 8.2 Digital micrometer

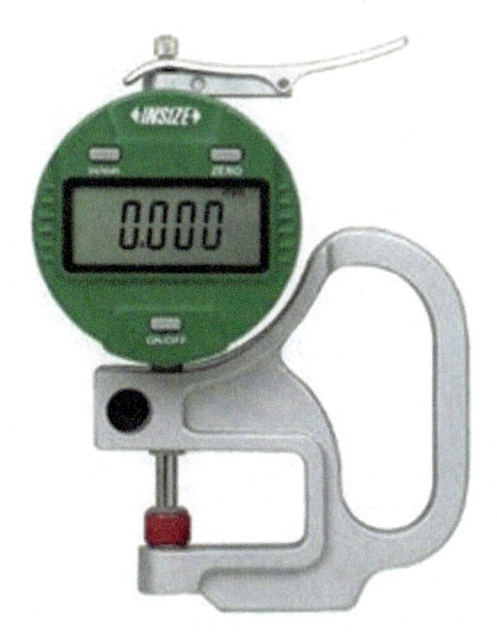

$$\text{Bulk} = \frac{\text{Thickness in micron}}{\text{Substance in}\left(\text{gms/mt}^2\right)}$$

8.2.2 Mechanical Tests

(i) **Bursting Strength**: Bursting strength is the resistance put up by a circular sheet of paper to rupture when pressure is applied from one side. It depends on the tensile strength and extendibility of paper. Bursting strength is commonly referred as Mullen.

The test is carried out according to IS: 1060 (Part-1). The test specimen is held between the clamps on a BS tester. It is then subjected to an increasing pressure by a rubber diaphragm. The diaphragm is expanded by hydraulic pressure at a controlled rate until the test specimen ruptures. The pressure reading at that point is recorded as bursting strength. The test should be performed on both the sides of the paper; top side as well as on the bottom side of the paper. It is expressed as kg per square centimeter [2].

(ii) **Burst factor**: This is conceptually a measure of the efficiency of the grammage of the paper in giving the paper its bursting Strength or the bursting strength per unit grammage. This can be understood by the fact that some papers have got long fibres and a lot of overlap of fibres, as they have high tensile strength even though grammage may be low and vice-versa. The calculation is as follows [2]:

$$\text{Burst factor(BF)} = (\text{BS} \times \text{gsm})/1000$$

(iii) **Tensile Strength**: The limiting resistance of a test piece of paper or board submitted to a breaking force applied to each of its ends under the conditions defined in the standard method of test. The tensile strength is generally expressed as breaking length where two ends of the strip of test specimen are clamped in the jaws and allowed to pull each other under tension in a specific speed till the paper breaks. Standard: IS: 1060 (part-1):1966 [2] (Fig. 8.3).

(iv) **Breaking Length**: The calculated limiting length of a strip of paper or board of uniform width, beyond which, if such a strip suspended by one end, it would break by its own weight [2].

$$\text{Breaking Length (in meter)} = \frac{\text{Tensile strength in kg cm}}{\text{Substance in gms per square metre}} \times 100{,}000$$

(v) **Folding Endurance**: The number of double folds in opposite directions, at the same place, which paper under specified tension, can stand up to the point when it ruptures. This test give information about certain properties of paper,

8.2 Paper and Paper Board

Fig. 8.3 Universal testing machine

such durability which cannot be obtained by other tests. A strip of continuously folded till it breaks, the number of double folds giving the folding endurance. Report the average, maximum and minimum of the number of double folds that the test pieces are able to sustain up to rupturing point in each direction separately. Standard: IS: 1060 (Part-1):1966 [2] (Fig. 8.4).

(vi) **Tearing resistance**: The average force required to tear a specimen of paper after an initial tear. The tearing resistance is usually greater in the cross direction than in machine direction. Hold outer tongues of the test piece in a fixed clamp and the centre tongue in the middle clamp. Release the pendulum and note the load necessary to continue the tear. This test can be carried out in a single piece of paper or a pack of two or more test piece so adjusted that the reading is not less than 25% and not more than 75% of the capacity of the instrument [2] (Fig. 8.5).

The tearing resistance shall be tested separately for machine and cross direction. Report the average, maximum and minimum of the readings in each direction separately and state the number of test pieces used for each direction. Standard: IS: 1060 (Part-1):1966 [2].

Fig. 8.4 Folding endurance tester

Fig. 8.5 Tearing resistance tester

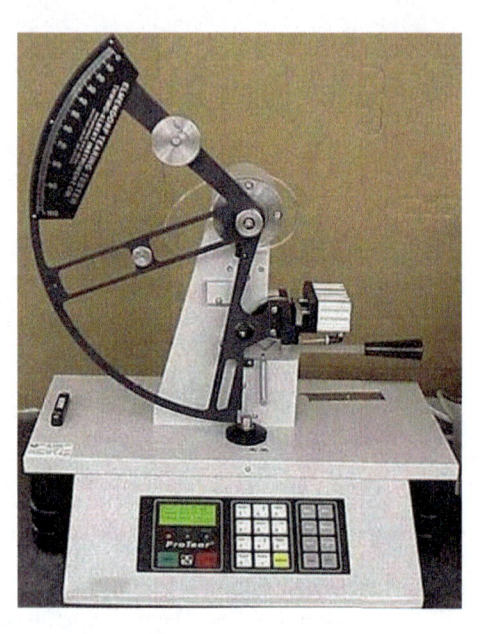

(vii) **Smoothness**: Smoothness is concerned with mechanical perfection of paper surface. Smoothness means freedom from lumps, wire and felt marks, fuzziness, foreign matter, inter fiber voids, crush, cockles, mechanical damage and incomprehensibility and other gross surface in perfections. It is measured in a microscope by using a micrometer focusing adjustments by which the microscope is focused on a small area of the paper surface and the micrometer reading is noted. The paper is further moved to a small determined distance, the microscope is focused and a new reading is taken [4].
Standard: TAPPI T 538. It is expressed as ml/s.

8.2 Paper and Paper Board

(viii) **Porosity**: Paper is highly porous material. The porosity of paper is measured in terms of time required for 100 cc of air under a standard pressure to pass through a 1.000 square inch area of the paper. Gurly Hill SPS tester is mostly used for the measurement of porosity. The test is carried out according to TAPPI T 460. Measurement of porosity is a control test in paper making since it indicates variation in structure and density. It is important factor for printing purpose. It is expressed as ml/100 cc [5] (Fig. 8.6).

(ix) **Stiffness**: This test is suitable for paper boards. The representative test specimen of 5 test pieces of having size of 15 mm wide and 50 mm long is cut and the both the ends of each piece of test specimen are clamped on the stiffness tester. The each sample is allowed to bend at an angle of 15° in both machine and cross direction. The force required to bend the sample is measured and expressed in Taber. This test is carried out according to Indian Standard IS:1060 (Part III);1969 The test value is also expressed as Stiffness factor by using this formula [6]:

$$\text{Stiffness Factor} = \frac{\text{Average stiffness}}{10\, t\, 3}$$

where, t = thickness in mm.

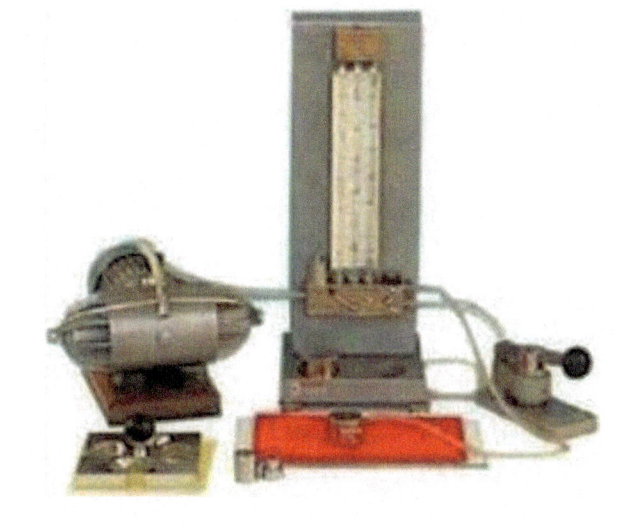

(x) **Ring crush test for kraft liner** (RCT): The RCT of paper is defined as the maximum vertically applied compressive force on the rim of a circular ring of the paper without the paper buckling. RCT is measured as per TAPPI T 882. The test strip is formed into a ring when it is inserted into a specially designed holder. A top down load is applied on this strip of paper till it buckles. The reading at this point gives the RCT of the paper. It is suggested to measure RCT-MD and RCT-CD of the paper. When the specimen is cut along the machine

Fig. 8.6 Smoothness and porosity tester

Fig. 8.7 Ring crush tester

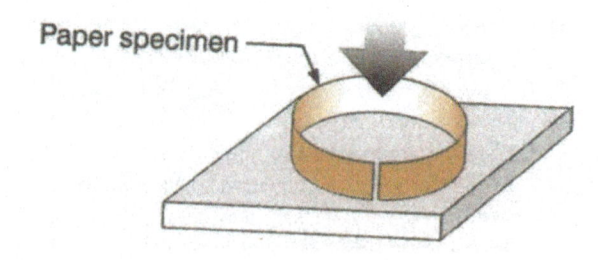

direction of the paper, it is known as CD specimen. This is because in the specimen, cross direction fibers are coming under load. Similarly, when the specimen is cut along the cross direction, it is called as MD specimen. The edges of the test specimen should be clearly cut, without tear or frays. For RCT test, a minimum of 10 specimen of each test unit of each sample. The test is extremely sensitive to moisture content, even slightly wet fingers can alter the readings. It is recommended to wipe hands or put plastic gloves [7] (Fig. 8.7).

(xi) **Corrugating Medium Test (CMT)**: CMT is a measure of the crushing resistance of a laboratory fluted strip of corrugating medium. Unit is Newton (N). It is also called as Concora Test. It measures the ability of recycled corrugating medium to keep two liners separated. The test is carried out according to TAPPI T 809. Single facer of a flute is prepared. Fluting medium of a specific size is then separated from the single facer. The specimen is kept on lower platen of compression tester keeping the flutes facing up and upper plate is allowed to approach lower plate at a constant speed till it crushes the fluting medium. The force at which the flute is crushed, is CMT value and can be read on the indicator [8].

8.2.3 Chemical Tests

(i) **Water absorptiveness of paper and paper board or Cobb value**

Water absorbency or Cobb test is a characteristic pertaining to the sheet's ability to resist water penetration and absorption.

The test is carried out according to TAPPI T 441. A sample of test specimen is weighted and then the specimen is placed on a rubber sheet over a metal plate and then clamped by means of a metal ring. Pour 100 ml of water and the paper to absorb for a period of 45 s and then remove the water. Wipe out the specimen for remaining water and weigh the test specimen within a period of 15 s to complete the test within 60 s or one minute.

This test is also carried for 30 s. The absorption of water by the test specimen is calculated by comparing the reading before and after absorption of water. It is expressed as gm per square meter [9].

Fig. 8.8 Hot air oven

(ii) **Moisture Content**

The conditioned specimen is weighed and heated to a constant weight to expel moisture. The difference between two weighing gives the moisture content. This method applied to all paper, paperboard and paper products except the containing significant quantities of water content that are volatile at 103° ± 2 °C. The test specimen for moisture Content shall not be conditioned.

About 2 gm of paper sample is cut into small pieces and placed on a wide mouth glass bottle with a glass cover in triplicate and then samples are exposed to a air oven by removing the cover and maintained the temperature of 103 ± 2 °C for a period of 2 h. Remove the glass bottle, cool it and the weigh. Expose it again for 15 min and repeat the process till the consistent value is obtained. Standard: IS: 1060 (Part 1):1966 [2] (Fig. 8.8).

$$\text{Moisture content, percent by weight} = \frac{W \times w}{W} \times 100$$

where

W = Original weight of the conditioned specimen before drying and

w = Weight of the specimen after drying.

(iii) **pH Value**: The following method is suitable for the regular run of commercial and industrial papers, the water extracts of which are normally acidic and usually buffered. It is not adequate for determining the pH of unbuffered and neutral papers, such as insulating papers, which require a more accurate method for eliminating error due to the absorption of carbon dioxide by the water extract during its preparation and testing. Electro metric PH meter is used to measure the pH value. Standard IS: 1060:1969 (Part-III) [6]

(iv) **Ash Content**: The percentage of ash is, therefore, taken to be an index of added mineral matter or loading. A known weight of the specimen is burnt, ash weighed and the percentage calculated [10].

Tear about I g of the specimen into small shreds and place in a previously weighed crucible. Again weigh. Heat carefully over a Bunsen burner to ensure that the paper burns quietly until it is completely charred. Transfer the crucible into a muffle furnace at $800 \pm 25°0$ and heat until all the carbonaceous matter is burnt off. Cool the crucible in a desiccator, weigh and repeat the operation till the weight is constant. Standard: TAPPI T 413.

Calculation

Calculate the ash as percentage on the original weight of the material as follows:

$$\text{Ash, percent by weight} = 100 \times \frac{w - X}{W - X}$$

where,

w = weight in g of the crucible and the ash,

X = weight in g of the crucible, and

W = weight in g of the crucible and the material.

8.2.4 Optical Tests

(i) **Gloss**: This test is performed to find the degree of specular reflectance of papers like super calendared, imitation art and art. Determine the reflectance of each specimen by the method appropriate to the instrument used, utilizing the table supplied with the instrument. Report the mean, minimum and maximum results for the top-side and wire-side separately. Standard IS: 1060-1960 (Part-II) [11].

(ii) **Opacity**: The method of test described below covers the procedure for measuring the opacity for all kinds of paper and paper products by determining the apparent light reflectance. The apparatus shall be capable of measuring the apparent light reflectance as prescribed in this method. It may measure the values separately or give directly the ratio of the apparent reflectances. Place the test piece first over the standard white backing. Then over the standard black backing and then measure the apparent reflectance of the light.

The ratio of reflectance over black backing to that over white backing, expressed as a percentage is the contrast ratio. Calculate the average contrast ratio from determinations on both sides of each test piece. Standard IS: 1060-1960 (Part-II) [11].

8.2 Paper and Paper Board

Fig. 8.9 Brightness tester

(iii) **Brightness**: This method is intended for determining the 45°–0° directional reflectance of uncoloured paper for blue light. In the paper trade, this quantity is known as 'brightness'. The test is carried out according to Indian standard IS:1060-1960 (Part II) where the sample is exposed to Tungsten lamp of the brightness tester and adjust the angle of light in such manner that the light will on the surface at that specified angle so that angle of incidence light and angle of reflectance light will same. The angle of reflectance light in percentage is measured. Standard IS:1060-1960 (Part-II) [11] (Fig. 8.9).

8.3 Corrugated Fibre Board

The important tests to be carried out for Corrugated fibre board are given below [12]:

(i) **Board grammage by ply separation method**

Cut at least 10 samples of 10 cm × 10 cm from the board to be tested and condition to equilibrium at 27 °C +− 1 °C for 24 h. Separate the components by soaking in warm water with flutes vertical until easily separated. Carefully remove the adhesive as completely as possible without disturbing the fibres, may be by gentle rubbing. The individual layers of separated paper sheets are dried in oven at 105 °C for 1 h. Recondition at 27 °C ± 1 °C for 24 h. Weigh the each specimen to determine the grammage of the liner. To obtain gm of the fluting medium, divide the weight of the specimen (in gm) by take up factor of that flute [13].

(ii) **Caliper**

The thickness of corrugated board is the distance in millimetres measured between the two parallel contact plates of a micrometer between which the test specimen is subjected to a pressure of 20 kPa. It is measured in millimeter [13].

(iii) **Bursting strength**

The CFB sample is tested by using Mullen Bursting equipment to measure the force required to burst the fibres of 3 ply or 5 ply or 7 ply board. This would indicate about the quality of paper is used for the fabrication of CFB. It is expressed as kg/cm^2 [13] (Fig. 8.10).

(iv) **Edgewise compression Test** (ECT)

ECT of a corrugated board is defined as the maximum vertically applied compressive force along the edge of the board without board buckling.

The test is carried out according to TAPPI T 811. The sample is kept under horizontal plates with the flutes vertical. The plates are then pressed down with a constant speed of 10–15 mm/min so that the load on the edge of the board gradually increases. The load at which the board buckles gives the ECT. SI unit is kN/m [14].

(v) **Flat crush Test (FCT)**

The FCT test is a measure of the resistance of the flutes in corrugated board to a crushing force applied perpendicular to the surface of the board. This test is satisfactory for single-faced or single wall corrugated board, not for double wall or triple wall corrugated board because of lateral motion of the central facing or facings. This test is carried out according to TAPPI T 825. In this method, after cutting the test specimen in circular shape of 50 mm diameter is placed on the lower platen of the instrument with the corrugations parallel to the front of the machine fail. Failure is defined as the maximum load sustained before complete collapse. Its unit is kg/sq.cm [15] (Fig. 8.11).

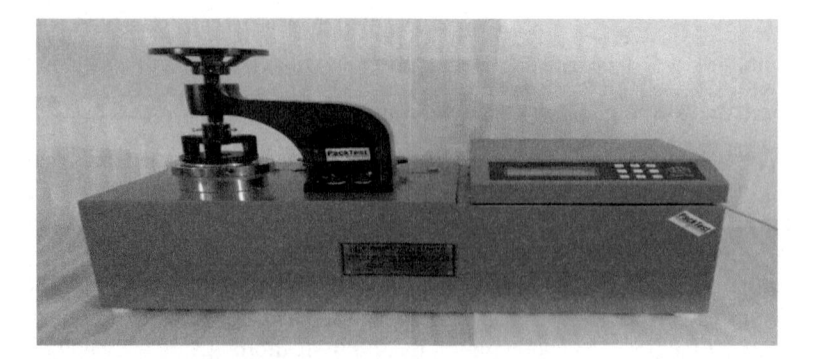

Fig. 8.10 Bursting strength tester

8.3 Corrugated Fibre Board

Fig. 8.11 Flat crush tester

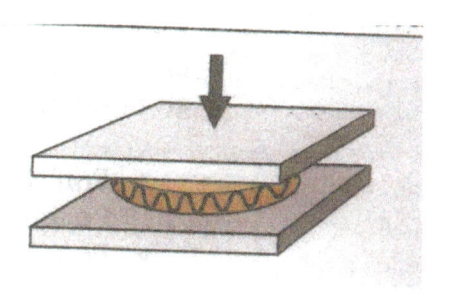

(vi) Puncture Resistance

Puncture is a highly visible effect but it is extremely difficult to quantify. In some respects, it is an indirect measure of other hazards. Puncture can result from a box falling on to a corner of another box or from the penetration of nails or hooks from a surface or being improperly handled in equipments like froklifts, hooks, gangropes etc. The purpose of puncture resistance is a measure of the energy needed to punch through a material.

The test is carried out according to TAPPI T 803. The machine works on the simple principles of "free fall under the gravitational pull". A pendulum with a pyramidical shaped head is released from a certain height. The free falling pendulum acquires kinetic energy and the head punctures the board. The required energy is recorded and it is expressed as Ounce inch/tear inch or kg-cm [16].

8.4 Corrugated Fibre Board Box

The following important test are carried out to assess the performance of Corrugated Fibre board Boxes.

8.4.1 Box Compression Strength (BCT)

Significance: This test is conducted to measure the ability of a corrugated fibre board box to take top-down loads. It also indicates to measure the stackability of the box and determines how much load can be stacked upon the box without the walls of the box buckling (Fig. 8.12).

Test Method: The testing is carried out by following the test method given in TAPPI T 804 or IS: 7028 (Part-1)-1987 There are two types of variations in the test [17].

(a) Dynamic crush resistance: The box is placed in a press between two parallel plates where the pressure is applied to the box at right angles to its flaps. The

Fig. 8.12 Compression strength tester of Corrugated fiber board box

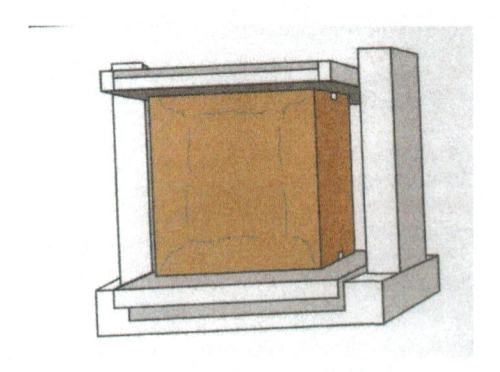

speed of the plates remain constant and fixed. This test is normally continued until the first sign of damage to the box appears. The crush resistance is expressed in kg.

(b) Static Crush resistance: Under this test, a plate bearing a given load is placed on the box. The test is carried out for a fixed period of time. The result is compared with the initial state of the box and the differential is noted.

8.4.2 Drop Test

Significance: This test is conducted with the aim to determine the ability of the package to withstand during rough handling, the degree of protection offered to the contents by the package and also compare the different types of packaging for the same product.

Test Method: This test is carried out by following the test method given in TAPPI T 802 or IS: 7028 (Part-4)-1987. The height of the drop test will depend upon the purpose of the test. In some case, the height and number of drops are prescribed in the specification. The drop height is specified by considering the gross weight of the box. The details of drop height and number of drops are given below:

Weight of the packaged product (kg)	Drop height (mm)
<9.1	762
9.1–18.1	610
18.1–27.2	457
27.2–45.2	305

In general, 10 sequential drops are carried out on an individual filled corrugated box and the box is dropped upon its corner adjacent to the joint of box, three edges adjacent to that corner and then all six faces of the box. At the end of the test, the observations are taken on the individual box for any kind of damage, press etc. on the outer surfaces and then after opening the box, internal observations are also taken for any kind of damages of the products which are packed into the box [18].

8.4.3 Vibration Test

Significance: This test is carried out to simulate the mechanical hazards which would occur on the package during transportation [19].

Test Method: The test is conducted as per TAPPI T 802 or IS: 7028 (Part-4)-1987. A filled corrugated box is placed on a vibration table and then adjusted the amplitude and frequency of vibration. The box is then vibrated by considering the following objectives:

(a) Until a determined level of damage has been reached.
(b) For a particular period of time, after which the effects are noted.

After the test is conducted for a specific period of time like 30 min or 60 min, the following observations are taken:

(a) any kind of damages occur on the box.
(b) any kind of damages of the product inside the box.

8.4.4 Stack Load Test

Significance: This test is carried out to simulate the Static load impact on the filled box during Storage at godown.

Test Method: This test is carried out by following the test method as per IS: 7028 (Part-I)-1987. The total load to be superimposed over the individual box is to be calculated by considering the stack height in the warehouse. A filled corrugated box containing the products inside the package is taken and then a metallic plate of having dimension more than the top surface of the box is placed over the box. A number of square shaped concrete blocks of having equivalent calculated load is superimposed over the box and then kept it for 24 h. The arrangement is to be made in such a manner that concrete blocks should not be fallen. After the specified period of time, the observations are taken about any kind of damages like crush or bulging of box etc. In addition, the observations are also to be taken from the products packed inside the package about any kind of damages due to the impact of static load [20].

8.4.5 Inclined Impact Test

Significance: This test is carried out to simulate with shunting in railways for bulk packages. The main objective of this test is to determine the ability of the package to stand up to rough handling, the degree of protection offered to the contents by the packages due to mechanical impact on the package during transportation [21].

Fig. 8.13 Inclined impact tester

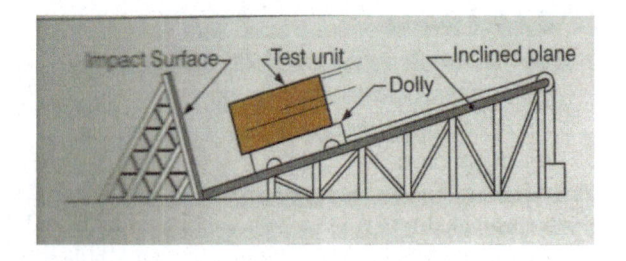

Test method: The test is carried out by following the Test method as per IS: 7028 (Part-3)-1987 TAPPI T 801. A full closed box is placed on a trolley. This trolley is supported on rails fixed to a tilted surface. The trolley is then released and crushes into a wall which is angles to the tilted surface. The impact may be either on a side or edges or at the corner of the box. After the test, the observations are taken to check the filled box for any kind of damages to the box as well as the product packed inside [22] (Fig. 8.13).

The distance travelled or the speed at the moment of impact is noted. The test is repeated for certain number of times until there is a sign of damage.

8.4.6 Rolling Test

Significance: This test is carried out to simulate the mechanical impact on the packages of having more than 45 kg which are normally handled by rolling on the floor.

Test method: This test is carried out as per the test method given in IS: 7028 (Part-2)-1987. A filled box of having weight more than 45 kg is tipped over on its side by rolling for a distance of 1000 cm from a point A to B and then repeat the same from B to A. However, the test is performed according to the shape and the centre of gravity of the package [23].

8.4.7 Water Spray Test

Significance: This test is carried out to simulate the condition while the boxes are exposed to rain water. The test is conducted to determine to what extent the package protects its components from the rain and also to assess the effect of rain on the performance of the package [24].

Test Method: The test is conducted by following the test method given in ISO: 2875 TAPPI: T805 A box is filled with dry sand and closed by means of pressure sensitive

8.4 Corrugated Fibre Board Box

plastic tapes at bottom and top of the box in such a manner that there should not be any gaps in between the flaps of the box. The filled box is then placed in a rain chamber on a flat form which is fixed over a plastic container. The water is sprayed over the box from a nozzles at a standard rate and the water not absorbed by the box is collected in containers which is fixed below the platform. In general, the outer liner of the box should have either water proof coating or very low cobb value to assess the performance of water absorption due to rain water [25] (Table 8.1).

8.5 Plastic Films and Laminates

The quality of Plastic films and laminates are assessed by considering the different types of tests to assess the following properties:

8.5.1 Physical Tests
8.5.2 Mechanical Tests
8.5.3 Optical Tests
8.5.4 Physico-Chemical Tests

8.5.1 Physical Tests

(i) **Thickness (mm)**

In the imperial system, film thickness is expressed in terms of mils: 1 mil = 0.001 inch. The term 'gauge' is also used to express the thickness of very thin films: 100 gauge = 1 mil.

For example, 0.00048 inch polyester film is called '48 gauge' film, in the metric system the thickness is measured in micrometers. That is, 1 mil = 25.4 m.

Test method: the test piece is cut and then clamped between the two Jaws of micrometer. The thickness of the film is seen directly translated and analog meter attached to it.

Standard: ASTM E 252-78 [26].

(ii) **Density**

Significance: This is chemical property of polymer. It is a direct function of chemical composition which depends on specific gravity of C-H(C-H) polymer that is about 1.0. Fewer hydrogen atoms in polymer more dense the material. Density is important as an indication of the film yield in inch sq./lb or cm sq./kg [27].

Principle: the density of a film sample film often measured by means of density gradient column which is mixture of two fluids different density, the proportions of which change uniformly from top to the bottom of the column. A series of calibrated glass floats covering the required density range is placed in column.

Table 8.1 A summary of different tests related to kraft paper and corrugated fibre board boxes

Packaging materials	Important tests	Related standards
(A) Kraft paper as liner materials	Grammage or substance	IS: 1060 (Part-I)-966
		TAPPI T 410
	Thickness or caliper	IS: 1060 (Part-I)
		TAPPI T 411
	Water absorbancy or cobb value	IS 4006 (Part-1)
		TAPPI T 441
	Burst index	IS: 4060 (Part-I)-1966
		IS: 2771 (Part-I)-1987
		TAPPI T 807
	Ring crush test (RCT)	TAPPI T 822
	Corrugating medium Test (CMT)	TAPPI T 809
	Tensile breaking strength	IS: 1060 (Part-I)-1966
		TAPPI T 494
	Internal moisture content	IS: 1060 (Part-I)-1966
		TAPPI T 412
(B) Adhesive	Solid content	IS: 2257-1987
	Viscosity	Cup method
	Board grammage by Ply separation method	IS: 7063 (Part-IV)-1987
		ISO: 536
		PEFCO: NO_2
		NF: Q03-001
	Caliper or thickness	IS: 1060 (Part-I)-1987
		ISO: 3034
		FEFCO: No. 3
		NF: Q03-030
	Bursting strength	IS: 1060 (Part-I)-1987
		IS: 2771 (Part-I)-1990
		TAPPI T 810
		ISO: 2759
		FEFCO: No. 4
		DIN: 53141
		NF: Q03-054
	Edge crush test	IS: 7063 (Part-II)-1990
		TAPPI: T 811/T823

(continued)

8.5 Plastic Films and Laminates

Table 8.1 (continued)

Packaging materials	Important tests	Related standards
		ISO: 3037, FFFCO: No. 8
		DIN: 53149
	Water absorption cobb value	NF: Q03-041
		IS: 4006 (Part-I)-1985 IS: 2771 (Part-I)-1990 ISO: 535 FEFCO: NO 7 TAPPI T 492/T 819 DIN: 53132 NF: Q03-035
	Types of flutes	IS: 2771 (Part-I)-1990
	Number of plys	IS: 2771 (Part-I)-1990
	Puncture Resistance	IS: 4006 (Part-2)-1987 TAPPI T 803
(C) Corrugated fibre board boxes	Internal dimension	IS: 2771 (Part-1)-1990 IS: 13228-1991
	Style of box	IS: 6481-1971
	Flap bending test	IS: 2771 (Part-I)-1990
	Compression strength	IS: 7028 (Part-6)-1987 TAPPI T 804
	Drop test	IS: 7028 (Part-4)-1987 TAPPI T 802
	Vibration test	IS: 7028 (Part-2) TAPPI T 817
	Stack load test	IS: 7028 (Part-I)-1987
	Inclined impact test	IS: 7028 (Part-3)-1987 TAPPI T 801
	Rolling test	IS: 7028 (Part-2)-1987 DIN: 55449
	Water spray test	ASTM B-113 ISO: 2875 TAPPI: T805 DIN: 55447

Procedure: film samples are cut and then immersed for 30 s. In for example ethanol or methanol. Then, they are placed in the column using tweezers, taking care not to introduce air bubbles. After 3 h measurement is made, against the scale of the heights of the floating samples and of markers against their respective densities and the best fitting curve drawn through the points. From this cut the densities of the film samples can be read.

(iii) **Yield Strength** (m sq./kg mm or inch sq./lb. Min)

Significance: This test tells that how much length of the film of particular thickness can be achieved for certain mass of material from amorphous or crystalline form of polymer that is, reproducibility of polymer. This is useful in prici t because different fns has different yield which depends on specific gravity (density) [28].

Formula for calculating inch²/lb is

$$\frac{27.69}{\text{Specific gravity} \times \text{Thickness}} = \text{inch}^2/\text{lb}\left(14.2\,\text{cm}^2/\text{kg}\right)$$

(iv) **Change in Linear Dimension** [29]

Significance: This is another technique that utilizes a property specific to plastic films which produces a tight wrap even on an uneven shaped article. The principle on which shrink wrapping is based in known as "Plastic Memory." In other words, a film that has been stretched during manufacture (at temperature above its softening point) and then cooled to "freeze-in", the consequent orientation of the molecule tends to return to its original dimension when reheated.

Application: Wrapping over a complete pallet for the Collation of cans, bottles or cartons. LDPE or LLDPE films are cured most widely because they are cheap, tough and waterproof.

8.5.2 Mechanical Tests

(i) **Tensile Strength and Elongation**

Tensile strength is the maximum stress that a material can sustain. Elongation is the elongation at breaking point and is important as a measure of films ability to stretch [30].

Principle: Tensile strength most commonly uses scripts out of which one grip grip is fixed and the other is my allowed to move at a constant speed the mean value of the speed of separation such that the initial strain rate on the test specimen per millimetre of the test specimen for minute. So, if one uses 50 mm GL, the speed should be 500 mm per minute the load range of the machine the breaking load of that pieces fall between 15 and 85%.

Procedure: the gauge length of the specimens shall be 50 plus or minus 1 mm and the width should be 10–25 mm from the out of 5 test pieces from the sample in the lengthwise direction and 5:00 in the crosswise direction the total length of the specimen should be at least 50 mm longer than the GL thickness of the fire should be measured by using micrometer condition the test pieces for not less than 1 h at a temperature 27 plus or minus 2 degree Celsius and 65+ or −5% RH. Then clamp their ends in the grips separated by 50 mm. Start the machine at a preadjusted speed or 500 mm/min and note the load and alongation at break.

Calculation and Report: the Tensile strength at break shall be recorded in MN/m sq. (kg/cm sq.) from the original area of cross section. The mean value of the five results shall be expressed for the lengthwise and crosswise sample. Elongation at break shall be expressed as the percentage of the original length between reference lines. The mean of the Five results shall be expressed in lengthwise and crosswise.

(ii) **Seal Strength and Heat Sealibility** Principle and procedure is Same as tensile strength testing machine. The test is carried out according to ASTM F 88-68 [31].

Significance: This test is used to determine ability of the plastics films getting stuck together in the presence of heat and exposure time. Where, there will not be true fusion weld the position of the package product will find their way out of the package through the seal this method is used to prevent leaking of the products from the packet eliminate the chances of seal wrinkles by keeping packaging material under tension in two directions while they are being sealed (Fig. 8.14).

The heat sealability of the packaging film is a very important property when considering its use in on wrapping or bag making equipment, and the integrity of the seal is important to the ultimate package.

(iii) **Tear Strength (gmF)**

Significance: the usual tests for measuring tear strength actually measure the energy absorbed by test specimen in propagating tear that has already been initiated by cutting a small nick in the test pieces by a razor blade [32].

Summary: Tear strength requirement mat be high or low according to the particular use. Where tear tapes are to be incorporated ease of tear propagation in one direction is avoided in shipping sacks which might be punctured during transit. In all tear test, test pieces should be cut from both machine and transverse directions because tear strengths can vary widely according to the direction of the tear.

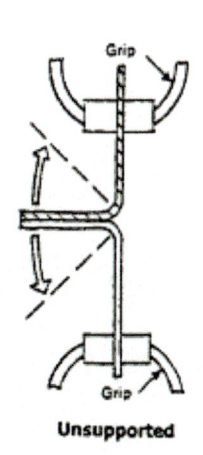

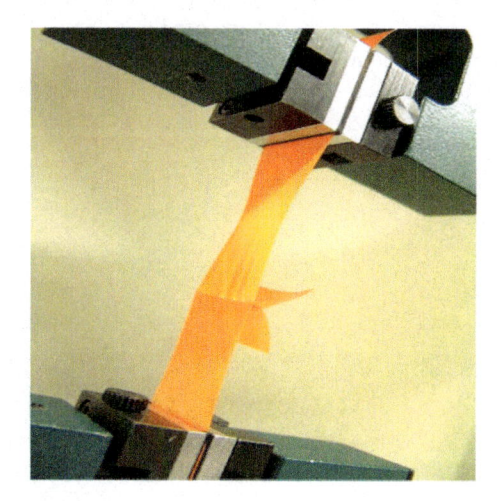

Fig. 8.14 Heat seal strength

(iv) **Impact Strength** (Dart)

Significance: An impact strength of a film is a measure of its ability to withstand shock loading.

Summary: One method of measuring this is falling Dart method. In one variation of this, the dart is dropped at a constant height and its weight is adjusted from a minimum (just too light to rupture in two test pieces) upto maximum (just heavy enough to rupture all test pieces). The weight at which 50% of the test pieces rupture multiplied by the height through which the dart falls is taken to be impact strength of the film [33].

(v) **Coefficient of Friction**

Significance: The frictional property of film are important when the film comes into close contact with the metal surface such as 'plough' formers, guides etc. which are used to perform tubes or to fold the wrapping material around the product. High slip (low coefficient of friction) is desirable in this situation. Friction is also a factor when film passes over free running roller [34].

A co-efficient of friction that is too low could level to slippage instead of a positive drive of the rollers; a coefficient that is too high could lead to wrap around of rollers that should break.

Test method: one method of Measuring the coefficient of friction utilizes a sled made from metal block wrapped around with a sheet of foam rubber. The film sample is taped to the rubber in order to give the sled base a smooth wrinkle free surface if the film/metal Coefficient is required this sled is placed on a smooth metal table. For measuring film/film coefficient another sample of film is taped to the metal table. Either the Sled or the table can be motor driven and the horizontal force of the Sled is measured using a spring balance or strain gauge. The original deflection of the load measuring device is noted and used to calculate the static coefficient of friction by dividing the weight on the Sled. The average deflection is also noted and used to calculate the dynamic sliding coefficient of friction [34] (Fig. 8.15).

Fig. 8.15 Coefficient of friction test

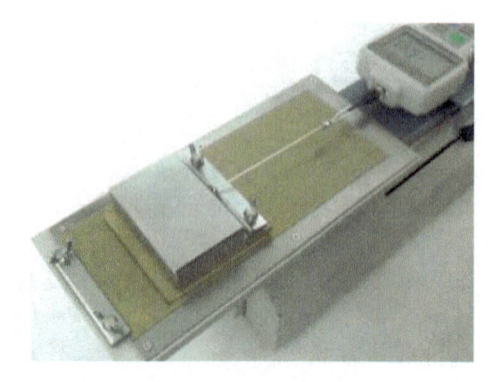

8.5 Plastic Films and Laminates

(vi) **Blocking**

Significance: Blocking is the tendency of the two adjacent layers of film to stick together particularly when left under pressure for some time as when film are stuck in cut sheets.

Testing method: Degree of blocking is determined by the force required to separate the two layers of blocked film when the force is applied perpendicular to the surface of the film. Blocking can also be assessed by measuring the force required per inch of film to draw a 1 by 4 inch diameter rod perpendicular to its Axis at a rate of 5 inch per minute between the adhering films there by separating them. This is determined by observing the adhesion/Cohesion between the contiguous forces of the specimen when they have been subjected to a pressure of 35 g per centimetre cube for 24 h [35].

(vii) **Melt Flow index**

Significance: This test is used to measure the flow ability of the plastics material when the material is subjected to extrusion through extruder at specific pressure at 190 °C+ or −0.5 °C.

Test method: The M. F. I tester is kept on the knob is set to 190°+ or −0.5 °C which one can verify by pressing the push button arrangement below the digital meter. After some time when the extruder attains that specified temperature which can be seen on the digital meter, the sample is cut in pieces and weighed (if not available in granular form) and poured into the hopper. Then the pressure of 2 kg or 5 kg is applied from the top by inserting rod with load in the hopper. The timer is then made on the extruded material coming out of the extruder collected on the glass plate at a certain interval and weight on digital balance. Then M. F. I is calculated and expressed as gm/10 min [36].

(viii) **Rubproofness**

Significance: This test is usually done to determine the loss prints from paper or paperboard when rubbed against each other at a certain pressure and revolution. This is usually desired in wrappers like soap packaging.

Test method: In this method when the surfaces come in contact their strike each other of which Appeal of the wrapper get lost losing some amount of pigments from the wrapper which has been deposited Dyeing printing or COATING. This test method involves one large disc and one small disc both discs are then mounted with the specimen which is already weighed and allowed to move a certain cycles and pressure already specified by the customer. The difference between the weight of the original sample and rubbed sample is calculated and reported as weight loss in grams. This is also expressed as abrasion loss and is defined as the number of milligrams of material removed per thousand revolutions calculated by multiplying 1000 and materials removed in milligram divided by a number of revolution of the turntable [4].

(ix) **Coefficient of Linear Thermal Expansion** [37]

Significance: Plastics are extremely sensitive to change in temperatures. Crystallinity also has a number of important effects upon the thermal properties of polymer. Molecular orientation has significant effect on the thermal property. The orientation has significant effect on the thermal property orientation tends to decrease the dimensional stability at higher temperature. Molecular weight of the polymer affects the low temperature flexibility and low temperature brittleness. So, thermal property plays important role in plastics.

Test method: This test method requires the use of fused Quartz tube dilatometer for measuring change in length (dial gorge LVDT) and liquid bath to control temperature. The Test is conducted by mounting a preconditioned specimen usually between 2 and 5 inches long into the dilatometer. The dilatometer along the measuring device is then placed below the liquid level of the bath. The temperature of both is varied as specified. The change in length is recorded.

The coefficient of linear thermal expansion is calculated as follows:

Change in length of the specimen due to heating or cooling

$$X = \frac{\Delta L}{L0 \Delta T}$$

X = Coefficient of linear thermal expansion 1 °C.

ΔL = Change in length of the specimen due to heating or cooling.

$L0$ = Length of specimen in room temperature.

ΔT = temperature difference in degree Celsius over which the change in length of the specimen is desired.

(x) **Change in Linear Dimension and Shrinkage (%)**

Test Method: ASTM E 831-19.

Significance: This is another technique that utilizes a property that is specific to plastic films which produces or tight wrap even on an awkwardly shaped article. The principle on which shrink wrapping is based is known as 'plastic memory'.

Film that has been stretched during manufacturing (at a temperature above is softening point) and then cooled to 'freeze in' the consequent orientation of the molecule tends to return to its original dimension when reheated.

Application: the collection of cans, bottles or cartons to the wrapping of whole palette loads. LDPE and LLDPE films are cured most widely because they are cheap tough and WATERPROOF. Where clear and more sparkling film is required then PP or PVC are the more likely choices. Heating is usually carried out by means of hot air tunnels.

8.5.3 Optical Tests

(i) **Gloss**: The specular gloss is defined as the luminous reflectance factor of a specimen at the specular direction. This method has been developed to correlate the visual observation of surface shininess made at roughly corresponding angles. The light beam is directed towards its specimen at a specified angle and the light reflected by the specimen is measured. The test is carried out at an angle of 20°, 60°, and 85°. The method is followed as per ASTM D 523 [38].

(ii) **Transparency**: It is defined as the ratio of transmitted light to the incident light and it is generally expressed in percentage.

(iii) **Haze**: It is the generally accepted that if the amount of light is deviated more than 2.5° from the incident beam, the light flux is considered to be haze and it is reported in percentage.

8.5.4 Phsico-chemical Properties [39]

(i) **Water Vapour Transmission rate (WVTR)**

Significance: One of the prime function of the packaging film is to act as barrier to gases and vapours. Cookies and snack items for example need to be kept dry, conversely cigarettes have to be protected from moisture. Many items particularly fatty foods have to be protected from oxygen pick up. The measure of permeability is therefore, important.

Summary: the standard tests for water vapour permeability consists of wax sealing test specimen over months of metal dishes cinataununv a desiccant, the dishes are weighed initially and then placed in a temperature and humidity control cabinet. Weighing are carried out at regular intervals and weight gains calculated. Permeability is calculated as weight if water vapour per unit area per unit time at given temperature and humidity [40].

(ii) **Oxygen Transmission Rate (OTR)**

Measure of gas-permeability is carried out under controlled temperature and relative humidity. The film is used as a position between. A test cell and an evacuated monomer. The pressure across the film is usually 1 atm (101.3 kPa) as the gas passes through the film test piece, the Mercury in the capillary leg of the monometer is depressed. After a constant transmission rate is attained, a plot of Mercury height against time gives straight line, the slope of which can be used to calculate gas transmission rate [41].

(iii) **Overall Migration Test**

Significance: This test is used to determine the amount of food particles which has been transferred to the packaging material. This arise due to the reaction between the plastic materials and the food products. This is very useful test when product is meant for storage.

Test method: a sample of material or article is brought into contact with the simulating solvent under specified test conditions (duration and temperature) which is chosen by table of specific food products for a given condition of use. At the end of the prescribed duration, the contact liquid is evaporated and the residue is weighed [42].

Test Method: As per IS: 9845-1987.

(iv) **Specific Migration Test**

Significance: This test is carried out to assess the leaching of restricted elements from packaging materials to food products which will have harmful effect to human health.

Test Method: IS: 9845-1987—to prepare the extractant & IS: 3025 (Part-2)—for the detection of heavy metal content [43].

8.6 Plastic Containers [44]

Plastic Containers are rigid in nature. These Containers can be made either by means of blow moulding, Injection moulding, Stretch blow moulding, rotational moulding process. The quality of plastic containers are evaluated by considering the following important tests:

(a) Qualitative tests
(b) Performance Tests

(a) **Qualitative Tests**: The qualitative tests are normally carried out for the empty containers as per the test method given in IS: 2798:1998

(i) Measurement of Dimensions: A micrometer height gauge is used to measure the different dimensions of the Containers.

- Overall Height—Place the Container on a surface plate and the measure the highest point on the Container using a micrometer height gauge at two positions as follows:
  - Close to but avoiding the part line
  - At 90° to the position specified at (a)

The height is recorded as the mean of the two readings and the accuracy or measurements shall be 0.1 mm.

- Diameter—A vernier micrometer or Circumference gauge is used. The container diameter shall be ascertained by either of the micrometer or circumference gauge method. The accuracy of the measurement shall be 0.1 mm.

8.6 Plastic Containers ...

- Neck Height: A micrometer depth gauge is used to do the measurements.
- Neck and thread Dimension: Micrometer or vernier with an accuracy of measurement of 0.02 mm is taken to measure the diameter of the neck. The diameter is recorded as the mean of the two diameters at right angles.
- Measurement of wall thickness: The wall thickness of container ascertain either by micrometer or dial caliper gauge method. The sample of the container is cut into three pieces (top, middle and bottom) with a pair of scissors or hacksaw blade. The wall thickness is measured by means of micrometer or screw gauge fitted with ball point tip, at four different places in each section. The average of all readings are considered to be the wall thickness.
- In case of dial gauge, the wall thickness is measured with the help of dial caliper fitted with spherical anvils. The measurement should be with an accuracy of 0.02 mm.
- Measurement of Fill point: A sample container is filled with water to its rated capacity and determine fill point by depth micrometer measurement from top sealing surface to surface of liquid.

(ii) Weight of the Container: This is a very simple test. This test is carried out to determine the quantity of polymeric materials used to make this Containers. It is expressed in "gm"

(iii) Brimful Capacity of the Container: The plastic container is weighted to an accuracy of 0.1 gm. The water at ambient temperature i.e. $27° \pm 2$ °C is filled into the container within approximately 3 mm of brim and then top-up carefully pouring water by a dropper and then the weight the filled container to an accuracy of 0.1 gm. The difference of weightings is the mass of the water recorded in gm and result is expressed to the nearest 0.1 g. The mass of the water in gm or volume of water is measured numerically equal to the brimful capacity of the container in milliliteres.

(iv) Environmental stress Crack Resistance (ESCR): This particular test is carried out as per test method given in Indian Standard IS: 8747. The test method consists of exposing the partly filled and sealed blow-moulded Container to the action of a stress cracking agent, within the Container at an elevated temperature. The test is important to determine the effect of plastic resin on the stress-crack resistance of the Container [44, 45].

Method

About 15 Containers are filled to one-third of overflow capacity (180 ml) with stress Cracking solution and then put approximately 0.1 ml dye solution in each containers. Placed these containers in the air oven at a temperature of 60 ± 1 °C in a vertical position with the finish up. Inspect the containers for environmental Stress Crack failure hourly for the first 8 h and thereafter at least once each 24 h. Remove the containers and records are taken about the failure of containers for cracking and the exposure time.

(iv) **Verticality Test**: The sample container is filled with water to get more stability and then place on its base on the flat smooth plate having a pillar bolted to it at right angles. Adjust the "V" block mounted on the pillar in such way that the bottle is tightly fitted.

(b) **Performance Tests**: The performance tests are carried out for the filled Containers. The important performance tests are given below:

(i) **Closure Leakage Test**: The plastic containers are filled with Colored water up to nominal capacity or the material to be packed at ambient temperature and the closely tight with the closure. The containers are then placed in inverted condition on a white blotting paper without any external support for at least 30 min. After that each and every containers will be checked for any leakage which would be evident from any visible stains on the blotting paper [44].

(ii) **Air Pressure Leakage**: The test is carried out by maintain the specified pressure in the Container and detecting any leakage with water or soap solution. A rubber plug is fixed at the end of air pressure line where air compressor is used for the test. The testing pressure can be regulated by an air pressure valve and read on the pressure gauge connected to the end of the air line. The testing air pressure is regulated at 35 kPa to an accuracy of $\pm 2\%$ and then immersed the container fully or partly in water reservoir and detect any leakage by the bubbles of air escaping through the water. For large Container, soap solution is applied at various parts of joints and to check the bubbles to detect the leakage [44].

(iii) **Drop Impact Test**: The drop test is used to measure the ability of the container to withstand rough handling while in a packed condition. The sample container is filled with water upto nominal capacity at an ambient condition. Close the containers properly with the closure with the inner seal heat sealed to the mouth. The containers are dropped under free fall condition from a height of 1.2 m upto 5 kg or 5 L capacity, for containers of 10 kg or 10 L, the drop height should be 1.0 m and the drop height would be 0.5 m for the Containers of 15 kg or 15 L capacity on a hard smooth surface. The containers shall be ruptures nor there should be any kind of Leakage from the walls of the containers. Slight deshaping of the container shall be acceptable if there is no leakage of water [44].

(iv) **Hydrostaic Pressure Test**: The plastic container shall be filled with water to exclude all air and then connected to the water supply which should be connected to a tapered rubber plug to seal the mouth of the Container. A suitably modified screw cap may be used instead of rubber plug. The pressure shall be increased upto 2 kg/cm^2 or as specified and then hold it for 5 min. There should not any sign of rupture or leakage from the Container other than from around the mouth or localized bulging of the containers shall be considered as failure the test [44].

(v) **Handle Pull Test**: The sample container is filled with water upto the nominal capacity and then close tightly with caps. The container is then fixed in inverted condition and hang it from the handle which would be attached with weight equal to double the nominal capacity of the container through a hook. Keep this container in this position for 24 h and then observations should be taken for any of damage to the handle from the hinges [44].

(vi) **Stack Load Test**: Four plastic containers are filled with water upto nominal capacity at an ambient temperature. The containers are then arranged in a block of 2 × 2 on a rigid level flat surface. A top load evenly distributed on a flat plate placed on the unsupported containers, equivalent in magnitude to the total weight of identical packages stacked on top to a minimum stack height of 3 m and kept it for a period of 24 h. Examine the containers after 24 h of test period and there shall not be any sign of breakage or cracks or permanent buckling which could reduce the strength or intends to cause leakage of water or reduction in effectiveness of the closure or cause instability in stack condition [44].

(vii) **Test for Ink adhesion for printed Containers**: The test is carried out to determine the adhesion quality of printing on the surface of the plastic containers. Two numbers of 25 mm wide transparent pressure sensitive tape or cello tape is applied to the printed area of the container; one piece down the length of the container and the other one round the circumference. Pressure is applied firmly on to the container and then leave it for 15 s. Tape is removed by pulling slowly at 1 cm/s from one end at about 90° to the container surface. There shall be no significant removal of the print from the surface of the container and the print shall be legible to the naked eye after the test.

(viii) **Test for Product Resistance of Printed Containers**: The printed Container is left to stand for at least 24 h after printing. Smear the Containers, or representative section cut-out from the printed area with the product at 40 ± 2 °C and then leave it for 1 h. Wash the container or its representative section with cold water. Rub each container or representative section firmly with hard paper tissue ten times. There shall be no significant removal of print from the surface of the Container and the print shall be legible in the naked eye after the test [44].

8.7 Metal Containers

The important tests related to physical, mechanical and chemical properties of tinplate are given below [46]:

8.7.1 Physical Properties

(i) Skid or Stillage: Tinplate sheets received from manufacturers of tinplate fabricators are either in a bundle or coil form. The external condition of the packet should be examined carefully in order to check the ability of tinplate to withstand deformations like bending, spinning and drawing operations during fabrications.

(ii) Size of Select: The actual dimensions of both sides of the sheets in terms of length and width are required to be checked by maintaining the tolerance level ±3 mm over the specified value.

(iii) Shape of Sheet: The sheets should be in flat conditions in both vertical and horizontal positions.

(iv) Thickness of Tinplate: A number of test methods or techniques like gravimetric, volumetric, coulometric and non-destructive X-ray fluorescence are used to determine the thickness of tinplate sheet and then expressed in either mm or an average.

(v) Surface Finish: Normally, the surface of tinplate is either bright finish or stone finish. Stone finish surface with lower tin coating is comparatively less resistant to corrosion as compared to the tinplate with bright finish. Surface finish should be checked for any spots, dust particles, rust particles, excess oil spots, scratches, wood grain effect and other grain defects which might deteriorate the quality of a particular grade.

(vi) Grain Direction: Grain direction plays an important role for making the 'can' body in order to have a quality product. The direction of grains along the circumference of 'can' body is called 'c' direction but if it is along the vertical axis is called 'h' direction. However, this does greatly affect the making of 'canends'.

8.7.2 Mechanical Properties

The properties are important to assess the impact strength of tinplate which effects fabrication of tinplate containers. The important tests under the category are as follows [47]:

(i) **Tensile Strength**: This test is carried out for the double reduced tinplate sheet where the proof stress value i.e. Rp is measured as per the international method. Rp is defined as the stress at which a non-proportional elongation equal to a given percentage of the original gauge length occurs. The test is carried out in a normal way by using wedge grip.

(ii) **Hardness Test**: This test has got more significance along with other tests like bend test, cupping test and spring back test etc. to determine the temper classification of different grades of tinplate sheet so as to describe a combination of inter–related mechanical properties such as hardness, ductility and springiness within tinplate technology. However, Rockwell Superficial Hardness test is used to assess the single reduced tinplate grades. This test is often performed after de tinning in Clerk's solution.

(iii) **Fluting Test**: This is an important test for fabricators for the manufacturing of three piece metal 'can' where the rectangular body blank is formed into cylindrical body and hence the sheet is formed into a cylinder of about 50 mm diameter by standard three roll forming machine. The formed cylinder is examined visually for evidence of flats or any irregularity in the circumference of the cylinder.

8.7.3 Chemical Properties

The tests pertaining to chemical properties have got more relevance to tinplate canners. Few of the important tests are explained below:

(i) **Thickness of Tin Coating**: This test is important in order to have resistance quality against corrosion. Thickness is measured after de-tinning of tinplate. Clerk's solution, prepared by dissolving 20 gm of antimony trioxide (Sb_2O_2) in one litre of concentrated Hydrochloric acid (con. HCl) is used as de-tinning solvent. Normally, the differential tin-coating is applied on the tinplate by means of electrolytic tin coating techniques and grammage of tin-coating for body and the ends have thickness as D 11.2/5.6 and E 5.6/5.6 respectively. On account of this, there is a deviation of thickness in tin-coating of tinplate. Gravimetric method is commonly used to determine the thickness of tin coating.

(ii) **Lacquer Adhesion Test**: This test is conducted for the tinplate which is coated with either Acid Resistant (AR) or Sulphur Resistance (SR) lacquers for different applications. Depending on the type of food, different lacquered tinplates are used for the fabrication of 'cans'. Lacquer adhesion test is performed by scotch tape test method.

(iii) **Lacquer Continuity Test**: This is an important test to determine the uniformity of lacquer coating over the tinplate sheet. In case of non-uniformity or uneven instances, the product comes in contact with the inner layer of the tin 'can' and these get exposed to the un-even coated surface resulting in deterioration of quality of product. This is a very simple test where cotton swab dipped in acetone is applied over the lacquer surface to determine the continuity of lacquer the other tests under mechanical and chemical properties are carried out in the laboratory to ensure the performance of tinplate and hence, these tests are termed as subjective or performance tests.

(iv) **Scotch Tape Test**: A simple test generally used to estimate the degree of lacquer adhesion is the Scotch Tape Test. A length of 25 mm wide pressure sensitive adhesive tape is firmly applied to a pattern of scratches penetrating the lacquer film and then pulled away with a sharp jerk carrying with it particles of lacquer from poorly adherent films. The scratch pattern should be made in repeatable form, lose spacing of the scratches helping to give a more definite result. A steel comb with 20 teeth spaced I mm apart used to make two sets of scratches at right angles to one another gives a suitable pattern.

To asses failure of adhesion during heat processing, scotch tape test may be made after test samples have been exposed either to boiling water for 30 min or to steam in a pressure Vessel. For the pressure vessel, it is common practice to immerse part of the test piece in water and after treatment for 30 min at 120 °C to inspect separately the parts of the lacquered surface exposed to water and steam.

(v) **Porosity or Iron Solution Value (ISV) Test**: The test is carried out to determine the micro porosity through tin coating and is a measure of how much iron is exposed. Lower the ISV, better the performance of the 'can'.

(vi) **Passivation Layer Test**: Passivation treatment of tinplate is done to improve the stabilising properties of the surface. It improves the corrosion resistance, prevents excessive growth in tin oxide, surface dis-colouration and improves lacquer adhesion properties. To assess the passivation, surface chromium on tinplate is determined by chemical analysis and then compared with the standard value.

(vii) **Pickle Lag Test**: A specimen is immersed in hydrochloric acid and the time needed to reach the steady rate of hydrogen evolution is recorded. This is a difficult test and is done on base steel surface after removing the tin.

(viii) **Alloy Tin Couple Test (ATC)**: The test is carried by evaluating the porosity in iron tin alloy layer. The basis of the test is the measurement of the current flowing between a sample of tinplate from which the free tin layer has been removed and a tin electrode of much larger size when both have been immersed, with external electrical connection in the test medium (grape fruit juice) for 20 h. This is a difficult test. The current density on the test piece expressed as micro amperes sq. cm is the ATC value.

8.7.4 Testing of Tinplate Containers or Cans

The tinplate containers or 'cans' are manufactured either in 3-piece or 2-piece and thus called as 3-piece or 2-piece containers. In case of 3 piece 'can', the edges of circular tinplate material are joined together to make the body of the 'can' where a different technique is followed like lock seam, spot welded or 'stitched' seam, adhesive joined seam, soldered seam and welded seam. These 'cans' are mostly used for the packaging of processed food products. Important tests carried out in the laboratory to assure the quality are given below.

8.7.4.1 Objective or Qualitative Tests [48]

(i) **Physical Observation**: The visual examination of the 'cans' for any defects in seaming, any spots or specs on the surface finish, print quality, number of beads etc. prior to packing the 'can' for dispatch.

(ii) **Thickness of Tin-Coating**: The canners do carry out the tests to evaluate the tin coating thickness to ensure that the tin 'can' would be suitable for the packaging of food products or non-food items so that the metal 'cans' would not be affected by corrosion. This test is carried out in the same way as mentioned earlier.

(iii) **Lacquer Adhesion and Lacquer Continuity Test**: These tests are also carried by the manufacturer of food or non-food items and the method of tinplate testing is as given here:

- Double Seam Analysis: In order to ensure proper double seaming, double seam portion of the 'can' is cut and then checked under a magnifying glass or through profile projector.
- Tare Weight: Tare weight of the empty 'can' is evaluated by gravimetric method. It is expressed in gm.

8.7.4.2 Performance or Subjective Tests

In order to ensure performance of the 'can' during handling, storage and transportation, tinplate containers are subjected to different tests which are given here:

(i) **Compression Strength**: The test is carried out to assess the withstanding ability of the tinplate container under compressive force where the container is subjected to compression testing equipment. The compressive force in kgf and the deflection in 'mm' are measured.

(ii) **Leakage Test**: The filled containers are also subjected to leakage test by immersing the same in water. The appearance of any bubble will indicate about leakage of the 'can'.

(iii) **Product-Package Compatibility Study**: A particular type of food product is packed in a metal 'can' made with a particular type of lacquer, either Acid Resistance (AR) or Sulphur Resistance (SR) depending on the requirement of the food product. Hermetically sealed containers are kept in the laboratory. Metal 'cans' are opened at fixed interval of time and the quality of the product is evaluated in terms of dis-colouration, change in taste, odour, texture etc. This particular test is carried out in the laboratory by sensory evaluation method where 5–10 panelists are engaged to assess the product quality by following hedonic scale scoring method. Similarly, changes in physicalproperties of the metal 'cans' are also checked to evaluate the quality.

(iv) **Oxygen Head Space Analysis**: Vacuum is created inside the metal 'can' above the top surface of filled products during the canning process by way of subjecting the 'cans' in the exhausting chamber so that there would not be any chance of multiplications of microbes due to non-availability of oxygen gas. However, it is also required to check the residual oxygen at the headspace of hermetically sealed filled metal containers. This particular test is carried out with the help of oxygen headspace analysis tester to ensure that the metal 'cans' are absolutely free from oxygen.

(v) **Drop Test**: The filled 'cans' are often subjected to drop test to asses the impact strength of metal 'cans'. Moreover, this test will also ensure leakage performance during storage, handling an transportation. This is a subjective test. The objective of this test is to determine that neither liquid media nor the product will come out due to drop impact However, there could be the appearance of dent on the 'can'.

(vi) **Salt Spray Test**: This test is not commonly carried out for the metal 'can' However, in some cases, depending on the customer requirement, the empty 'can' are subjected to salt spray chamber where the carton is exposed to 5% salt solution vapour under 38 °C temperature for 50 h At the end of exposure time, the metal 'cans' are observed for the appearance of any spots of corrosion on the 'can' body or at the joint. This is also a subjective test to assess the quality of tinplate used for the fabrication of metal 'cans.

8.8 Glass Containers

Glass is very transparent and therefore its contents can be clearly seen for any impurities or suspended particles, but the transparency of the glass Containers sometimes becomes detrimental to its contents, as some of the products tend to deteriorate on exposure to sunlight. In such cases, amber colour glass or green colour glass is used. These colours filter the detrimental wave length of sun light. Another important characteristics of the glass Containers is its top to bottom compression resistance. Packaging of products like beer in glass bottles is a good example where the use of good compression resistance of glass bottle is made [49].

The quality of glass Containers are evaluated by conducting the following tests in the laboratory:

(a) Physical Tests.
(b) Mechanical tests.
(c) Chemical Tests.

(a) **Physical Tests**:

(i) **Visual surface defects**: The bottles are visually Checked for any different kind of defects like offset finish, split finish, dirty finish, rough finish, chocked neck, presence of bubbles, spots, Specks, hair line cracks on the body surface, stones, thin body, thick bottom etc.

(ii) **Dimensions**: These refers to height, body diameter, neck diameter etc. These are critical as they affect productivity on the high speed machine. The dial gauge thickness tester is used to measure the dimensions and the values are expressed in mm.

(iii) **Finish Dimensions**; The term "Finish" refers to that part of the glass Container which takes on the closure. For efficient sealing, dimensions of the finish should strictly conform to the tolerance of dimensions.

(iv) **Capacity**: Capacity of the bottle is defined either as the brimful capacity or capacity upto a filling height agreed upon. The brimful capacity is determined by considering the difference in weight between the empty bottle and the bottle filled with water upto the brim. The temperature of water used for filling should be constant, perfectly 25 °C. Corrections are to be applied for water density variation which is standardized at 4 °C. While measuring the capacity to a filling point, a depth gauge may be used inside the neck of the bottle to verify the distance from the centre of the liquid meniscus to the top of the bottle.

(v) **Verticality**: This test determines not only the deviation of the whole body from the vertical but also the combined effect of various deformations which may also be present. The verticality test can be done by following the test method as per IS:2091-1983 [50].

Method: Fill the bottle with water in order to provide more stability and place it on its base on the flat plate having a shaft bolted to it at right angles. Adjust "V" block mounted on the shaft in such a manner that it is in contact with the outer diameter of the bottle at about the middle. Adjust the dial indicator fitted to the shaft so that its measuring point comes with the outer edge of the neck of the bottle. Rotate the bottle, keeping the body always in contact with the "V" block and note down the minimum deflection on the indicator. The verticality is expressed as half the difference between the maximum and minimum readings of the dial indicator obtained during rotation of the bottle through 360° (Fig. 8.16).

(vi) **Ovality**: Ovality is described as the extent of deviation of a cylindrical bottle from a perfectly circular cross section. The ovality is determined by measuring the maximum and minimum leading horizontal dimension along the circumference, using a vernier caliper. Provided the star wheel could accommodate, ovality is not a very critical defect, excepting where the bottles are labeled by rolling on their side.

Fig. 8.16 Verticality tester

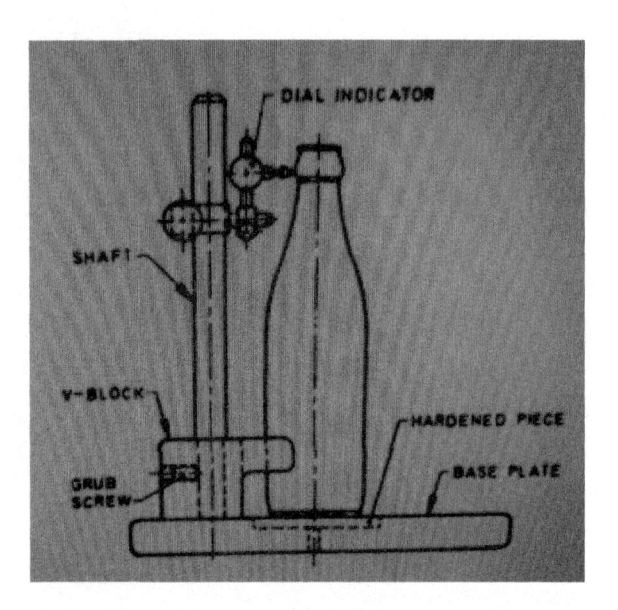

(b) Mechanical Tests:

(i) **Hydrostatic pressure Test**: This test is important for the glass bottle where the liquid products are filled under pressure like soda water, Carbonated beverages etc. In order to carry out the test, a suitable testing equipment is to be designed where the bottle to be tested is required to be held by split collar so designated that the bottle is not clamped but is suspended from the bead of the finish. A seal shall be provided which shall be water-tight at the pressure to be applied. The apparatus should also have the provision to pump the water into the bottle through non-return valve and the pressure indicated on a suitable pressure recorder. A automatically controlled timing mechanism shall be built into the apparatus so that pressure is applied for 1 min ± 2 s (Fig. 8.17).

Test method: Fill the bottle with water a room temperature. Hold it by split collar and then seal. Apply a pressure of 1 MPa (10 kgf/cm^2) for a period of 1 min ± 2 s. The sample bottle has to withstand that pressure and there should neither any deformation nor damage or breakage to the bottle.

(ii) **Annealing Test**

Fig. 8.17 Hydrostatic tester

8.8 Glass Containers

Annealing is a process of slowly cooling hot glass objects after they have been formed, to relieve residual internal stresses introduced during manufacture. Once the bottles are annealed, it is necessary to check whether the bottles are properly annealed. The testing is done by viewing the annealed bottle through a polorioscope where stained glass shows interference of colours when viewed through polarized light. Well annealed bottles should not show any play of colours.

(iii) **Thermal Shock Test**

This is a test wherein glass articles are subjected to sudden temperature differences to which the bottles are likely to be subjected in actual filling and using. The testing is carried out as per test method given in IS:6506-1972 [51].

Method:

A wire-net busket is to be taken to hold the glass samples for immersing into hot water bath and then transferring them into cold water bath. Both the water bath is to be filled with water and then adjusted the temperature like $L1 \pm 1$ °C by considering L2 is the temperature for tape water and (L2-L1) is the required temperature in which the glass sample need to be exposed. The temperature at cold water bath can be adjusted by using ice water. The sample bottles are taken in wire net busket and then immersed into hot water bath and allow to soak for 15 min by maintaining the temperature at $L1 \pm 1$ °C and then transfer each of the samples, filled with water, to the bath at temperature, completing the process of transference in 10 ± 2 s for each article, and immerse almost completely in the bath, for a specified period, which shall not exceed two minutes, to enter them. Finally examine each of the articles for any kind of damage.

(iv) **Pedulum test**: This test is carried to assess the mechanical strength of the bottles. The test consists of dropping a steel ball of 400 gm to fall suddenly from a height of 10 cm high on the bottle held rigidly. In case of milk bottle, the bottle should not crack. Another pendulum test is carried out where the steel ball swings and strikes at the bottle held rigidly. Bottles which have to withstand impacts, are generally made with thicker wall distribution and carefully annealed (Fig. 8.18).

(c) **Chemical Tests**

(i) **Density Test**: This test is important to get an idea about the ingredients of glass particles instead of conducting chemical analysis of Glass composition. Dr. Preston has developed the "Perison density Comparator" with a help of which density measurements are made. Density test is very sensitive test and density of glass changes even if there is slight variation in the composition of glass or even in the degree of annealing.

(ii) **Alkalinity test**: This test is important to determine the leaching of alkali from different types of glasses like Type I, II, and III. The chemicals like Alcohol (95%), Methyl Red, 0.05 N Sodium Hydroxide, 0.02 N Sodium Hydroxide, 0.02 N Hydrochloric Acid, 1 N Potassium Dichromate Solution

Fig. 8.18 Pendulam tester

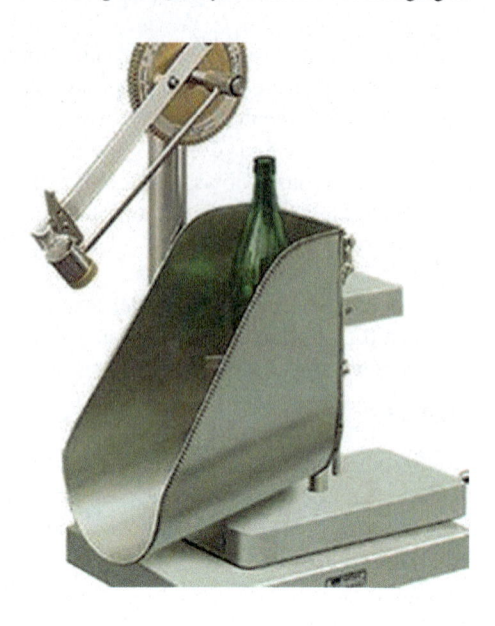

Test method

About six Glass containers are required to conduct this test and all the container should pass the test.

Fill the containers completely with warm water, empty and allow to drain for thirty seconds, removing the last drops of water by touching the inverted rim with filter paper No. 41 (12.5 cm) Whatt-man. Repeat the washing three times. Fill the containers to their prescribed capacity with Acid Methyl Red Solution and seal on Ampoule Sealing Machine. Heat in an autoclave at a temperature of 121 °C for 30 min. Cool and examine the color of the solution.

When the containers of are of colored glass, remove the solution for examination on a thoroughly washed while glazed tile. The container passes the test, if the color of the test solution has not changed form pink to the full yellow color of Methyl Red as indicated by comparing it with the color of a solution prepared by adding 0.1 ml of 0.05 N Sodium Hydroxide to 10 ml of Acid Methyl Red Solution. The containers, which have once passed the test, may fail to do so after being stored. A samples from a batch of containers which fails to pass the test after storage may be washed internally with a 5% v/v solution of Glacial Acetic Acid followed by three washing with water, and resubmitted to the test. If the sample then passes the test each container of the batch is similarly washed before being used.

8.9 Composite Containers [52]

The important tests are as follows:

8.9 Composite Containers [52]

8.9.1 Weight of Empty Container

Five numbers of empty Containers are taken at random from the production line and then take the weight in gm in an Weighing balance with an accuracy of 0.1. Average of five readings will be considered as the weight of Composite Containers.

8.9.2 Adhesive Bonding

By using a sharp knife, cut the ends off the composite Container body as close to the end seams as possible. Split the container body lengthwise so that it can be flattended out as a rectangle. Peel apart the liner overlap, liner, body plies, labels and label overwrap. Measure any area where there is no fibre tearing adhesive bond and calculate the percentage of the total area involved to determine that 80 percent of minimum fibre tearing bonds are available.

8.9.3 Body Wall Thickness

Body wall thickness shall be measured with a round head micrometer with at least count of 0.02 mm. The thickness shall be measured at four points diametrically opposite and the mean of the four readings shall be taken as the thickness of the wall.

8.9.4 Drop Test

The filled Composite Containers are subjected to the drop test at predetermined drop height to asses the quality of materials used for the fabrication of containers in terms of withstanding Capacity.

Test Method:

Ten samples of filled Containers shall be taken at random from the filling line by maintaining the filling temperature within ±2 °C. The drawn samples shall be dropped from a height of 1.2 m on a flat surface. Each Conter shall be dropped four times in the following sequence:

(a) On the top ends.
(b) On the bottom ends.
(c) On flat in lengthwise and
(d) On opposite side in lengthwise.

The samples shall be deemed to have passed the test if not more than 2 containers burst due to impact during drop test.

8.10 Plastic Sqeezable Tube

Plastic based laminated squeezables are subjected to the following tests to asses the quality [53].

8.10.1 Physical Tests

8.10.1.1 Visual Defects

The finished tubes shall be examined visually for the following defects:

(a) **Critical Defects**

    (i)   Incorrect print design or missing colour;
    (ii)   Insect, hair, or other gross contamination on/in tube or cap;
    (iii) Open seam;
    (iv) Open shoulder weld;
    (v)   Hole or slit in tube or sleeve;
    (vi) Incomplete head;
    (vii) Blocked orifice, and
    (viii) Stripped threads.

(b) Other Defects:

    (i)   Delamination;
    (ii)   Seam poly flow/white line barely visible;
    (iii) Seam melted on outside to point where the outer layers of polyethylene arc pulled off or wrinkled;
    (iv) Damaged ends;
    (v)   Missing, cracked or badly chipped insert;
    (vi) Flashing or pigtails in the orifice;
    (vii) Poly strings inside or attached to tube;
    (viii) Inside contamination; and
    (ix) Outside contamination.

8.10.1.2 Dimensions

- Outside diameter of tube

Test Method: This method accurately determines the outside tube diameter using a Go/No-Go gauge. Push sample through the correct hole diameter on the gauge. The tube should fit through the gauge with a small amount of effort. A no fit or extreme effort fit indicates that the tube does not meet specification.

8.10 Plastic Sqeezable Tube

- Inside diameter of tube

Test Method:

A mandrel of appropriate diameter for the tube is taken for measuring the diameters. Push the tube or sleeve over the mandrel and note the amount of force or pressure necessary. The observation is also taken whether the tube or sleeves fits too tight or loose over the mandrel. **Results**: For laminated tubes or sleeves, there should be some frictional resistance, but not enough to buckle the wall. The fitment should not be loose.

8.10.2 Mechanical Tests

8.10.2.1 Shoulder Bond Strength

Test Method: Cut the body/sleeve into approximately 15 mm strips terminating the cut at the shoulder. Holding the head in one hand and a strip in the other separate the strip from the head. Angle of 90° should be maintained between strip and the axis of the tube.

Rate bonds according to the following,

(a) **Excellent**: hard to pull, laminate/tube body material tears
(b) **Good**: easy to pull but slight tearing of laminate/tube body material
(c) **Poor**: easy to pull, no tearing of laminate/tube body material.

8.10.2.2 Bursting Strength

Test Method: Seal the open end of the empty tube. Remove the cap and connect the nozzle with appropriate adopter, matching tube neck and thread fitment, to air supply line. Apply air pressure to the tube at the rate of 70 kPa/5 s and continue raising the pressure till the pressure reaches 140 kPa. Maintain this pressure for 15 s. The tube or side seam shall not show any bursting at the tube body or side seam. Raise the pressure further till the tube bursts. Observe the rupture of the tube. The tube passes the test if the body ruptures before the seal fails.

8.10.2.3 End Seal Integrity Test

Test method: Cut the sealed empty tube atleast 50 mm below the seal. Make two parallel cuts, perpendicular to the seal at the corner of the seal. Pull both strips until the sample fails. If the seal is good it will fail by fracturing the laminate, while the sealed interface remains unaffected. If the seal opens up by the pull the sealing is improper. A break should occur at the base of the seal next to the tube body or elsewhere in the body to ensure a good seal.

8.10.2.4 Torque Test of Caps

Test method: Torque-meter, standard cap holders is used to conduct this test. Take the tube under inspection. Insert tube mandrel of same diameter as that of tube diameter into torque meter for holding the tube. Insert the cap of auto-capped tube into the cap holder at bottom portion of the torque meter. Holding head portion of the torque-meter in one hand and the tube in other hand rotate head portion of the torque meter in anti-clock wise direction to unscrew the cap from tube nozzle. Note the reading observed in the display of torque meter. Reading is expressed in N/cm (Range: 15–45 N-cm).

8.10.3 Physico-chemical Test

8.10.3.1 Overall Migration

The tube shall comply with the overall migration limits of 60 mg/l, maximum of simulants and 10 mg/dm^2, maximum of the surface of the material or article when tested by the method prescribed in IS 9845 [42].

8.10.3.2 Specific Migration Test

Significance: This test is carried out to asses the leaching of restricted elements from packaging materials to food products which will have harmful effect to human health.

Test Method: IS: 9845-1987—to prepare the extractant & IS: 3025 (Part-2)—for the detection of heavy metal content [43].

8.11 Plastic Pouches

The following tests are conducted in the testing laboratory to asses the performance of Plastic Pouches used for food products [54].

8.11 Plastic Pouches

8.11.1 Plastic Pouch Compression Test

Significance: The plastic based flexible pouches are finally packed into Corrugated fibre board boxes or shipper Containers. During storage and transporation, if the shipper containers are having adequate load bearing capacity, in such case, there is a probability that load will be transmitted into the plastic pouches leading to the leakage of pouches. The pouch compression or burst strength would indicate about the load bearing capacity of the individual pouches.

Test Method: There are 2 basic methods of test.

8.11.2 Static Compression Test

Where a package is loaded upto a predetermined level, held there for a predetermined time period and then relaxed. This may or may not lead to a burst.

8.11.2.1 Dynamic Compression Test

Where a package is loaded incrementally till it cannot bear the load any longer and eventually bursts open.

8.11.2.2 Static Compression Test

Place a sealed package or plastic flexible pouch filled with its original contents or water on the bottom plate of the compression machine. The top plate of the machine is brought down and allowed to touch the pouch. Then load is applied by allowing the top plate to compress the pouch with the static load and hold it for 60 s. Remove and examine the pouch for any kind of leakages and seal integrity. The recommended load for static compression test is as follows (Table 8.2).

Table 8.2 Weight Package and static load

Weight/volume of package (g)	Static load (kg)
<100	20
100–400	40
400–2000	60
>2000	80

8.11.2.3 Dynamic Compression Test

Place a sealed flexible package filled with its original contents or water and then placed on the bottom mplate of the compression machine. The top platen will be allowed to move downwards at the rate of 10 mm/min and apply the load till the pouch ruptures either from seal area or from the body and the contents come out. Pouches should be strong enough to withstand the forces that will be subjected to them in the event of its outer package (shipper) failure. The higher Dynamic Compression Strength, the better is the quality of the pouch.

8.11.3 Pouch Burst Test [55]

Significance: This test in principle is similar to the Pouch Compression Test, except that the pouches are subjected to internal air pressure in lieu of external force. The internal burst test is considered to be an important test in order to asses the seal quality of flexible pouches while the pouches are exposed to mechanical hazards during storage and transportation. This test will also indicate about the quality of flexible materials either single or multilayer materials used for making the flexible pouches.

Test Method: The flexible pouch sample is mounted in a burst test which internally and increasingly pressurizes a package until the package around the seal area get "bursts" and opened up due to increased pressure. The pouches are examined visually and noted the type of failure and the pressure at which failure occur. The test is conducted as per the test method of ASTM F 2054-07.

8.11.4 Vacuum Leakage Test (ASTM F2338-09(2013))

Significance: This test method detects the leakage of plastic pouches by measuring the rise in pressure (vacuum loss) in an enclosed evacuated test chamber containing the test package. Vacuum loss results from leakage of test package headspace gases and/or volatilization of test package liquid contents located in or near the leak. When testing for leaks that may be partially or completely plugged with the package's liquid contents, the test chamber is evacuated to a pressure below the liquid's vaporization pressure [56].

Test Method: The methods require a Vacuum Dessicator chamber to contain the test package and a leak detection system designed with one or more pressure transducers. This method is ideal for packages whose contents have headspace or some amount of air or gas within the package. Placing the package inside the vacuum dessicator, closing the lid, and submerging it under the water, vacuuming the chamber. If air bubbles appear, then the package is compromised or faulty.

8.11.5 Drop Test of Pouches [57]

The filled Plastic pouches are subjected to the drop test at predetermined drop height to asses the quality of plastic pouches in terms of withstanding Capacity.

Test Method:

Sixteen samples of filled pouches shall be taken at random from the filling line by maintaining the filling temperature within ± 2 °C. After 15 min, the drawn samples from the filling line shall be dropped from a height of 1.2 m on a flat surface. Each pouch shall be dropped four times in the following sequence:

(a) On the Side.
(b) On the opposite
(c) On flat longer edge and
(d) On opposite longer edge.

The samples shall be deemed to have passed the test if not more than 2 pouch burst due to impact during drop test.

References

1. Export Quality Management, ISO 9000 Quality Management Systems, 1993, Published by Federation of Indian Chambers of Commerce & Industry
2. Indian Standard IS: 1060-1966 (Part I) Methods of Sampling and Test for paper and allied products, published by Bureau of Indian standard
3. The Technical Association of the Pulp and paper Industry (TAPPI T 410): Grammage of paper and paper Board (weight per unit area)
4. The Technical Association of the Pulp and paper Industry (TAPPI T 538): Roughness of Paper and paperboard (Sheffield Method)
5. The Technical Association of the Pulp and paper Industry (TAPPI T 460 om-02): Air resistance of paper (Gurley method)
6. Indian Standard IS: 1060-1969 (Part III) Methods of Sampling and Test for paper and allied products, published by Bureau of Indian standard
7. The Technical Association of the Pulp and paper Industry (TAPPI T 822 om-02): Ring crush of paperboard (Rigid Support Method)
8. The Technical Association of the Pulp and paper Industry (TAPPI T 809 om-99): Flat crush of corrugating medium (FCT test)
9. The Technical Association of the Pulp and paper Industry (TAPPI T 441 om-09): Water absorptive of sized (Non-bibulous) paper, paper board and Corrugated fibreboard (Cobb test)
10. The Technical Association of the Pulp and paper Industry (TAPPI T 413): Ash in wood, pulp, Paper and Paperbaord Combustion at 900 °C
11. Indian Standard IS: 1060-1960 (Part II) Methods of Sampling and Test for paper and allied products, published by Bureau of Indian standard

12. Quality and Standards of Corrugated fibre board boxes, N.C. Saha, Packaging India Journal, April-May Issue, 2010, published by Indian Institute of Packaging
13. Corrugated Box Handbook, Published by Core Emballage Limited, p 124, 1261999
14. The Technical Association of the Pulp and paper Industry (TAPPI T 811om-17): Edgewise Compressive Strength of Corrugated Fibrebaord (Short Column Test)
15. The Technical Association of the Pulp and paper Industry (TAPPI T 825): Flat Crush Test of Corrugated Board- Rigid Support Method
16. The Technical Association of the Pulp and paper Industry (TAPPI T 803 om-10): Puncture test of Container board test Method
17. The Technical Association of the Pulp and paper Industry (TAPPI T 804): Compression Test for Fibre Board Shipping Containers
18. Indian Standard IS: 7028 (Part-4): Performance Tests for Complete, filled Transport packages for Vertical drop Test
19. Indian Standard IS: 7028 (Part-2): Performance Tests for Complete, Filled Transport packages for Vibration
20. Indian Standard IS: 7028 (Part-1): Performance Tests for Complete, filled Transport packages for Stacking Tests under Static Load
21. Indian Standard IS: 7028 (Part-11): Performance Tests for Complete, filled Transport packages for Inclined Impact Test
22. The Technical Association of the Pulp and paper Industry (TAPPI T 801om-06): Impact Resistance of Fibreboard Shipping Containers, Test Method
23. Indian Standard IS: 7028 (Part-5): Performance Tests for Complete, filled Transport packages for Rolling Test
24. Indian Standard IS: 7028 (Part-8): Performance Tests for Complete, filled Transport packages for Water spray Test
25. The Technical Association of the Pulp and paper Industry (TAPPI T 805): Water Resistance of Fibreboard Shipping Containers (Spray Method)
26. American Society for Testing and Materials (ASTM E 252-05): Standard Test Method for Thin Foil, Sheet and Film by Mass Measurement
27. American Society for Testing and Materials (ASTM D 792–20): Standard Test Method for Density and Specific Gravity
28. American Society for Testing and Materials (ASTM D-4321-99): Standard Test Method for Package Yield of Plastic Film
29. American Society for Testing and Materials (ASTM D: 1204): Standard Test Method for Linear Dimensional Changes of Non rigid Thermoplastic Sheeting or Film at Elevated Temperature
30. American Society for Testing and Materials (ASTM D-882-18): Standard Test Method for Tensile Properties of Thin Plastic Sheeting
31. American Society for Testing and Materials (ASTM F 88-68): Seal Strength Testing of Flexible Barrier Materials
32. American Society for Testing and Materials (ASTM D-1004-09): Standard Test Method for Tear Resistance (Graves Tear) of Plastic Films and sheeting
33. Indian Standard IS: 2508: 2020 (Amendment-3)—Polyethylene Films and Sheets-Specification
34. American Society for Testing and Materials (ASTM D-1894-09): Standard Test Method used for determining Static (μ_s) and Kinetic (μ_k) Coefficients of Friction of Plastic Film and Sheeting
35. American Society for Testing and Materials (ASTM D-3354-15): Standard Test Method for Blocking Load of Plastic Film by the Parallel Plate Method
36. American Society for Testing and Materials (ASTM D-1238): Standard Test Method for Melt Flow Index Test of Plastic Granules
37. American Society For Testing and materials (ASTM E831-19): Standard Test Method for Linear Thermal Expansion of Solid Materials by Thermomechanical Analysis
38. American Society for Testing and Materials (ASTM D-523): Standard Test Method for Specular Gloss

References

39. Physico-Chemical properties of Plastic Film and Laminates-Tests and their significances, N.C.Saha, Packaging India Journal, June–July issue, 2004
40. American Society for Testing and Materials (ASTM F-1249): Standard Test Method for Water Vapour Transmission Rate
41. American Society for Testing and Materials (ASTM D-3985-17): Standard Test Method for Oxygen Gas Transmission Rate
42. Indian standard IS:9845-1998: Determination of Overall Migration of Constitutents of Plastics Materials and articles intended to come in contact with Food staffs
43. Indian standard IS:3025 (Part-2)-2004: Determination of 33 elements by Inductively Coupled Plasma Atomic Emission Spectroscopy
44. Indian Standard, IS: 2798: 1998: Methods of Test for Plastics Containers
45. Indian Standard IS: 8747:1977: Methods of Test for Environmental Stress-Crack Resistance of Blow moulded Polyethylene Containers, Published by Bureau of Indian Standard
46. Quality Control and Testing of Tinplate Containers, N.C. Saha, Packaging India Journal, Oct–Nov 2009, Published by Indian Institute of Packaging
47. Packaging Technology educational volumes (Set-2), pp 229–240, published by Indian Institute of Packaging, 2015
48. Tinplate- a versatile and eco-Friendly packaging Medium, published by Indian Institute of packaging and Tin Plate Promotion Council, 2009
49. Glass Containers-Properties, Significance & Tests, published in Technical Bulletin, V.R. Krishnamurthi, 1996
50. Indian Standard IS: 2091-1983, Specification for Multi-trip Glass Beer Bottle, Published by Bureau of Indian standard
51. Indian standard IS: 6506-1972 Method for Thermal Shock Tests on Glassware, published by Bureau of Indian standards
52. Indian standard IS: 11357: 1985: Specification for Composite Container for dry Products
53. Indian Standard IS:17480:2020—High Density Polyethylene Multi-Sqeezable Tube for Packaging-Specification
54. American Society for Testing and Materials (ASTM F 2338-09 (2013) Standard Test Method for Nondestructive Detection of Leaks in Packages by Vacuum Decay Method
55. American Society For Testing and Materials (ASTM F2054–07): Standard Test Method for Burst Testing of Flexible Package Seals Using internal Air Pressurization within Restraining Plates
56. American Society for Testing and Materials (ASTM F2054-07) Standard Test Method for Burst Testing of Flexible Package Seals Using Internal Air Pressurization Within Restraining Plates
57. Indian standard IS:11805: 2007: Polyethylene Pouches for Packaging Liquid Milk-Specification

Chapter 9
Sustainable and Green Packaging-Environmental Issues

9.1 Introduction

In the last 5 years or so, the two words that have attracted the most attention in world-wide industry and economic planning are "sustainable packaging". Almost everybody is trying to prove to the world what a good citizen he or she is by targeting sustainable development as the most important objective in everything that they do; this has now become the primary goal for almost all corporate activity or economic planning.

There is absolutely no doubt that every package used has to be as sustainable as it can be and this consideration has to be an integral part of all packaging design.

One has to first understand what sustainability is all about.

9.2 Defining Sustainability

We start by trying to define what "sustainability" really means in today's context. According to the dictionary, the word sustainability comes from the verb "to sustain", which has two important meanings:

- to strengthen or support.
- to keep something going over time or continuously.

Sustainability is now associated with evaluation of a process or technology and, as it happens, it actually encompasses both of the definitions listed above. On the one hand, it must strengthen or support the total system we live in (environment, habitats, living species, plant life, resources etc.) and, on the other, it must keep it going over time or, preferably, continuously. To put it in another and more practical way, sustainability of a technology or process is the concept of using it to meet the

---

With contributions by "S. Chidambar".

needs of 'today' without in any way compromising the interests of future generations or the 'tomorrow'. It is the latter part of this statement that has to be treated as the inviolable bottom line.

Based on this, we can lay down the two essential requirements of a sustainable technology or process as follows:

- It must not impose any burdens on the environment or on material resources, whether renewable or non-renewable
- It must not cause any deterioration in their quality or availability.

9.3 Origins of Sustainability

In 1987, the United Nations constituted the World Commission On Environment and Development which released a path-breaking report entitled Our Common Future (popularly known as the Bruntland Report). This report is probably the starting point of most modern concepts and formal structured thinking on Sustainable Development [1].

Sustainability originally started off by focussing on ecological measures that were needed to sustain the environment. Down the line came the realisation that there were other things that needed to be addressed—particularly the issue of conserving natural resources that were either not basically renewable or renewable in the short to medium term (like fossil fuels), thereby rendering them also non-renewable for all practical purposes. The sustainability umbrella was, therefore, enlarged to address all these other concerns as well.

In 2015, the United Nations General Assembly formally adopted the "universal, integrated and transformative" 2030 Agenda for Sustainable Development, a set of 17 Sustainable Development Goals (SDGs). The goals are to be implemented and achieved in every country from the year 2016 to 2030. The Official Agenda for Sustainable Development adopted on 25 September 2015 outlined the following 17 Sustainable Development Goals and laid out its associated 169 targets [2].

- Poverty—End poverty in all its forms everywhere
- Food—End hunger, achieve food security and improved nutrition and promote sustainable agriculture
- Health—Ensure healthy lives and promote well-being for all at all ages
- Education—Ensure inclusive and equitable quality education and promote lifelong learning opportunities for all
- Women—Achieve gender equality and empower all women and girls
- Water—Ensure availability and sustainable management of water and sanitation for all
- Energy—Ensure access to affordable, reliable, sustainable and modern energy for all
- Economy—Promote sustained, inclusive and sustainable economic growth, full and productive employment and decent work for all

9.3 Origins of Sustainability

- Infrastructure—Build resilient infrastructure, promote inclusive and sustainable industrialization and foster innovation
- Inequality—Reduce inequality within and among countries
- Habitation—Make cities and human settlements inclusive, safe, resilient and sustainable
- Consumption—Ensure sustainable consumption and production patterns
- Climate—Take urgent action to combat climate change and its impacts
- Marine-ecosystems—Conserve and sustainably use the oceans, seas and marine resources for sustainable development
- Ecosystems—Protect, restore and promote sustainable use of terrestrial ecosystems, sustainably manage forests, combat desertification, and halt and reverse land degradation and halt biodiversity loss
- Institutions—Promote peaceful and inclusive societies for sustainable development, provide access to justice for all and build effective, accountable and inclusive institutions at all levels
- Sustainability—Strengthen the means of implementation and revitalize the global partnership for sustainable development

The Sustainable Development Goals (SDGs) replaced the eight Millennium Development Goals (MDGs), which expired at the end of 2015. The MDGs had been established in 2000 following the Millennium Summit of the United Nations. Adopted by the 189 United Nations member states at the time and more than twenty international organizations, these goals were advanced to help achieve the following sustainable development standards by 2015.

- To eradicate extreme poverty and hunger
- To achieve universal primary education
- To promote gender equality and empower women
- To reduce child mortality
- To improve maternal health
- To combat HIV/AIDS, malaria, and other diseases
- To ensure environmental sustainability (one of the targets in this goal focuses on increasing sustainable access to safe drinking water and basic sanitation)
- To develop a global partnership for development.

Let us now examine how sustainability impacts Packaging.

9.4 Addressing the Concerns

We need to first list the major concerns that have to be addressed. Many of the concerns are inter-dependent. The first group of concerns have to do with natural resources and the fact that some of the key ones are either non-renewable or, for all practical purposes, rendered thus because the renewal process takes up inordinate time. It is no longer a question of whether we will run out of resources but when this will happen. The clock is already ticking.

These resources include:

- Fossil fuels (coal, petroleum and natural gas)—The reserves of these resources are finite and conservation of these need conscious and immediate attention. While the world is not going to run out of natural gas as soon as it does of the other resources, the overall situation is extremely grim. It is absolutely essential to see that there is no depletion of these resources. Any resources used have to be recovered or regenerated for use as fresh inputs at the end of the life-cycle of all products and processes. Thus, thinking has to follow a "cradle to cradle" approach rather than the current "cradle to grave" approach (where post-consumer waste is addressed by merely by collection/treatment and valuable resources are disposed of—mostly through landfills).
- Various minerals (e.g. metallic ores, rare metals and industrial raw materials).

Many other man-made resources are also concerns either because they are scarce or in short supply (like energy) or because existing infrastructure is inadequate or under strain. The latter usually becomes more of a constraint in poor and developing countries, which unfortunately constitute a major part of the world.

There are a number of concerns that relate to the environment, as we know it. The important ones are:

- Quality of the environment and the need to protect it from pollution (atmosphere, water, soil).
- Water—especially fresh water and ground water tables. Many people do not appreciate the acute scarcity of this resource. Although 71% of the earth is covered by water, only 2.5% is fresh water and most of this is locked up in the Antarctic ice sheet. The remaining freshwater is found in glaciers, lakes, rivers, wetlands, the soil, aquifers and the atmosphere.
- Due to the water cycle, fresh water supply is continually replenished by precipitation. However, it is still a limited amount and management of this resource is crucial. Awareness of the global importance of preserving water for ecosystem services has only recently emerged. During the twentieth century, more than half the world's wetlands have been lost along with their valuable environmental services. Increasing urbanization pollutes clean water supplies and much of the world still does not have access to clean, safe water.
- Forest cover—Since the Neolithic Revolution (which saw wide-scale transition of many human cultures from a lifestyle of hunting and gathering to one of agriculture and settlement), about 47% of the world's forests have been lost to human use. Present-day forests occupy only about a quarter of the world's ice-free land with about half of these occurring in the tropics. While forest cover In temperate and boreal regions is gradually increasing, deforestation in the tropics is a major concern.
- Land use—Over the years, land use has changed for the worse due to agriculture, urbanisation and industrialisation. Feeding more than seven billion people

takes a heavy toll on the Earth's resources (about 38% of the Earth's land surface). Added to this are the resource-hungry activities of industrial agribusiness—everything from the need for irrigation water, synthetic fertilizers and pesticides to the resource costs of food packaging, transport and retail.
- Global warming and climate change—Apart from greenhouse gases that cause global warming, ocean circulation patterns also have a strong influence on climate/weather and, in turn, the food supply of both humans and other organisms. A sudden alteration in circulation patterns of ocean currents could drastically alter the climate in some regions of the globe. Almost ten per cent of the world's population—about 600 million people—live in low-lying areas vulnerable to sea level rise.

Ozone depletion—use of chemicals like chloro-fluorocarbons in packaging and some industrial processes is a serious problem.

Preservation of all living species and plant life and protection of their habitats and natural food chains. Accumulation of plastics in oceans and marine habitats is a major problem.

- Energy—All the most widely used processes for generating electricity create warming and/or release of greenhouse gases. Usage of energy also creates warming, whether for domestic purposes or industrial activity. Generation of energy also uses up scarce
- non-renewable resources like coal or other fossil fuels. Sources of renewable energy need to be actively pursued and used more widely.
- There are also some human activities that impact on the environment. In the context of packaging, the most important of these is the generation of post-consumer waste and the management of its collection and disposal.

9.5 Some Guidelines

Given that these concerns need hardly be emphasised often enough, the simplest approach to sustainability is to try and lay down some basic guidelines which will form the basis for formulating an ideal and workable strategy. Some of these may be simplistic but, nevertheless, need to be stated so that nothing is overlooked. In no particular order of priority, we could list these guidelines as follows:

- Preserve the quality of the environment at all costs.
- Prevent depletion of resources, both natural and man-made.
- Conserve and/or reduce energy usage and constantly work at improving energy efficiency.
- Reduce waste, both during productive processes and post-consumer.
- Use renewable and recyclable or reusable materials/systems.
- Use cleaner and greener processes.
- Safely recover all materials. This includes biological means, if necessary.
- Avoid or reduce greenhouse gas emissions.

One can see that, unfortunately, some of these objectives are in conflict with others. Let us take an example to illustrate this, drawing upon a standard packaging operation viz. printing. World-wide, there is a move towards using water-based inks in place of solvent-based systems because they are greener, safer to handle and cause no VOC emissions into the atmosphere. The downside of this is that water-based inks are much more difficult to dry and need much higher energy for the purpose and take a longer time. There is, thus, a clear conflict of interest. Another thing that has to be kept in mind, and this is unfortunately a vital criterion in the practical world, is that what one does has to be economically viable and cannot end up being less cost-effective than the process one is trying to replace or modify. This is why we need to devise a scientific, analytical and quantitative approach to sustainability instead of going by notions or instincts. As we will see later, this is the reason for developing a 'score-card' system that accounts for both the positives and the negatives of a process with the ultimate objective that all technologies or processes must be burden neutral.

Having come this far, let us now see if it is possible to generate a definition of what constitutes Sustainable Packaging. A lot of work is still being done to arrive at such a definition but the most definitive work on this has been carried out by the Sustainable Packaging Coalition, an industry-wide coalition in the USA. They have developed a set of criteria that could form the basis of a universal acceptance of what is sustainable packaging. According to the SPC, sustainable packaging:

- Is beneficial, safe and healthy for individuals and communities throughout its life cycle.
- Meets market criteria for performance and cost.
- Is sourced, manufactured, transported and recycled using renewable energy.
- Maximises the use of renewable recycled source materials.
- Is manufactured using clean production technologies and best practices.
- Is made from materials healthy in all probable end of life scenarios.
- Is physically designed to optimise materials and energy.
- Is effectively recovered and utilised in biological and/or industrial cradle to cradle cycles.

9.6 Hierarchy of Options

People have tried to develop a hierarchy of options in packaging design for sustainability and this keeps changing and evolving all the time. For a long time, people used the 3-R (reduce, reuse, recycle) system as the guiding principle and it seemed good enough. However, over the last few years, people have realised that there is more to it than just the three R's. This has now been modified to 5 R's. The highest position in the hierarchy is now "remove".

In other words, the first priority is to remove any unnecessary or extraneous components of the packaging and cut down on parts that are not really required, e.g. secondary or tertiary "layers" that are not really needed. The fifth R stands for "renew or recover". This means that, in case it is not possible to either reuse or recycle the packaging, then it must be renewed or the inputs recovered to either make it suitable for use in a new application or to ensure that some of the input "burden" is negated to the extent possible and some value recovered. This latter option includes incineration to recover energy although one would exercise this option only as a last resort when absolutely nothing else can be done with the waste.

Walmart, one of the leading drivers in the sustainable packaging movement, go even further. They follow what they call their 7-R philosophy, in which they add two more R's to the five listed above—"revenue" and "read". The first implies that not only does sustainable packaging help protect the environment and conserve scarce resources, it makes outstanding economic sense as all such measures lead to only one thing—money saved! This is as powerful an incentive as any. They also stress "read" as the way to keep oneself abreast and to educate and share/exchange information with system partners and other people to keep taking the process forward.

9.7 Evaluating Sustainability

There is no gainsaying the need for evaluating sustainability using a sound scientific and quantitative method. The most basic reason for evaluation is the fact that, while implementing measures to improve sustainability, there is quite often a conflict between the various basic objectives, as explained earlier. Therefore, it is essential to know whether a particular exercise has had a net positive contribution or not when compared with a reference system or with other alternative systems. The evaluation is usually carried out by conducting what is called a Life Cycle Analysis (LCA). Originally, the LCA was based on a "cradle-tograve" approach that studies the system through its whole life cycle from birth to the stage where it "dies" or exhausts its usefulness at disposal. However, the latest concept is that LCA must follow a "cradle-to-cradle" approach because, at the end of a system's life-cycle, all its components must either be reconverted into a usable form or, if this is not feasible, as much of the inputs must be recovered as possible for reuse.

An LCA studies and evaluates the net burden of the whole process or system—materials, resources, energy, emissions and waste—and arrives at the net burden throughout the life cycle. The life cycle itself is broken up into 5 phases as follows:

Phase 1 Production of raw material
Phase 2 Manufacture of containers or converted packaging materials
Phase 3 Transportation
Phase 1 Waste management
Phase 1 Reuse/recycling/renewal/recovery

All systems must ideally be burden-neutral or, if possible, burden-negative.

Many case-studies have established that the one phase in the LCA that generally has the most profound effect on the bottom line is Phase 3 because it usually accounts for the maximum amount of energy expended and emissions generated in the entire life cycle.

This phase consists of two parts—transportation of packaging materials (commonly referred to as "packaging miles") and transportation of packaged product (commonly referred to as "product miles"). This is why factors like the packaging materials, the package form, the weight per package, the distance from which the packaging is sourced, the distance over which the packaged product is distributed and cube utilisation during storage and transportation all play a vital role in the whole process. This is also why it has been generally found that systems like flexible packaging are the most desirable because they are lightweight, offer the best cube utilisation and are eminently suitable for in-house form-fill-seal operations; all this drastically reduces the impact of both packaging and product miles.

It can be emphatically stated that light-weighting is practically everything and far outweighs any other consideration in calculating net burdens. Lighter-weight materials also produce less post-consumer solid waste. Another very important criterion is energy consumption and this can be directly related to the conservation of the environment. By and large, the processes that generate energy or involve its consumption are, by their very nature, sources of warming. Therefore, more energy consumed directly equates to more generation of heat or generation of greenhouse gases. It is very important to appreciate the importance of these two criteria to accurately analyse and comparatively rate competing systems.

The Walmart scorecard, which was developed and announced in 2007, is used by them to evaluate vendors and reward suppliers who have helped them to achieve their avowed objective of reducing their packaging. This is a good example of how sustainability of packaging can be measured. In this scorecard, Walmart have ascribed weightages to the various factors involved as follows:

- 15% Greenhouse gas or CO_2 per MT of production
- 15% Material value
- 15% Product/package ratio
- 15% Cube utilisation
- 10% Transportation
- 10% Recycled content
- 10% Recovery value
- 5% Renewable energy
- 5% Innovation

As we can see, the two major contributors to Phase 3 of LCA, i.e. cube utilisation and transportation, together make up as much as 25% of the total weightage.

9.8 Debunking Some Myths

There are some common myths and mistaken notions about sustainability that are, unfortunately, widely prevalent. The first of these is that all packaging made from natural and renewable sources (agricultural materials like corn, starch or wood) are always more environment-friendly. Based on existing knowledge and, when viewed scientifically, this is largely incorrect. While it is true that useful packaging materials have been developed from natural and renewable inputs, e.g. paper, cellulosic films and biopolymers like polylactide (PLA) resins, these generally have two basic handicaps:

- they usually have a higher specific gravity than those of materials like synthetic polymers like commodity plastics, and
- they also have poorer barrier properties, which means we need to use them in higher thicknesses as compared to synthetic polymers to obtain equivalent performance.

In effect, we could actually end up using higher-weight materials and this could mean a higher net environmental burden. There is no denying the fact that, being made from renewable agricultural resources, they greatly help in conserving scarce petroleum or natural gas reserves but that is another issue altogether.

Use of paper/board (which is sourced from natural resources) has also been touted as more sustainable than other materials but this is usually not true. Every tonne of paper produced uses up three tonnes of water, which the paper-making process converts into polluting or toxic effluents that need to be effectively treated before release. Then, again, recycled printed paper/board is not recommended for direct food contact because printing inks used on paper are usually based on mineral oils that migrate into food products and are toxic. (The government of India has recently banned the use of printed paper/newsprint to directly wrapped foodstuff.) Hence, a total review of the system needs to be carried out before evaluating whether it is more sustainable.

9.9 Biodegradability

Myth number two is that all bio-degradable packaging is always more environment-friendly. The issue of bio-degradability is very complex and the perspective has undergone a drastic change ever since we have established that arguably the most serious and pressing problem that the environment is faced with is that of global warming. This warming is due to generation or emission of "greenhouse" gases like CO_2, methane, So_x, NO_x and sulphur hexafluoride. When substances break down chemically or biologically in nature (called aerobic degradation), the decomposition generates carbon compounds. But, in real life, most organic waste like vegetation and food scraps and degradable packaging materials actually land up in land-fills where they undergo anaerobic degradation and they end up generating methane, a greenhouse gas that is some 20 times as potent as CO_2 in terms of warming.

When we consider the fact that the decomposition process of degradable materials in landfills is extremely rapid in the presence of wet organic waste like food scraps with the emission of deadly methane, it becomes imperative that they be kept out of land-fills. (Landfills are the most dangerous man-made sources of methane.) Ideally, degradable materials should be disposed of only in closely managed composting programmes (called industrial composting) that are designed to provide optimal oxygen, moisture and temperatures (usually much higher than those encountered in backyard composting) with controls on emissions to deliver a carbon–neutral process. Such industrial composting facilities are extremely scarce even in places like the USA. There is also now a serious shortage of land for land-fills and, consequently, land-fill charges are becoming exorbitant. Therefore, the latest initiatives in the advanced countries are now actually targeted at keeping organic waste out of land-fills at any cost and the mantra is—resort to land-fills only if no other reuse/recycling/recovery or incineration option is available.

9.10 Some New Concepts on Resource Utilisation

The latest concepts in Sustainability are all centred around building Circular Economies.

It all started off with the realization that the a major focus of Sustainability has to be the conservation and preservation of our natural resources—especially non-renewable resources and those that are not renewable in the near term (like coal or fossil fuels)—from emphasis on just the protection of the environment and eco-systems. While the latter is certainly important to sustain the quality of life and to prevent deterioration of living conditions, a very real threat is that we are going to run out of resources some time in the future.

Indeed, the World have had an excellent track record of technology development over the last four decades in alleviating the ecological situation. For example, when the global energy crisis hit us in 1974, the countries were actually flaring every bit of natural gas instead of putting it to use or focusing on developing technologies to make its use as a feedstock commercially viable. Since then, everyone have been able to see this happen and, if anything, natural gas has become the preferred feedstock for petro-chemicals and heating/cooking (both domestic and industrial). Likewise, electrical and hybrid vehicles are soon expected to overtake conventional vehicles that use traditional IC engines based on fossil fuels. Industry is also constantly working on efficiency in usage of energy and natural resources. While all this will certainly mitigate the situation in the short to medium term, this will only help in postponing the inevitable depletion and eventual drying up of reserves of non-renewable resources.

This is where Circular Economies come in. The world need to rapidly work towards development of Circular Economies to replace our Linear Economies which overwhelmingly rule the roost today.

9.11 Linear Economies

Today, the world is totally dominated by Linear Economies where the linear "take, make, dispose" economic model rules supreme. Economic development and manufacture and consumption of products has evolved solely on linear designs. This model relies on large quantities of cheap, easily accessible materials and energy. The sequence of all manufacturing activity and consumption is:

- Manufacture/acquire all required resources
- Convert these resources into products
- Use these products
- Dispose of residual product waste after use

The following diagram sums up the approach of Linear Economies (Fig. 9.1).

While this model has been highly successful in generating industrial development and engendering high rates of growth, no deliberate attention is paid to conservation of resources. This has necessitated rethinking on use of energy and resources. Even though the design philosophy of product cycles that evolved was a "cradle to grave" concept which does focus a lot of attention on handling and disposal of post-use waste as part of the overall life cycle, there was no conscious effort to see that resources used were either re used/recycled or regenerated to enable their fresh use as inputs for new product cycles. While the focus on reduction of resource use and reduction of fossil energy per unit of economic output in the twentieth and twenty-first centuries has been noteworthy, this has only succeeded in postponing but not avoiding the impending resource crunch, albeit in the longer term future.

Even the most advanced linear economies are surprisingly wasteful in their models of value creation. In Europe, material recycling and waste-based energy recovery

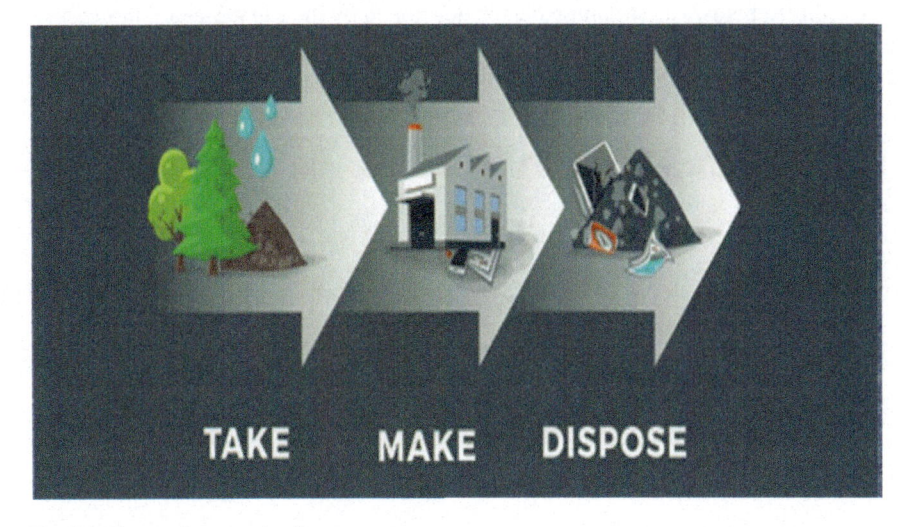

Fig. 9.1 Types of economies-linear economy

captures only 5% of the original raw material value. Analysis has also found significant structural waste in sectors that many would consider mature and optimized. For example, in Europe, the average car is parked 92% of the time and 31% of food is wasted along the value chain. Rapid urbanisation is only going to make things worse. (Overall demographic growth is projected to bring the global proportion of people living in cities to 66% by 2050.)

All this led to the "cradle to cradle" concept that gave rise to thinking of Circular Economies as the way forward. The rationale for transitioning to a circular model is increasingly gaining ground.

9.12 Circular Economies

What is a Circular Economy? A circular economy is one that is restorative and regenerative by conscious design. It aims to keep products, components, and materials at their highest utility and value at all times, distinguishing between technical and biological cycles. A circular economy addresses and focuses completely on optimal usage of resources. The accent is to try and use renewable resources wherever possible; even if non-renewable resources are used, they have to be restored or regenerated at the end of the product life cycle so that they can be used once more in new product cycles so that there is no drain on finite reserves of these non-renewable resources. This new economic model seeks to ultimately decouple global economic development from finite resource consumption. While optimally managing the flow of renewable resources. The Circular Economy deliberately designs out wastage from the system (Fig. 9.2).

The fundamental design of this model is based on circular loops and loops within loops. Given below is a structural design developed by the Ellen Macarthur Foundation (the pioneers and prime movers of the Circular Economy concept) and McKinsey together with the World Economic Forum [3] (Fig. 9.3).

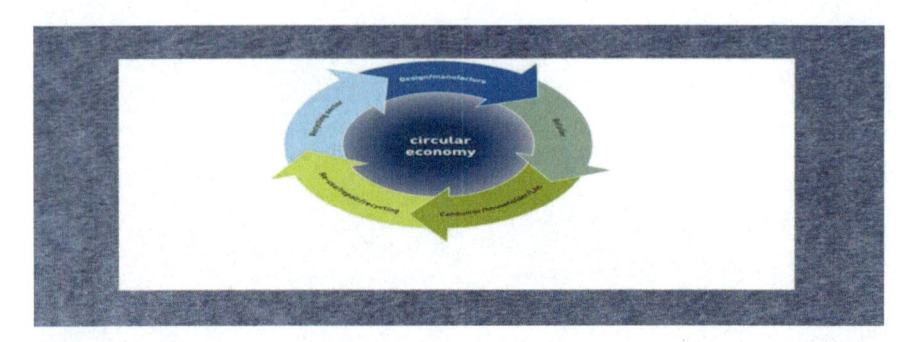

Fig. 9.2 Types of economy-circular economy

9.12 Circular Economies

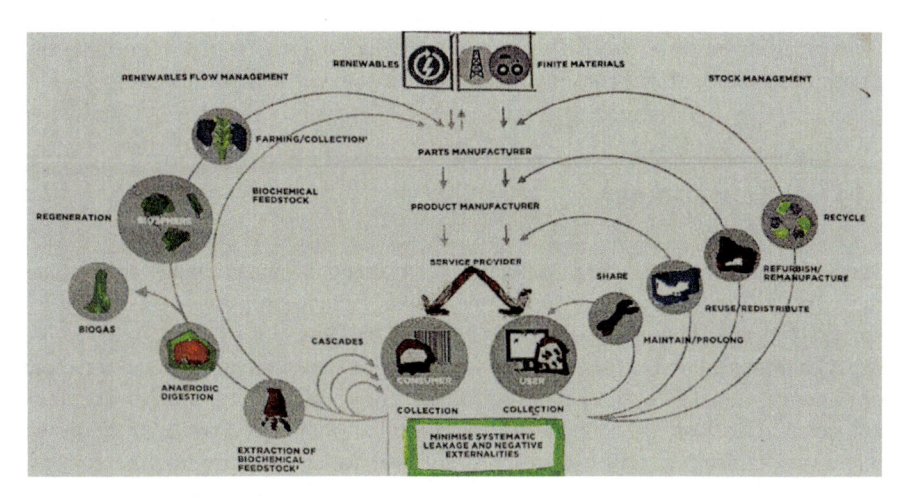

Fig. 9.3 Loop diagram on economy

One of the sectors that has been identified as key to the success of implementing Circular Economies is Plastics—especially Plastic Packaging. (The Packaging sector accounts for about 40% of global Plastics usage.)

While the concept of Circular Economies is truly futuristic and not so easy to implement in the short term, some countries like Japan are well ahead of the curve and really need to be complimented for their efforts. As far back as 2000, Japan developed and enacted their Law For The Promotion of Efficient Utilization Of Resources. The law treats materials as vital resources and mandates the use of disassembly plants aimed at recovery and restoration of materials. This has led to significant recovery of many materials like metals that were hitherto deemed as waste and consigned to disposal systems like landfills.

9.13 Thinking Out of the Box

There are some interesting examples of out-of-the-box thinking that has contributed to making systems more sustainable.

The first of these is a move by the government of Australia. Some years ago, they completely banned the use of conventional filament-based light fittings and mandated that all lighting has to be based on CFL fittings or fluorescent tubes. This immediately saved an immense amount of energy and considerably reduced warming.

Although there was an immediate increase cost by way of initial investment, there was actually a saving in cost in the long term due to the longer life of CFL fittings and savings in energy consumption.

The second example is a move by Walmart. Now, Walmart is a massive organisation with millions of employees world-wide. They decided to cut down on the size of their calling cards by 30%. This immediately led to a saving of several thousand tonnes of paperboard and conservation of a critical material input/resource.

9.14 Greenwashing

One undesirable result of the Sustainability initiative is what is termed "greenwashing". This is the misleading of opinion by falsely claiming positive sustainability benefits and reporting claims not entirely based on facts. For example, when some FMCG companies first tweaked formulations of cleaners/detergents by making them more concentrated, they initially made fancy claims about how much water they had saved. This was not truly valid because, while they saved water in their own operations, consumers still had to use additional water while mixing the concentrates for use in their homes. Thus, there was no real saving of water in the total system. However, to their credit, what they did achieve was a significant saving of packaging materials and systems because the number of bottles/intermediate packaging/shipping cartons needed came down; this should have been correctly reported as such. Many LCA's are also cited to clam improved sustainability but, in fact, the terms of reference of LCA's have to be properly studied before such claims can be termed acceptable.

There are now strict guide-lines on reporting and the powers that be frown on misreporting and sustainability claims not backed by facts or scientific evidence.

9.15 Statutory Mandates

The Government of India has enacted two major acts on Plastic Waste Management, especially Packaging Waste.

The first one, notified on the 18th of March, 2016, was called the Plastic Waste Management Rules 2016 (PWMR 2016) [4], and tried to lay down rules in accordance with concepts enshrined in the principles of Circular Economies. It mandated that systems for the management of Plastic Waste were part of extended producers' responsibility.

The rules very categorically stated under Responsibility of Producers, Importers and Brand Owners that:

- The *producers*, within a period of six months from the publication of these rules, *shall work out modalities for waste collection system based on Extended Producers Responsibility and involving State Urban Development Departments, either individually or collectively, through their own distribution channel or through the local body concerned.*
- *Primary responsibility for collection of multi-layered plastic sachets or pouches or packaging is of Producers, Importers and Brand Owners who introduce the*

products in the market. They need to establish a system for collecting back the plastic waste generated due to their products.
- ***Manufacture and use of non-recyclable multi-layered plastic shall be phased out in Two years time.***

The rules further specified that *"No producer shall on and after the expiry of a period of Six Months from the date of publication of these rules in the Official Gazette manufacture or use any plastic or multilayered packaging for packaging of commodities without registration from the concerned State Pollution Control Board or the Pollution Control Committees."*

While the intentions of the Government to push the economy towards a Circular Economy were noteworthy, the provisions of this act were extremely stringent and totally impractical. In fact, they were quite impossible for industry to implement. A large proportion of the Packaging Systems in use were based on multilayered plastic structures or combinations with metallised films/metallic foils for adequate barrier/shelf-life and were no longer permissible, if the act was to be strictly followed. This would create Constraint in packaging industry as the industries are not equipped with proper R&D for alternative materials.

Based on feedback from various quarters, the Government realised the problem of fully implementing the act and they, therefore, replaced this act with a new one. In a major relief for Industry and consumers, the Government issued new rules for Plastic Waste Management that superseded the PWMR 2016 and allowed the continued use of Multilayered Plastic Structures. The new rules—called the Plastic Waste Management Rules, 2018—were notified on the 27th of March 2018 [5].

In a major change of thinking, the Government introduced the concept of a new 'alternate use' process viz. 'energy recovery', which it defined as "energy recovery from waste that is conversion of waste material into usable heat, electricity or fuel through a variety of processes including combustion, gasification, pyrolisation, anaerobic digestion and land fill gas recovery". It now banned the use of only "multi-layered plastic which is non-recyclable or non-energy recoverable or with no alternate use". In effect, almost all existing packaging structures in use could continue to be used. However, the extended producers' responsibilities of brand owners and manufacturers continued to be in force. How effectively these will be fulfilled can only be seen and evaluated over the next couple of years depending on what brand owners/producers do to take on these responsibilities.

9.16 Final Recommendations

There is no doubt that all corporate organisations need to take a triple bottom line approach for monitoring and reporting results (economic performance, social performance and environmental performance) for integrated sustainability and clearly identify it as an overriding priority. The commitment has to be a key corporate objective and driven top down.

Here are some pointers on how to manage the sustainable packaging process:

- Design life-cycles, not packages.
- Evaluate scientifically, not by instinct.
- Always study the whole 'system'.
- More energy used = more warming.
- Less is more: light-weighting is everything.
- Removal of unnecessary packaging and/or reduction in packaging is the most effective option.
- Use recycled materials and facilitate recycling (e.g. on a PET bottle, use a PETG label and not a paper one).
- Phase 3 of LCA requires the most focus.
- "Plastics" is not a dirty word. In fact, they are the lowest-burden options.
- "Natural" is not always better. Source of materials is not very relevant.
- Agricultural sources should be thoroughly evaluated for availability and the impact their diversion will have on the food chain. (See what is happening when corn is diverted for bio-ethanol for automotive fuel.)
- Think out of the box (e.g. higher product concentrations for liquid detergents/chemicals, higher product/package ratios, packs that can be reused for other applications, refill concepts etc.).
- Dissemination of knowledge within the organisation and to system partners (suppliers, consumers, designers etc.) plays a vital part in the process.

Finally, here is the most powerful reason to go "green"—the icing on the cake is that all the measures that need to be taken viz. source and material reduction, energy conservation, savings on transportation costs, recovery of materials etc. all lead to significant economic savings and healthier bottom-lines.

Art the end of the day, Sustainable Packaging actually creates more wealth for oneself and the world thinks the world of you.

References

1. United Nations General Assembly (20 March 1987) Report of the World Commission on Environment and Development
2. United Nations General Assembly—The Sustainable Development Goals
3. Ellenmarcarthur circular economy diagram, 2019. https://www.ellenmacarthurfoundation.org/circular-economy/concept/infographic
4. Govt. of India—Plastic Waste Management Rules 2016
5. Govt. of India—Plastic Waste Management Rules 2018

Chapter 10
Life Cycle Analysis of Packaging Material

Anup K. Ghosh and Susmita Dey Sadhu

10.1 Life Cycle Analysis (LCA)

Life cycle analysis is one of the most crucial parts of commercial application of any material. Any product starting from a small candy to any luxury car or anything has certain impact on lifestyle, environment, society and economics of that area. Environment, society and economics are interconnected with each other. If one of them gets adversely affected, the other two will automatically respond to that change. So, to maintain a balance, we need to consider the economic, social and environmental aspects of products, such as food packaging in various forms coming to the market. When the balance is struck among the economy, environment and society, then only the packaging material as well as the product becomes sustainable.

Sustainable development is defined in literature as "development that meets the needs of the present generation without compromising the ability of future generations to meet their own needs" [1].

To produce a sustainable product, the impact of that product during its life cycle needs to be quantified. A product or service is designed because it is required by the people in the society. Sometimes there are a few applications and ideas which have brought revolution to the society. For example, when mineral water in plastic bottles was introduced in India, anyone hardly imagined that it will grow as such a big industry. In fact when the first plastic was invented, there was hardly any idea that within a century plastic will find its application almost in all spheres of life. After completing the lifetime, the product either goes to the grave or again recycled to produce a new product. For sustainable usage of the product, the resources should be recyclable and economical. The relation between sustainability, environment, society and economy is depicted in Fig. 10.1. LCA provides the tools and the methodology to quantify its effect on all the three spheres during its life cycle.

Sustainability takes into account each and every small detail of a product at micro or macrolevel. Life cycle may be defined as "compilation and evaluation of the inputs, outputs and the potential environmental impacts of a product system throughout its life cycle" [ISO. Environmental management-life cycle assessment—principles and

Fig. 10.1 Components of sustainability

framework (ISO 14040). Geneva: ISO; 2006]. LCA considers the beginning of "life" of a product at the stage of designing of the product followed by the resource and raw material procurement, product manufacturing, storage, transportation, consumption, collection of wastes and disposal/recycling to complete destruction of the material or production of the new product again. Some items are recycled completely, so that a cradle to cradle approach may be adopted. The LCA approach for the products which undergoes complete destruction after its lifetime is completely different from the earlier one. But both the products will have an impact on its surrounding as the resources are taken from the environment.

LCA may use various methodologies to calculate a products impact in terms of energy, emission or water. LCA estimates and assesses the impacts that attribute to environmental changes like climate change, global warming, land, water and air pollution and depletion of resources and consequently gives a better representation of the impact of the product. A product may have a better performance compared to its opponent interns of emissions but may pollute the soil much more adversely on disposal and degradation. Hence, LCA of emission may not always give a clear picture of the environmental impact of the better or preferred product [2].

For LCA estimation, the data collection is the most important and most challenging step. In most of the cases, the product manufacturer and the LCA analysts both do not have any clear idea on the amount of energy, water and emissions involved in every step of the product life cycle. A large number of data need to be collected and documented to quantify the impact of a product on various aspects of life. Moreover, in LCA, the determination of the starting point and ending point is one of the trickiest parts. For example, if the LCA of a PET bottle needs to be calculated, the energy calculation may start at the point where the PET granules are purchased. Another calculation in which the production of PET from its raw material and the transportation of the raw material to the production site is also involved may change the final outcome in a completely different manner. Again, as PET is recycled to produce many other important commodities, it changes the energy calculations once again. LCA calculation of the a multiproducts system is another challenge as the way of partitioning of all the contributing factors for a particular product will decide the final fate of the product.

Life cycle analysis is still in its nascent stage but plays an essential and important tool to estimate the effect and impact in various sphere of life during its lifetime by virtue of its presence at the interphase of different aspects.

10.2 Definition

LCA is an environment tool, which analyzes the complete environmental impact of a product during its lifetime. It can also be named as the "cradle to grave" analysis [1]. LCA analysis includes the impact of the raw materials during (1) synthesis or extraction (2) during the production stage (3) at the time the product is being and (4) during the disposal. Figure 10.2 shows phases of life cycle analysis.

Different types of energy consumed during the whole life cycle (mainly electricity, energy derived from fossil fuels, etc.), the consumed resources whether renewable or non-renewable (water, nutrients, chemicals, fuel, etc.), liquid effluents contaminated with chemicals or solid suspended particles (hydrocarbons, nitrates, heavy metals, sulfates, etc.) and biochemical oxygen demand, gaseous emissions (CO_2, NO_x, SO_2, etc.) are considered collectively for the analysis. The health hazards related to processing of jute and its adverse effect on human health and chemical pollution in case of paper production have also been taken into account.

10.3 Methodology

Primarily four different stages were described for any LCA calculation.

a. *Goal and Scope Definition*

Goal and scope defines a boundary for the LCA calculation of the product system for one unit of the product. The unit of the product is important not only for measurement purposes but for comparison with other product system. Further for the manufacturing of a unit amount of a material or product, the goods, services, packaging material, manpower, transport, energy and establishment cost may be calculated. When two or more products are manufactured in parallel, then the partition of all resources and impacts may also be done for each unit of product.

b. *Inventory Analysis*

Since the life cycle analysis is done based on the amount used and emissions and waste generations due to the product, the results may vary as follows:

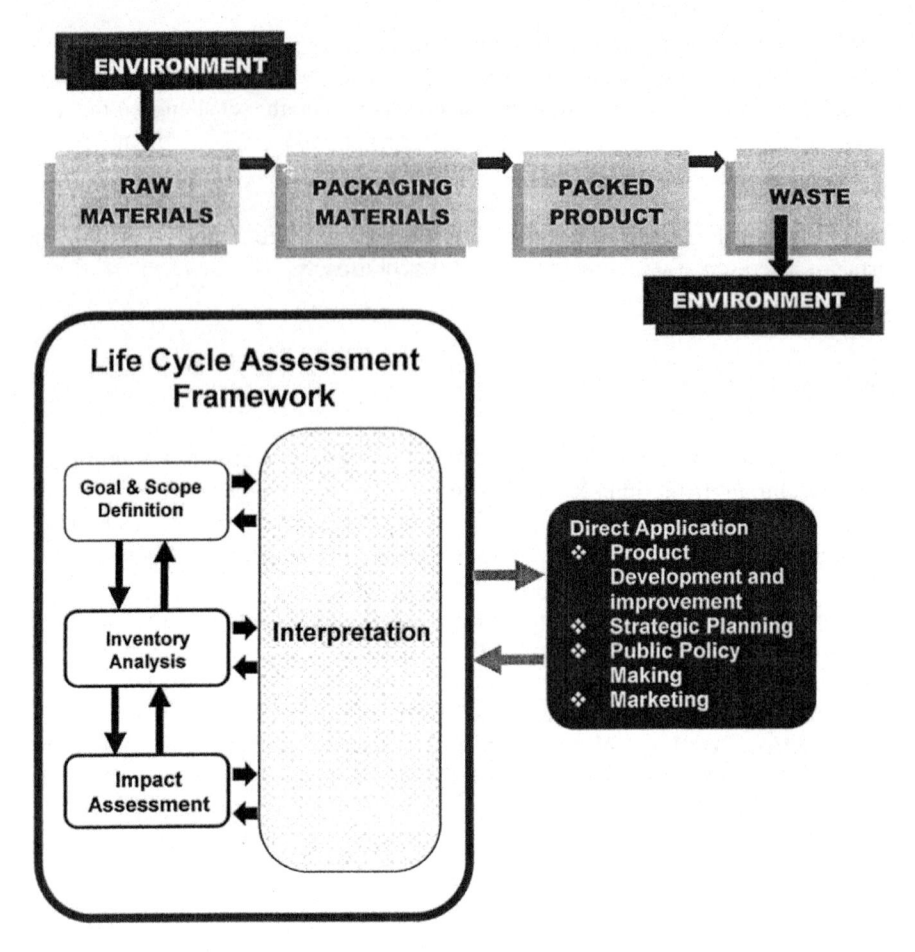

Fig. 10.2 Phases of life cycle analysis [3]

I. From one region to another

The cost of manufacturing depends on things like labor cost, availability in that area, transportation cost, distance of travel, etc., and thus, the LCA changes from place to place.

II. Over a time period

When the environment impact is measured, for example, based on the calculation of the CO_2 emission, the same time period should be considered for comparison of various products. The total amount of CO_2 emitted will vary if the time of CO_2 collection varies.

III. In different time period

10.3 Methodology

LCA measured during a particular time period does not remain same over a different time period. The cost of things, environmental aspects, etc., keep on changing which will also affect the LCA.

c. *Impact Assessment*

Impact assessment focuses on providing data and indicates the estimated impacts on various aspects like human life, society and environment. It gives a quantitative as well as qualitative evaluation data of a product life cycle.

d. *Life Cycle Improvement*

Life cycle improvement is the analysis of LCA data of two or more products to suggest and make changes in the present LCA of a product. This gives an opportunity to estimate and understand the impact of any change made to a particular product line. For example, if an alternate route is suggested for manufacturing a chemical, then all sorts of impacts of both the routes may be calculated and compared. This will lead to a better understanding of the best suited path for the manufacturing of the product. Thus, the LCA measurement and its result will vary depending on factors like number of phases which are considered for the calculation, with change in geographical area, over a time period and with changed time periods.

As the life cycle analysis is gaining its popularity and importance, a need was felt to have a systematic method for calculating the life cycle analysis. In view of this, ISO standards are made for providing a framework for LCA:

- International Standard ISO 14040 (1997) on principle and framework
- International Standard ISO 14041 (1998) on goal and scope definition and inventory analysis
- International Standard ISO 14042 (2000) on life cycle impact analysis
- International Standard ISO 14043 (2000) on life cycle interpretation.

When the data for LCA calculation is collected, the data should be tabulated for the consumption of energy, water, emission and any other resources at every stage of the product life cycle. This includes the raw material collection and extraction, energy usage, transportation, packaging, storage, recycling and final disposal or recovery from that waste. Depending on the nature of the product, either cradle to cradle or cradle to grave approach may be adopted. A model of resource and pollution flows is shown in Fig. 10.3 for complete recycle of plastic product.

10.4 Life Cycle Impact Assessment

The importance of probable environmental influences is analyzed by the life cycle impact assessment (LCIA) phase [1]. The LCI results are used to analyze inventory data as well as a number of environmental impact and environmental category indicators. The most important factors in the LCIA phase analysis are (ISO 14040:2006) (i)

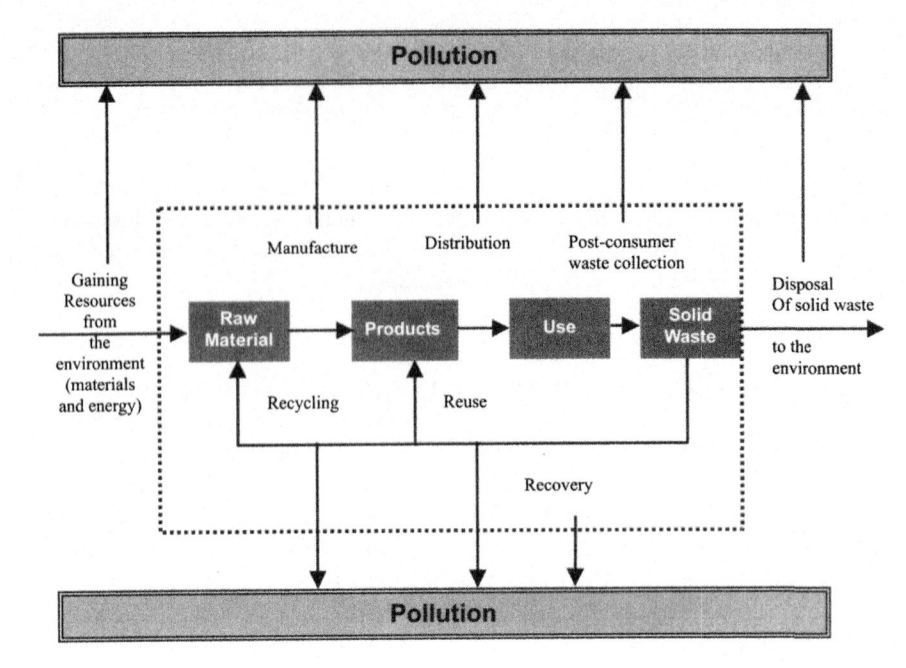

Fig. 10.3 Model of resource and pollution flow in the environment, for the complete life cycle of a product

selection of impact categories, (ii) category indicators, (iii) characterization models, (iv) assignment of LCI results (classification) and (v) calculation of category indicator results (characterization). Furthermore, normalization, grouping and weighting can be carried out. The assessment includes the following impact categories:

1. **Climate change** is brought about by the increase in the amount of greenhouse gases and results in warmer climates, globally.
2. **Acidification** is caused by emissions of sulfur dioxide and nitrogen compounds in the atmosphere. It negatively impacts the green cover and the pH balance of aquatic ecosystems.
3. **Eutrophication** is because of the infiltration of waterways by phosphorus and nitrogen. It eventually leads to a rapid growth rate of aquatic organisms.
4. **Photo-oxidation** is caused due to the generation of chemicals (such as ozone) from hydrocarbons and nitrous oxides under bright sunlight. High concentrations of ozone in the upper atmosphere have a detrimental effect on living organisms.
5. **Particulate matter formation**. Industrial activities, energy generation and transport activities lead to generation of small particles. Minute particles have the ability to infiltrate into the human respiratory system and cause disorders.
6. **Fossil fuel depletion**. Recent human activity has led to reduction of non-renewable resources including fossils, mineral as well as certain metals.

10.4 Life Cycle Impact Assessment

7. **Human toxicity**. There are three major routes through which human health is exposed to harmful substances: through drinking water and food, through the air we breathe and through the skin.

The LCI components are classified and characterized into impact category indicator results in the first step of LCI analysis. It is followed by standardization of the impact category indicator results. Weighing may be done to collate the impact category results into a single impact value with the description of the product systems with overall impacts on the surrounding.

10.5 Carbon Footprint

The measurement of carbon footprint involves life cycle and life cycle assessment (LCA). Carbon footprint is the amount of greenhouse gases (GHGs) generated throughout the entire life cycle of a product. Carbon dioxide equivalent of the GHG emissions is calculated using the data of 100-year global warming records. In the carbon footprint calculations, all the greenhouse gas emissions are considered. However, the carbon footprint of a fiber-based print product majorly consists of methane and nitrous oxide, in addition to carbon dioxide. The global warming potentials of these three GHGs are carbon dioxide (CO_2) = 1; methane (CH_4) = 25; nitrous oxide (N_2O) = 298. Despite the standards ISO 14040 and 14,044 being used for LCA methodology, the calculation of carbon footprint requires different guidelines for carbon-specific features.

10.6 Significance of LCA Study in Packaging Material

When the case of packaging material is considered for LCA, the basic purpose is to compare different products with same application to find out the most suitable packaging material with minimum adversity to environment. During this type of calculations, different types of materials like polymers, glass or metals are compared with each other. When two different materials are compared, the cost and energy requirement in preparation and extraction of raw materials, manufacturing the article, transportation, the efficiency as packaging material, whether it requires any extra safety component or not, its storage and distribution and finally disposal and recycling are considered. Apart from cost and energy requirement, other resources like water and air are also considered along with the carbon footprint that it generates and emits in the environment.

When LCA of a package is considered made from same material, even its design plays a very important role. For example, comparison may be done between an oil package of 5 lit and 5 packages of 1 lit oil pouch. Same material may be used in both the cases. But difference in LCA will arise in the amount of polymer used, the energy consumption for manufacturing the packages, then transport and storage mode, the shelf life of the eatable within those packs and the disposal and recycling modes of the products.

Life cycle analysis using "cradle to grave" approach is the only way to assess and compare the benefits for bulk packaging material by identifying inputs and outputs in the different phases of the life cycle analysis which involves two stages: The first stage consists of an process that accounts and produces an inventory of all the inputs and outputs of energy, material and emission in the life cycle of a product or package. The evaluation of the way the environment gets affected by inventory is discussed in the second. The basis of this study has been considered as one million ton (1 Mt) of bulk commodities in keeping with the view of the consumption in order of magnitude.

To perform LCA of any product, the whole life cycle starting from its birth to death or in other words from "cradle to grave" is to be taken into consideration. Life cycle of bulk packaging materials is divided into four phases: (I) Raw materials manufacture (II) Production of sacks (III) Usage (packaging and transportation) (IV) Waste management. Each phase of the evaluation takes into consideration the energy and water requirement for the operation and the emissions of harmful chemicals and gases. Other effects of these processes on the environment, such as depletion of natural resources and greenhouse effect, are also evaluated. A typical flowchart of product life cycle is shown in Fig. 10.4.

Fig. 10.4 Various phases of life cycle analysis

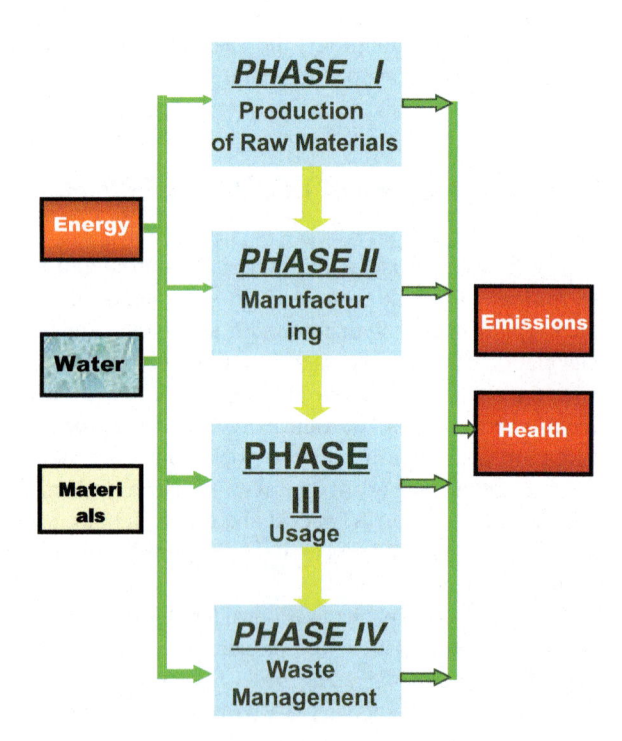

10.6 Significance of LCA Study in Packaging Material

LCA of Plastics in Packaging

10.6.1 Plastic Pouches Versus Glass Bottles in Milk Packaging

In this study, comparison of use of glass bottles and plastic pouches for milk packaging is done as per the LCA methodology [4]. Cradle to grave approach has been taken as consideration in this case. Comparing the relative weight of packaging material, it is seen the value for glass is very high compared with that of plastic pouches. According to the study, for producing plastic pouches for packing of one lakh liters of milk, only 0.40 MT of plastic raw material is required, but for glass packaging of the same, the raw material required is 45.4 MT of glass. Considering all the aspects of phase I and phase II, the energy consumption of glass bottle packaging is very high compared to plastic pouches. Manufacturing and recycling of glass bottles also require significantly high amount of water which is negligible when compared with the production of plastic pouches. There is a considerable amount of energy consumption in glass recycling than that compared with the plastic pouches. Pollution due to emissions of CO, CO_2 and NO_2 during transportation is also high in case of glass bottles as compared to plastic pouches (Fig. 10.5).

1. Energy saving during production of milk pouches is 1165.41 GJ/lakh L of milk which is 32 times lesser per unit packaging as compared to that of glass packaging.
2. Biological oxygen demand (BOD) and chemical oxygen demand (COD) values are 15–20 times higher in the case of glass bottles compared to plastic pouches which lead to severe environmental impact apart from health hazards.
3. Production of glass bottles may be held responsible for the maximum consumption of water which is about six times higher than that of plastic pouch for milk packaging.
4. The energy consumption for 95% reuse of glass bottles is found to be double than that consumed in making new plastic pouches.

Fig. 10.5 Milk packed in glass bottles vis-à-vis in plastic pouches

5. Also with respect to hygiene, it has been found that improper washing of bottles leads to contamination of milk.
6. Incineration of plastic pouches gives energy equivalent to about 16 MJ/kg for plastic pouches, whereas there is no incineration for glass bottles (Table 10.1).

Table 10.1 Life cycle analysis for packaging one lakh liters of "milk" [4]

	Glass		Plastic pouch	
Material required (Mt)	45.4		0.40	
	Energy[a]	*Water*[a]	*Energy*[a]	*Water*[a]
Phase I: production of raw material	671.92	1608.0	32.22	25.6
Phase II: production of bottles/pouches	530.27		4.56	
Total	1202.09	1608.0	36.78	25.6
Phase III: filling and distribution	*Fuel*[a]	*Energy*[a] *Single [Return]*	*Fuel*[a]	*Energy*[a] *Single [Return]*
	2049	114.75 [213.43]	1120	62.73 [106.64]
Phase IV: waste management	*Energy Consumption*[a]		*Energy Consumption*[a]	
Recycling percent				
100%	501.67		4.56	
80%	401.34		3.65	
60%	301.00		2.74	
50%	250.83		2.28	
Reuse (Including Transportation)	*Energy Consumption*[a]	*Water Consumption*[a]	*Energy Consumption*[a]	*Water Consumption*[a]
95%	277.8	509.1	143.4 (New plastic pouches)	26.6 (New plastic pouches)
80%	457.5	675.4		
60%	697.0	897.2		
Incineration	*Energy recovered*[a]		*Energy recovered*[a]	
100%	Not applicable		20.73	
80%			16.58	

[a]Units: Energy (GJ), water (thousand liters), fuel (liters)

Fig. 10.6 Oil packed in metal cans versus plastic containers

10.6.2 Plastics Versus Tin for Oil Packaging

A comparative study has been done on use of tin cans and plastic cans as packing material for oil using LCA methodology. The relative weight of packaging material itself amounts to 86,207 Mt for tin when compared with 63,218 Mt of plastic for packaging one million tonne of lube oil. Energy consumption in case of plastic in phase I is greater than that for the tin cans due to high feed stock energy values of polyethylene used to make plastic cans; production of tin cans is found to be highly energy intensive and in this respect, plastic cans consume less energy. The release of chemicals into water is high for the production of tin cans as indicated by the high value of COD. On the energy front during transportation, there is considerable savings in the use of plastic cans for lube oil packaging as plastic cans are lighter in weight. Reuse of tin cans and plastic cans is not viable option as this will lead to mixing of lube oil particles with the repacked materials. Incineration of plastic cans leads to considerable energy consumption (Fig. 10.6).

1. A 5 L oil plastic can weighs 275 gas against 375 g for a tin can.
2. Tin can consumes about 15% more energy than a plastic cans during production.
3. Tin can consumes 2550 L extra fuel for transporting plastic cans for an equivalent amount of oil.
4. Recycling plastics save about 25% more energy in comparison with tin cans.
5. Incineration of plastics can recovers about 50 GJ/ton of energy, whereas no energy can be recovered from tin cans.
6. High value of COD and BOD compared to 2% in case of plastic cans (Table 10.2).

10.6.3 PP-HDPE Woven Sacs Versus Jute/Paper Sacks

Food grains, fertilizer, cement and such other essential commodities need to be transported in bulk from manufacturer to the consumer in large bags known as bulk packaging bags. In India, the preferred size of such bags has carrying capacity of

Table 10.2 Life cycle data for different materials used for packaging one million ton of oil

	Tin cans		HDPE cans	
Material required (Mt)	86,207		63,218	
	Energy (thousand GJ)		Energy (thousand GJ)	
Phase I: production of raw materials	3846.02		5052.87	
Phase II: production of cans and liners	3638.54		1472.99	
Total	7484.55		6525.86	
Phase III: distribution	Fuel (Tons)	Energy (GJ)	Fuel	Energy
	83,770.49	4691.1	Taken as Basis	
Phase IV: waste management Recycling percent	Energy savings (thousand GJ/86207 ton)		Energy savings (thousand GJ/63218 ton)	
100%	1602.58		1602.28	
80%	1282.06		1296.23	
Incineration	Energy recovered		Energy recovered (thousand GJ/63218 ton)	
100%	Not applicable		3276.61	
80%			2621.29	

50 kg, which could be made either from jute or plastic raw material such as PP-HDPE. Paper could also be used if the capacity of the bag is reduced to 25 kg. Jute is most commonly used as bulk packaging medium in India. Other materials like multiwall paper sacks have also emerged as an alternative to jute. In recent years, plastics woven sacks have emerged as the true cost-effective performance orientated alternative to jute bags due to the restricted and erratic supply of jute, coupled with its drawbacks like poor moisture resistance, lower strength and volatile prices. Study done on the basis of general scheme of life cycle analysis is shown in Fig. 10.7.

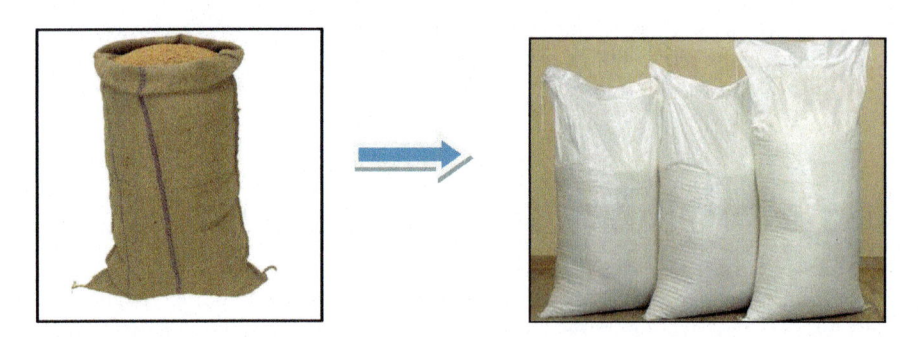

Fig. 10.7 PP-HDPE woven sacs versus jute/paper sacks

10.6 Significance of LCA Study in Packaging Material

Study reveals that PP-HDPE consumes significantly less raw material, energy, water and chemicals than jute and paper [5]. As a result, from the stage of production to transportation, PP-HDPE is more benign to nature than jute and paper. For packaging one million metric ton of bulk cereals with PP-HDPE woven sacks, only 2200 kg of plastic will be required vis-a-vis 13000 kg of jute and 7100 kg of paper. Such huge requirements of raw material deplete and obviously have an adverse impact on the environment for considerable period. The energy, water and chemical consumption, for producing the same quantity of packaging, is phenomenally large in jute and paper in comparison with PP-HDPE. Jute will consume 509.6 GJ of energy, 22.9 lakh L of water and chemicals to the tune of 265 tons/million metric ton of jute (MT). Paper needs 518.3 GJ of energy, 17.75 lakh L of water and 4260 tons chemicals of about 0.013 ton/MT of PP-HDPE. Jute farming also involves a lot of health hazards like respiratory diseases, skin disorders and cancers-arising from NO_2 and CH_4 produced during cultivation. Besides, workers also need to remain for 6 to 10 h in waist-deep water for retting of jute. Moreover, additives like the jute batching oil remain in final jute bags as residue. These residual additives when transported into food substances packed in jute bags cause several illnesses such as dizziness, headache, nausea, and vomiting. The toxic chemicals released by paper industry have high potential to harm earth's life forms. The reuse of PP-HDPE woven sacks in primary form or after minimum physical alteration is to extent of about 51% of domestic consumption of PP-HDPE woven sacks. The reuses are in the areas such as food grain storage, restitched shopping bags, hutment covering, covering for food grains, cattle sheds, poultry sheds, packaging for marble chips, lime stone etc. Non-reusable waste of PP-HDPE woven sack is mostly recycled through conventional routes and converted into products such as ropes, box, strapping, niwar putti, injection moldings, films, kudams, and nearly 46% of the domestic consumption of PP-HDPE sacks is reported to be recycled. The extent of non-usable PP-HDPE woven sacks discarded in the municipal solid waste stream is as low as of 3%.

1. The relative weight of only packaging material itself amounts to a very high value for paper and still higher value for jute when compared with that of PP-HDPE woven sacks.
2. Considering phase I and II together, the energy requirement for paper sacks manufacturing remains extremely high while those for jute bag and PP-HDPE woven sack are comparable.
3. The water requirement for both paper sacks and jute bags is significantly (more than 10 times) higher compared to that for PP-HDPE woven sacks.
4. There is a huge requirement of chemicals in paper bag manufacturing followed by jute bag manufacturing. The requirement of chemicals in the case of PP-HDPE woven sack is negligible.
5. The based jute batching oil (JBO) which is a hydrocarbon can be a potential health hazard.
6. The pollution of water related to the production of PP-HDPE woven sacks is negligible while it is extremely high in case of jute and paper sacks. In terms of BOD and COD, the wastewater from the production of jute and paper

Table 10.3 Net energy consumption considering two end-of-life cases

	NET energy consumption (GJ)		
	Jute	PP-HDPE	Paper
Fresh production	333,000	227,000	670,000
With recovery	333,000	132,000	500,000
With recycling	333,000	180,000	638,000

is highly polluted compared to the pollution in the production of PP-HDPE woven sacks.

7. High emission of methane (CH_4) is recorded during production of jute bags, and significantly excess generation of CO, CO2 and NO_x is noticed in transportation of these bags.
8. In the transportation front, there is considerable saving of energy in the use of PP-HDPE woven sack for bulk packaging as these sacks are lighter in weight than jute bags and the vehicles transporting the commodities have to make lesser number of trips for moving the same quantity of material, thereby reducing the fuel consumption.
9. PP-HDPE woven sacks are fully reusable and recyclable. Non-reusable waste of PP-HDPE woven sacks is mostly recycled through conventional routes and converted into different products.
10. There is a considerable amount of energy recovered when PP-HDPE woven sack goes through either recycling or waste-to-energy process. The recycling in the case of jute bags is generally not practiced. Recycling of paper results in very less energy saving as it has to go through the pulping process which is one of the steps involving major consumption of energy, water and chemicals. Also, because of the low calorific value of paper, the waste-to-energy process results in generation of lower energy compared to PP-HDPE woven sacks (Tables 10.3 and 10.4).

10.7 Conclusion

Though polymer-based packaging is a vehicle for sustainable development, the inappropriate ways of its disposal are fomenting wrong perceptions that it damages the health of environment. It is true that the bags made from these polymers prevent percolation of water in soil, if they are dumped in a landfill. However, the value of same bags could be utilized if they are recycled again. The reuse would consume less energy, save fossil fuels and would not accelerate global warming. There is more to gain from using and reusing plastics. Only we need to learn to manage its wastes carefully, by segregating them from decomposable wastes and channeling them to proper bins, as the developed countries have long been doing. After all, these polymers perform dutifully the role of a carrier from the doors of producer to consumer.

10.7 Conclusion

Table 10.4 Life cycle data for packaging of one million tonne of bulk commodities [5]

Material required (tonne)	Jute 12290			PP-HDPE 2310			Paper 7200		
	Energy (thousand GJ)	Water (thousand lakh liters)	Chemical (tonne)	Energy (thousand GJ)	Water (thousand lakh liters)	Chemical (tonne)	Energy (thousand GJ)	Water (thousand lakh liters)	Chemical (tonne)
Phase I (Production of Raw Material)	153.6	12.0	258.5	178.3	1.4	0.014	612.0	18.0	4647
Phase II (Production of Sacks)	179.4	9.7	Negligible	48.5	1.0	Negligible	57.6	Negligible	Negligible
Total	333.0	21.7	258.5	226.8	2.4	0.014	669.6	18.0	4647
Phase III Usage (Transportation per 100 km distance, 9 tonne truckload and 3.05 km/l fuel consumption)	Excess Fuel (thousand liters)	Excess Energy (GJ)		Excess Fuel (thousand liters)	Excess Energy (GJ)		Excess Fuel (thousand liters)	Excess Energy (GJ)	
	36.3	2035.9		Taken as basis (Zero Consumption)			16.6	927.9	
Phase IV Waste Management	Recycling (Energy saving[a] thousand GJ)	Incineration (Energy recovery thousand GJ)		Recycling (Energy saving[a] thousand GJ)	Ircineration (Energy recovery thousand GJ)		Recycling (Energy saving[a] thousand GJ)	Incineration (Energy recovery thousand GJ)	
	–	–		46.75	95.31		32.26	169.11	

[a] Compared to production of goods from virgin materials

References

1. Klopffer W, Grahl B (2014) Life cycle assessment (LCA): a guide to best practice. Wiley-VCH Verlag GmbH & Co. KGaA, Weinheim
2. Ciambrone DF (1997) Environmental life cycle analysis. CRC Press, Florida
3. Curran MA (ed) (2012) Life cycle assessment handbook: a guide for environmentally sustainable products. Wiley, New Jersey
4. Ghosh AK, Maiti SN, Saroop M, Tyagi S, Dey Roy S (2005) Life cycle analysis of plastics in packaging. Thompson Press, New Delhi
5. Ghosh AK, Tyagi S, Dey Roy S, Kulshreshtha B (2012) Life cycle analysis of PP-HDPE woven sacks vis-à-vis jute/paper sacks in terms of environmental studies

Appendix
Relevant Indian Standards of Packaging Materials for Food Products

S. No.	Indian Standards	Description
1	IS: 6622-1972	Grease proof Paper
2	IS: 9988-1981	Waxed Paper for Bread and biscuits
3	IS: 1397-1990	Kraft paper (second revision)
4	IS: 1398-1982	Packing paper, water-proof, bitumen laminated (second revision)
5	IS: 4658-1988	Coated paper and board (art and Chromo) (First revision)
6	IS: 8460-1977	Wrapping Tissue paper
7	IS: 3962-1967	Waxed paper for general packaging
8	IS: 3263-1981	Waxed paper for Confectionery (First revision)
9	IS: 9588-1990	Kraft paper (first revision)
10	IS: 1776-1989	Folding box board, uncoated (first revision)
11	IS: 2617-2006	Millboard, Grey board and Straw board (First revision)
12	IS: 12999-1990	Folding box board, coated
13	IS: 2771 (Part-1)-1975	Corrugated Fibre Board Boxes
14	IS: 1060 (Part-I)-1966	Methods of Sampling and Test for Paper and Allied Products, Part-1
15	IS: 1060 (Part-II)-1960 (Reaffirmed 1997)	Methods of Sampling and Test for Paper and Allied Products, Part-II
16	IS: 1060 (Part-III)-1969 (Reaffirmed 2004)	Methods of Sampling and Test for Paper and Allied Products, Part-III
17	IS: 10146-1982	Specification of Polyethylene for it safe use in Contact with food stuff and drinking water

(continued)

(continued)

S. No.	Indian Standards	Description
18	IS: 10151-1982	Poly Vinyl Chloride (PVC) and its Co-polymers for its safe use in Contact with Foodstuffs, pharmaceuticals and drinking water
19	IS: 7019-1982	Glossary of terms used in Plastics and flexible packaging, excluding Paper
20	IS: 2508-1984	Low density Polyethylene films
21	IS: 11352-1994	Flexible packaging Materials for the packing of Vanaspati in 100 g, 200 g 500 g, 1 kg, 2 kg and 5 kg Packs-Specifications
22	IS: 14129-1994	Flexible packaging Materials for the packing of Vanaspati in 10 kg and 15 kg packs-Specifications
23	IS: 12265-1987	Flexible packs for packing edible oils
24	IS: 12724-1989	Flexible packaging materials for packaging of refined edible oils
25	IS: 11434-1985	Ionomer resins for its safe use in Contact with foodstuff, Pharmaceuticals and drinking water
26	IS: 11705-1986	Ethylene acrylic acid (EAA) Co-polymers in contact with Foodstuffs, pharmaceuticals and drinking water
27	IS: 12247-1988	Nylon-6 Polymer for its safe use in contact with foodstuff, Pharmaceuticals and drinking water
28	IS: 12252-1987	PET (Polyethylene Terephthalate)/PBT (Polybutylene Terephthalate) for safe use in Contact with food stuff and drinking water
29	IS: 12724-1989	Flexible packaging Materials for the packaging of refined edible Oil-Specifications
30	IS: 11805-2007	Polyethylene Pouches for packaging Liquid Milk-Specifications
31	IS: 12883-1994	Poly Vinyl Chloride (PVC) bottles for edible Oil specification
32	IS: 12887-1989	Polyethylene Terephthalate (PET) Bottles for Packaging of Edible oil-Specification
33	IS: 13576-1992	EMA for its safe use in contact with foodstuffs, pharmaceuticals and drinking water

(continued)

Appendix: Relevant Indian Standards of Packaging Materials for Food Products

(continued)

S. No.	Indian Standards	Description
34	IS: 13601-1993	Ethylene Vinyl Acetate (EVA) Co-polymers for its safe use in contact with foodstuffs, pharmaceuticals and drinking water
35	IS: 10840 -1994	Blow moulded HDPE Containers for packing of Vanaspati-Specification
36	IS: 7408 (Part-1)	Blow moulded Polyolefin Containers-over 5 L capacity
37	IS: 7408 (Part-2)	Blow Moulded Polyolefin Containers-over 5 up to 30 L capacity
38	IS: 7408 (Part-3)	Blow Moulded Polyolefin Containers-closed head containers over 30 L, up to and including 200 L capacity
39	IS: 14543-1998	Packaged drinking Water (other than natural mineral water)
40	IS: 10840-1984 (Part 1)	Blow moulded HDPE Containers for Vanaspati
41	IS: 14537-1998	PET bottles for packaging of alcoholic liquors
42	IS: 8688-2004	Plastics Bottles for Potable water-Specification
43	IS: 1613-1960	Milk bottle Crates
44	IS: 9907-1981	HDPE Crates for milk bottles
45	IS: 2528-1964	Glossary of terms used in Plastic industry
46	IS: 2798-1998	Methods of Test for Plastic Containers
47	IS: 8747-1977	Methods of Test for Environmental Stress-Crack Resistance of Blow Moulded Polyethylene Containers
48	IS: 12100-1987	HDPE Woven sacks for packing flour
49	IS: 14887-2014	Textiles-High Density Polyethylene (HDPE)/Polypropylene (PP) Woven sacks for Packaging of 50 kg food grains-specification
50	IS: 16206-2015	Textiles-High Density Polyethylene (HDPE)/Polypropylene (PP) Woven sacks for Packaging of 10 kg, 15 kg,20 kg,25 kg and 30 kg food grains-specification
51	IS: 14968 -2015	Textiles-High Density Polyethylene (HDPE)/Polypropylene (PP) Woven sacks for Packaging of 50 kg/25 kg Sugar-Specification

(continued)

(continued)

S. No.	Indian Standards	Description
52	IS: 17427-2020	Wooden (Timber) Pallets for Packaging, Storage and Transportation – Specification
53	IS: 11357-1985	Specification for Composite Containers for Dry Products
54	IS 2034: 2012	Round Open Top Sanitary Cans for Butter and Cheese – Specification
55	IS 4638: 1981	Seamless Rectangular Fish Tins
56	IS 9976: 1981	Method of sampling of open top sanitary OTS cans
57	IS 9396 (Part 1): 1987	Round Open Top Sanitary Cans for Foods and Drinks—Part 1 Tinplate
58	IS: 10325-1989	Square Tins-15kgs/Litre for ghee, Vanaspati, edible oil and bakery shortenings-specification
59	IS 9991: 1981	Condensed milk Cans
60	IS 11104: 2012	Glossary of terms relating to open top sanitary cans
61	IS 11078: 2012	Round Open Top Sanitary Cans for Milk Powder-Specification
62	IS: 916-2000	Rectangular Tins for Liquids-Specification (Fourth Revision)
63	IS: 2471-1963	Methods of Test for metal Containers
64	IS: 1406-1995	Square Tins for Solid Products-Specification (Fourth Revision)
65	IS: 10325-2000	Square Tins-15 kg/litre for Ghee, Vanaspati, Edible oils and Bakery Shortenings-Specification (Second revision)
66	IS: 10339-2000	Ghee, Vanaspati, Edible Oil Tins up to 10 kg/litre Capacity-Specification (Second revision)
67	IS: 14407-1996	Aluminium Cans for Beverages-Specification
68	IS 9781: 1989	Glass jars for jams, jellies and marmalades-Specification (First Revision)
69	IS 9780: 1992	Glass bottles for tomato ketchup specification (First Revision)
70	IS 9154: 2020	Method for Determination of Alkali Resistance of Glass (First Revision)
71	IS 2091: 1983	Specification for multi-Trip glass beer bottles (Second Revision)

(continued)

Appendix: Relevant Indian Standards of Packaging Materials for Food Products   345

(continued)

S. No.	Indian Standards	Description
72	IS 1662: 1974	Specification for glass liquor bottles (Second Revision)
73	IS 1392: 1999	Glass milk bottles-Specification (Fourth Revision)
74	IS 13650: 1993	Glass containers for proteinase food
75	IS 12581: 1989	Glass containers for domestic fruit preserving-Specification
76	IS 11985: 1987	Specification for glass jars for pickles
77	IS 11369: 1985	Specification for glass honey jars
78	IS 10497: 2018	Method of test for the determination of brim-full capacity of glass containers by gravimetric method (First Revision)
79	IS 10516: 2018	Method of test for internal pressure resistance of glass containers (First Revision)
80	IS: 6219-1989	Methods of test for general purpose flat pallets for through transit of goods (Second revision)
81	IS: 7028 (Part-1)-2002	Performance tests for Complete filled transport packages: Part-1 Stacking Tests using Static load (second Revision)
82	IS: 7028 (Part-2) -2002	Performance tests for Complete filled transport packages: Part-2, Vibration Test at fixed low frequency (Second revision)
83	IS: 7028 (Part-3)-2002	Performance tests for Complete filled transport packages: Part-3, Horizontal Impact Tests (Horizontal or inclined test, pendulum test (Second revision)
84	IS: 7028 (Part-4)-2002	Performance tests for Complete filled transport packages: Part-4, Vertical Impact Drop Tests (First revision)
85	IS: 7028 (Part-5)-2002	Performance tests for Complete filled transport packages: Part-5, Rolling Tests (First revision)
86	IS: 7028 (Part-6)-2002	Performance tests for Complete filled transport packages: Part-6, Compression Tests (First revision)
87	IS: 7028 (Part-7)-2002	Performance tests for Complete filled transport packages: Part-7, Low pressure Tests (Second revision)
88	IS: 7028 (Part-8)-2002	Performance tests for Complete filled transport packages: Part-8, Water Spray Tests (Second revision)

(continued)

(continued)

S. No.	Indian Standards	Description
89	IS: 7028 (Part-9)-1987	Performance tests for Complete filled transport packages: Part-9, Stacking Test using Compression Tester (First revision)
90	IS: 7028 (Part-10)-1986	Performance tests for Complete filled transport packages: Part-10, Water immersion Test
91	IS: 7028 (Part-11)-1986	Performance tests for Complete filled transport packages: Part-11, Toppling Test
92	IS: 7028 (Part-12)-2002	Performance tests for Complete filled transport packages: Part-12, Vibration Tests using a sinusoidal variable frequency (Second revision)
93	IS: 7031-2002	Method of Conditioning for testing of Complete, filled transport packages (second revision)
94	IS: 7226-1989	Non-expandable general purpose, flat pallets for through transit of goods-Specification (Second Revision)
95	IS: 9111-1979	Glossary of Terms and Classification pertaining to transport packaging
96	IS: 9340-1993	Expandable Pallets-Specifications (first revision)
97	IS: 10106 (Part-1/Sec 1) 1990	Packaging code: Part 1, Product packaging, Section Food Stuffs and perishables
98	IS: 10106 (Part-2/Sec 6) 1990	Packaging Code: Part-2, Packaging materials, Section-6, Flexible Laminates
99	IS: 10106 (Part-2/Sec 1) 1985	Packaging Code: Part-2, Packaging materials, Section-1, Metals
100	IS: 10106 (Part-2/Sec 2) 1983	Packaging Code: Part-2, Packaging materials, Section-2, Paper and Paper board
101	IS: 10106 (Part-2/Sec 3) 1984	Packaging Code: Part-2, Packaging materials, Section-3, Plastics Materials
102	IS: 10106 (Part-4/Sec 1) 1982	Packaging Code: Part-4, Packages, Section-1, Metal Containers
103	IS: 10106 (Part-4/Sec 2) 1984	Packaging Code: Part-4 Packaging materials, Section-2, Paper and paper based packages
104	IS: 10106 (Part-4/Sec 3) 1984	Packaging Code: Part-4, Packaging materials, Section-3, Plastics Containers
105	IS: 10106 (Part-4/Sec 4) 1984	Packaging Code: Part-4, Packaging materials, Section-4, Glass Containers

(continued)

Appendix: Relevant Indian Standards of Packaging Materials for Food Products 347

(continued)

S. No.	Indian Standards	Description
106	IS: 10106 (Part-4/Sec 5) 1986	Packaging Code: Part-4, Packaging materials, Section-5, Wood based Containers
107	IS: 13714-1993	Dunnage pallets-Ware housing-Specification
108	IS: 16058-2013	Dunnage Pallets made of recycled Plastic wastes for warehousing application-Specification
109	IS: 13664-1993	Poly Pallets for Bag storage God owns-Specification
110	IS: 1260 (Part-2)-1999	Pictorial Marking for handling and labelling of goods: Part-2 General Goods (Third revision)

Index

A
Active modified atmosphere packaging, 176
Active packaging, 167, 173, 183–185, 188
Adhesive bonding, 299
Agriculture, 1, 5
Air pressure leakage, 288
Alkalinity test, 297
Aluminium, 94–103, 108, 114, 118–121, 123, 141
Annealing, 296, 297
Antimicrobial packing, 188, 189
Apportionment, 19
Aseptic packaging, 198, 199, 202, 207, 209
Ash content, 270

B
Barcode, 231
Biodegradadibility, 317
Blocking, 283
Blow moulding, 127, 133–139
Blow process, 122, 123
Board grammage, 271, 278
Boiling water bath, 196
Branding, 16, 18, 19, 23, 24, 39, 227, 230, 232, 233
Brand integrity, 230, 232, 256
Breaking length, 264
Breathable films, 167–170
Bulk, 263, 264, 275
Burnt deposit, 130
Burr, 130
Burst factor, 264
Bursting strength, 264, 272, 278, 301

C
Caliper, 263, 272, 278, 287, 295
Canning, 195–197, 200, 206
Carbon dioxide absorbers/Emitters, 185
Carbon footprint, 331
Categories of processed food products, 194
Chemical composition, 121, 146
Chemical tests, 268, 294, 297
Chemical treatments, 200
Chill marks, 108
Circular economy, 320, 323
Climate change, 311, 313, 326, 330
Closure, 20, 23
Cloth, 7, 21
Coating/Lacquering, 115
Cobb value, 268, 277–279
Coefficient of friction, 282
Coefficient of linear thermal expansion, 284
Co-extrusion process, 101
Collapsible crates, 143, 144, 148, 150
Communication, 3, 16, 18, 19, 28
Compensated vacuum, 177, 178
Components, 5, 10, 17, 20, 22, 24, 29
Composite, 89, 94–99
Conditioning, 262
Controlled atmosphere packaging, 167, 181, 182
Convolute method, 98
Corporate Social Responsibility (CSR), 231, 232
Corrugated fiber board boxes, 113, 140, 151, 155–158, 162
Corrugating adhesives, 153, 154
Corrugating medium test, 278
Corrugating or Fluting medium, 153, 154
Crystallinity, 127–129, 134

D

Density, 263, 267, 277, 280, 292, 295, 297
Digital printing, 245
Direct heating systems, 208
Drop test, 274, 279, 288, 293, 299, 305
Dynamic compression test, 303, 304

E

Ecosystem, 311, 312, 318
Edgewise compression test, 272
Energy, 326, 329–339
Environmental stress crack resistance, 287
Ethanol emitters, 185
Ethylene absorbents, 185
Eutrophication, 330
Expanded PolyStyrene (EPS) Containers, 172
Extended Producer Requirement (EPR), 231, 256
Extrusion blow moulding, 127, 133–135, 137, 138

F

Flash cooling, 208
Flat crush test, 272, 273
Flexible can, 204
Flexography, 244, 245
FMOT, 228, 229
Folding endurance, 264–266
Food, 1–8, 10–12, 14–17, 20, 23–25, 28–30, 33, 35–39, 43
Food contact safety, 238
Forming, 170, 173, 174
Form seal fill, 96
Fossil fuel, 310, 312, 313, 318, 327, 330, 338
Freeze drying, 214–217
Freshness indicators, 187
Functions of packaging, 18, 29

G

Gable top Carton, 201, 203
Gases in MAP, 175
Gas flushing, 174, 177, 178, 182
Glass, 113, 121–127, 137, 144, 331, 333, 334
Globalization, 1, 5
Global warming, 313, 317
Graphic design, 239, 240
Gravure printing, 244, 245
Greenhouse, 313, 316, 317
Green washing, 322

H

Handle pull test, 288
Heat, 200, 201, 204–206, 208, 209, 214–216, 218, 219, 222
High pressure processing, 210, 212
History of packaging, 6, 9
Holding pressure, 129, 130
Homogenization, 207, 208
Hot air, 200
Human toxicity, 331
Hydrogen peroxide, 200, 201
Hydrostatic pressure test, 288

I

Identification, 4, 9, 16, 19, 22, 24, 25
Impact assessment, 329
Inclined impact test, 275, 276, 279
Incorrect dimension, 130
Indirect heating systems, 208
Individual quick freezing, 212
Injection moulding, 126–129, 131–133, 137, 141
Injection stretch blow moulding, 133, 137, 138
Ink-jet printing, 245
Intelligent packaging, 187
Intermediate packs, 20, 21
Inventory analysis, 327, 329
Irradiation, 199, 201, 220, 221, 223
ISO, 325, 326, 329, 331

J

Jute, 327, 335–339

K

Key line drawing, 247, 251, 252
Kraft liners, 153, 154

L

Labels, 7, 9, 20, 23–28
Lamination process, 100
Lami tube, 99
Letterpress, 245
Liberalization, 1, 2
Life cycle, 325–332, 334, 336, 339
Life Cycle Analysis (LCA), 315, 325–329, 331–336
Life Cycle Impact Assessment (LCIA), 329
Linear economy, 319
Lithography, 244
Logistics, 4, 13, 19, 231, 232, 255

M

Manufacturing, 113, 116–119, 122, 127, 133, 140, 141, 147, 151, 155, 156
Marketing, 3, 13, 14, 16–19, 21, 23–25, 28, 29
Material, 113–118, 120, 122, 123, 126–128, 130, 132, 134–136, 139, 140, 145–147, 151, 152, 154, 155, 158, 162
Maturing, 172
Mechanical tests, 262, 264, 277, 280, 294, 296, 301
Melt flow index, 128, 135, 283
Metal, 113–122, 126, 132, 137, 141, 143
Microwave heating, 217, 218
Mixing, 170
Modified atmosphere packaging, 167, 174, 176
Moisture content, 262, 268, 269, 278
Molecular weight, 127–129, 135
Molecular weight distribution, 127, 128, 135
Moulded pulp trays, 171
Moulding, 172
Multilayer, 89, 94–97, 99–102

N

Neck finish, 235

O

Odour absorbents, 185
Optical tests, 262, 270, 277, 285
Overall migration test, 285
Oxygen scavengers, 184, 185
Oxygen transmission rate, 285

P

Packaging, 1, 3–26, 28–40, 42–44
Paper, 327, 335–339
Passive modified atmosphere packaging, 176
Pedulum test, 297
Peracetic Acid (PAA), 201
Performance, 261, 273, 276, 277, 286, 288, 291–294, 302
Phases of life cycle, 327, 328, 332
Phases of life cycle analysis, 327, 328, 332
Photo-oxidation, 330
pH value, 269
Physical tests, 262, 277, 294, 300
Plastic, 91, 94–96, 99–108, 113, 126, 127, 129–134, 136, 137, 139–145, 325, 329, 333–338
Plastic crates, 113, 139–141, 144, 145
Plastic sqeezable tube, 300
Plastic woven sack, 73, 74, 77
Plug assist, 107
Polyethylene Terephthalate (PET), 60, 61
Polypropylene (PP), 55, 57, 59–61, 65, 74, 75, 78, 84
Polyvinyl Alcohol (PVA), 55, 57, 64, 69
Porosity, 263, 267, 292
Pouch burst test, 304
PP-HDPE, 335, 337–339
Pre-expansion, 172
Presentation, 3, 4, 16, 18–20, 28
Preservation, 47, 48
Pressing, 171
Processing parameter, 129, 135
Production, 113–115, 117, 118, 126, 127, 129, 133, 134, 137, 139, 147
Promotional information, 19
Pulsed light, 199
Pulse electric field, 222
Puncture resistance, 273, 279

Q

QR code, 231
Quality, 261, 262, 272, 277, 286, 289–294, 299, 300, 304, 305
Quality control, 67

R

Radio frequency identification, 188
Reinforcement, 20, 23, 41
Requirements of food packaging, 48
Retort pouch, 204–207
Rigid, 47–49, 55, 56, 59, 61, 63, 80
Rigid packaging, 113, 126
Ring crush test, 267, 268, 278
Rolling test, 276, 279
Rub proofness, 289

S

Sack paper, 53
Saturated steam, 200
Seal strength, 281
Self-adhesive labels, 188
Semi rigid, 49
Sensation transference, 233, 234
Shelf life study, 238

Sink marks, 130
Smoothness, 266, 267
SMOT, 228
Spiral method, 98
Stability, 48, 54, 57, 60, 62, 67
Stacking crates, 141, 142
Stack load test, 275, 279, 289
180° stack-nest crate, 143
Stack-nest crates, 141–143
Steam pressure canner, 196
Stiffness, 267
Sulphite paper, 52
Superheated steam, 200
Supply chain, 231, 232, 255
Sustainability, 5, 19, 231, 232, 309–311, 313–316, 318, 322, 323
Sustainable development, 325, 338

T
Tamper-evident, 48
Tape plant, 75
Tearing resistance, 265, 266
Tensile strength, 264, 280, 281, 290
Test liners, 152–154
Tetra Brick Carton, 201
Thermal ink transfer, 245
Thermal shock test, 297
Thermoform, 89, 107
Thickness, 263, 264, 267, 272, 277–280, 287, 290–292, 294, 299
Time–temperature indicators, 187, 188
Tin, 114, 115, 119, 120, 123, 126, 335, 336
Tissue paper, 52
Trademarks, 9
Trimming, 171

U
Ultra-heat treatment, 207
Unit pack, 20–22, 25, 27
Urbanisation, 1, 2, 5
UV-C radiation, 200
UV sterilization, 223

V
Vacuum forming, 109
Vacuum leakage test, 304
Verticality, 288, 295
Vibration test, 275, 279
Visual hierarchy, 240–242

W
Warpage, 107, 128, 130
Water spray test, 276, 279
Water vapour transmission rate, 285
Wax paper, 53
Web stock, 100, 101
Weld marks, 131
Wooden crates, 140, 145, 146, 148, 150, 151

X
Xerography, 245

Y
Yield strength, 279

Z
ZMOT, 228

Printed in Great Britain
by Amazon